MW00843992

Introduction to Discrete Dynamical Systems and Chaos

WILEY-INTERSCIENCE
SERIES IN DISCRETE MATHEMATICS AND OPTIMIZATION

ADVISORY EDITORS

A complete list of titles in this series appears at the end of this volume.

Introduction to Discrete Dynamical Systems and Chaos

MARIO MARTELLI
California State University
Fullerton

A Wiley-Interscience Publication
JOHN WILEY & SONS, INC.
New York • Chichester • Weinheim • Brisbane • Singapore • Toronto

This text is printed on acid-free paper. ♾

For ordering and customer service, call 1-800-CALL-WILEY.

Library of Congress Cataloging-in-Publication Data:

Martelli, M. (Mario), 1937–
Introduction to discrete dynamical systems and chaos / Mario Martelli.
p. cm. — (Wiley-Interscience series in discrete mathematics and optimization)
"A Wiley-Interscience publication."
Includes bibliographical references and index.
ISBN 0-471-31975-9 (alk. paper)
1. Differentiable dynamical systems. 2. Chaotic behavior in systems. I. Title. II. Series.
QA614.8.M285 1999
515'.352—dc21 99-25865

Printed in the United States of America

10 9 8 7 6 5 4 3 2

To my wife and children

CONTENTS

PREFACE

The purpose of this book is to bring the fundamental ideas on discrete dynamical systems and chaos at the level of those undergraduates, usually in their junior year, who have completed the standard Calculus sequence, with the inclusion of functions of several variables and linear algebra. At this stage, students are in the best position for being exposed, during their college training, to the new ideas and developments generated in the last thirty years by the theory of discrete dynamical systems and chaos. The students' degree of sophistication permits the presentation of a broad range of topics and a fairly deep analysis of some nontrivial and historically interesting models. The importance and relevance of this exposure can hardly be described with better words than the ones used by R. Devaney (see [Devaney, 1989]). He writes: "*The field of dynamical systems and especially the study of chaotic systems has been hailed as one of the important breakthroughs in science in this century.*"

The book is divided into seven chapters and three appendices. Its content can be comfortably covered during a one-semester course, particularly if the teacher is satisfied with providing detailed proofs of only some fundamental results. As the title itself suggests, the topics of the book are limited to discrete dynamical systems. Several reasons have dictated this choice. The inclusion of both continuous and discrete systems would have created too large a body of material, with an inevitable loss of any in depth analysis. Moreover, a good understanding of continuous systems is hard to achieve without proper training in ordinary differential equations. Thus, their inclusion would have increased the prerequisites for the course. Another consideration that played an important role in the choice is the difficulty of establishing on theoretical grounds that a continuous system is chaotic. Chaos is one feature of dynamical systems that the book wants to present and analyze. It was considered awkward not to be in a position to prove that any continuous system is chaotic. A brief description of all chapters follows.

In Chapter 1 we present definitions and general ideas about discrete dynamical systems, together with some examples of significant interest derived from the recent research literature. In Section 1 we start with some examples of discrete dynamical systems and discuss the definition of discrete dynamical systems and the goals of the book. In Section 2 we introduce the standard definitions of fixed points, periodic orbits, and stability. In Section 3 we talk about limit points and aperiodic orbits, and we present a preliminary description of chaotic behavior. In Section 4 we give examples of systems, such as the system proposed by E.N. Lorenz to model atmospheric changes and the system proposed by J.J. Hopfield to model neural networks, which are later (Chapter 7) studied using the theory developed in the course.

Chapter 2 contains an extensive analysis of one-dimensional dynamical systems depending on one parameter. In the first section we introduce the cobweb method and the idea of conjugacy. In the second section we study the stability and instability of fixed points and periodic orbits. In Section 3 we present a result on global stability of fixed points. In Section 4 we introduce bifurcation, and we analyze this phenomenon both through examples and theoretically. The last section explores the implications of conjugacy and Li-Yorke chaos. The purpose of this chapter is to give the students some material to work on at the outset. Its only prerequisites are a few results from calculus of one variable, which are listed whenever needed, without proofs.

Chapter 3 contains an overview of those results of linear algebra and calculus of several variables which are likely to receive less attention during standard undergraduate courses. In the first section one finds something about the topology of $\mathbf{R}^q$ and of its structure as a normed vector space, the definition of continuity, and the equivalence of all norms. Section 2 deals with the operator norm of a matrix, the differentiability, first order-approximation, and the mean value inequality. The topics of this chapter are needed in Chapters 4 and 5.

In Chapter 4 we analyze discrete linear dynamical systems. Our study is based on three fundamental tools: the spectrum, a fundamental property of the spectral radius with respect to the operator norm of a linear map, and the spectral decomposition theorem. The first section explores the idea of representing the orbits of a linear system using eigenvectors. In Section 2 we study the case when the spectral radius is smaller than 1, and the case when all eigenvalues have modulus larger than 1. In Section 3 we present the spectral decomposition theorem, dividing the treatment into three cases: (1) when all eigenvalues are real and semisimple, (2) real but not semisimple, and (3) possibly complex. In Section 4 we investigate the saddle case, namely the case when some eigenvalues have modulus smaller than 1 and the others have modulus larger than 1. In Section 5 we analyze the case when at least one of the eigenvalues has modulus 1. Finally, in Section 6 we study affine systems, both in the case when 1 is not an eigenvalue and when it is.

With Chapter 5 we enter into the more challenging part of the book: the study of nonlinear systems in dimension higher than 1. In the first section we analyze systems having bounded invariant sets. Three types of maps are studied: contractive, dissipative, and quasi-bounded. We show here that the map proposed by Lorenz to model atmospheric behavior is dissipative and the one used by Hopfield to describe neural networks is quasi-bounded. Section 2 is devoted to maps having a unique fixed point that is a global attractor. Three classes of such maps are presented: contractions, triangular maps, and gradient maps. The third section deals with fixed points and periodic orbits that are sinks. In the fourth section we present repellers and saddles, with a brief excursion on stable and unstable manifolds. In Section 5 we discuss two fundamental results on bifurcation, including the Hopf case.

Chapter 6 is devoted to chaotic behavior. The first section opens the chapter with the definition of attractor and with a discussion of its relation to stability. In Section 2 we present a definition of chaotic dynamical systems based on the presence of a dense orbit and of its instability. Sensitivity with respect to initial conditions and other alternative definitions of chaos are also presented. In Section 3 we analyze the attractors of a chaotic system from the point of view of their dimension. Two types of dimension are discussed, the capacity and the correlation dimension. In the last section Lyapunov exponents are discussed together with their relation to stability and sensitivity with respect to initial conditions.

In Chapter 7 we present an extensive, although not complete analysis of the models introduced in Section 4 of Chapter 1, namely a blood-cell population model, predator-prey models for competition between two species, the model proposed by Lorenz as an approximation to the dynamics of atmospheric changes, and the Hopfield model of a neural network.

Appendix 1 contains an extensive collection of Mathematica programs that can be used to study discrete dynamical systems. They are referred to frequently throughout the book. The students and the teacher are free to use other symbolic manipulators such as Maple, or others. The ones presented here are simply for illustration purposes. An extensive analysis of a dynamical system is rarely possible without the significant help that can be provided by a computer, particularly when combined with at least a working knowledge of a powerful symbolic manipulator. It is hard, and perhaps even impossible, to get hands-on experience without using the enormous computing capability of a machine. Students are urged to learn how to use at least one of the many programs available. Some are designed strictly for the study of dynamical systems. Others address a much broader range of topics, and have quite a few features that can be exploited for a successful numerical study of discrete dynamical systems.

Appendix 2 has a list of references and possible team projects.

In Appendix 3 the reader finds short answers to selected problems. Many of the assigned problems can be solved without using a computing device. The answers to others are simpler to find with the aid of a symbolic manipulator. A few cannot be done without a computer, or a programmable calculator. A manual with detailed solutions of all problems is available from Wiley upon request.

The starred sections can be omitted without compromising the continuity of the presentation. They are clearly marked in the table of contents. The starred problems are easier to solve using the Mathematica programs listed in Appendix 1.

Wiley is providing a web site with some Mathematica programs which can be dowloaded and used in the investigation di dynamical systems, particularly in dimensions one and two. Please go to

ftp://ftp.wiley.com/public/sci_tech_med/dynamical_systems

Many scientists have written extensively about the importance of the topics presented in this book. It is hard to add something new to the things they have said so beautifully and appropriately. In particular, I believe that the number of mathematicians who feel that these topics should be made accessible to undergraduates largely exceeds the number of those who are still reluctant to "follow the trend." I belong to the first group, and this book is my attempt to provide one more tool for reaching the goal. Many friends have helped me in different ways. I am thankful to all of them. In particular I would like to mention Alfonso Albano, the late Stavros Busenberg, Courtney Coleman, Annalisa Crannell, Massimo Furi, and William Gearhart. When I was uncertain about the presentation of certain topics, or the most appropriate examples for clarifying definitions and theorems, they provided invaluable help. I am indebted to my students, to my daughter Monica, and to my son Teddy for finding numerous errors and misprints. I am thankful to my department, particularly to the chairman Dr. Jim Friel, for encouraging me in this enterprise by making a course on discrete dynamical systems and chaos mandatory for all math majors at CSUF.

Last, but not least, I would like to express my appreciation to the publisher for making the results of my effort available to the community of scientists and teachers. I sincerely hope that they will find this book useful.

Mario Martelli

Claremont, May 1999

CHAPTER 1

DISCRETE DYNAMICAL SYSTEMS

SUMMARY

The chapter opens with some simple examples of discrete dynamical systems and continues with a formal definition of discrete dynamical systems and an outline of the goals of our study. Definitions of stationary state, periodic orbit, and the concept of stability are introduced in Section 2. The third section provides an informal description of chaotic behavior by means of aperiodic and unstable orbits. Some interesting examples of discrete dynamical systems derived from the current literature are presented in Section 4. A detailed analysis of these systems is provided in Chapter 7 using the theory developed in the previous chapters.

Section 1. DISCRETE DYNAMICAL SYSTEMS: DEFINITION

1. Examples of Discrete Dynamical Systems

We start our study of discrete dynamical systems with three simple examples.

Example 1.1.1 After searching the interest rates offered by several banks and savings and loans in our area, we have decided to invest \$5,000 with Everest Savings, which offers a 6.5% interest rate compounded monthly. The teller explains that at the end of every month the new principal P_{new} will be equal to the principal of the preceding month multiplied by 1+(.065/12). In other words,

$$P_{new} = \left(1 + \frac{.065}{12}\right)P_{old}.$$

Denoting by P_0 our original investment of \$5,000, we obtain that after 1,2,...,n months P_0 has grown to

$$P_1 = \left(1 + \frac{.065}{12}\right)P_0,\ P_2 = \left(1 + \frac{.065}{12}\right)P_1 = \left(1 + \frac{.065}{12}\right)^2 P_0,\ \ldots\ ,$$

$$P_n = \left(1 + \frac{.065}{12}\right)P_{n-1} = \left(1 + \frac{.065}{12}\right)^n P_0.$$

The formula for P_n above can be generalized to every interest i and to every compounding period. For example, we learned that Mercury Savings is offering an interest rate of 6.8% compounded every 4 months. After m periods of 4 months, our investment has grown to

$$P_m = \left(1 + \frac{.068}{3}\right)P_{m-1} = \left(1 + \frac{.068}{3}\right)^m P_0.$$

Hence, 5 years with Everest Savings give a balance of $P_{60}=(1+.065/12)^{60}P_0=$ \$6914.09, while with Mercury Savings the balance is $P_{15}=(1+.068/3)^{15}P_0=$ \$6998.12. After 5 years we are slightly better off with Mercury Savings.

Let us use x_0, i, and m to indicate the initial investment, the interest and the number of compounding periods in a year, respectively. After n + 1 compounding periods, the amount x_{n+1} available to us is given by

$$x_{n+1} = (1 + i/m)x_n = (1 + i/m)^{n+1}x_0. \tag{1.1.1}$$

We have here a simple example of a discrete dynamical system. Let F(*x*) be the function

$$F(x) = (1 + i/m)x.$$

The system is governed by F and (1.1.1) can be rewritten in the form

$$x_{n+1} = F(x_n) = F^{n+1}(x_0), \tag{1.1.2}$$

where $F^{n+1}(x)$ represents the (n+1)th iterate of F. For example,

$$F^2(x) = F(F(x)) = (1 + i/m)F(x) = (1 + i/m)(1 + i/m)x = (1 + i/m)^2x.$$

In other words, $F^2(x)$ is evaluated by replacing x with $F(x)$ in the formula that defines F. To be more accurate we should write $F(i,m,x)=(1+i/m)x$, since both i and m play a role in the growth of the investment. However, we are interested mainly in the growth of x once i and m are fixed. Hence, we can still write $F(x)=(1+i/m)x$. The principal x is the **state variable** of the system. The interest i and the number m of compounding periods in a year are the **control parameters**.

The goals we are pursuing can be summarized as follows:

- Find the growth of any initial investment x_0, given i and m. In other words, look at the **evolution** of the state variable once the control parameters are **fixed**.
- Investigate how **changing** the control parameters (either i or m or both) affects the growth of the investment.

Frequently, both aspects are considered simultaneously.

We list some of the problems the reader may like to study:

- Given an interest i and a length m of the compounding periods, find how long it will take for an investment to double its original value.
- Find how much the interest should be for the investment to double in a certain period of time, assuming that the length of the compounding periods is known.
- Compare the combination of different interest rates and different compounding periods to see which bank offers the best deal.

Example 1.1.2 Our friend Ann, who is a biology major, is investigating the evolution of a colony of bacteria in the laboratory. In discussing the experiment we learn that it would be nice to have a formula that gives week by week the number N of bacteria per square inch. Ann tells us that in recent weeks there appears to be some kind of periodic behavior in the number of bacteria. One week their total appears to be higher than the number that can be supported by the laboratory environment (approximately 2.5×10^6 bacteria per square inch), and the week after the number appears to be lower.

To avoid dealing with very large numbers we let $x=N/10^6$ and we tell Ann that the proposed problem can be solved if we find the form of d_n and p_n in the discrete dynamical system:

$$x_{n+1} = x_n - d_n + p_n. \tag{1.1.3}$$

We explain that x_{n+1} represents the number of bacteria (divided by 10^6) at the beginning of the (n+1)th week of the experiment, x_n represents the same number the week before, and d_n, p_n are respectively the bacteria that died and were produced during the nth week. Ann tells us that experimental observations made so far suggest that about two-thirds of the bacteria die in any given week. Consequently, we suggest that $d_n=.7x_n$. Finding the form of p_n proves to be more challenging. She informs us that the growth of the colony was slow but steady at the beginning, and the periodic oscillations have been observed in the last few weeks. After some discussion and computer investigation we find that an acceptable form for p_n is $p_n=12x_n/(1+x_n{}^4)$. We rewrite (1.1.3) in the form

$$x_{n+1} = x_n - .7x_n + \frac{12x_n}{1 + x_n{}^4} = .3x_n + \frac{12x_n}{1 + x_n{}^4}. \tag{1.1.4}$$

We explain to Ann that this is a discrete dynamical system governed by the function $F(x)=.3x+12x/(1+x^4)$, where x is the state variable of our system. Using F, we write (1.1.4) in the more compact form

$$x_{n+1} = F(x_n). \tag{1.1.5}$$

We also remark that a more general version of (1.1.4) could be

$$x_{n+1} = x_n - ax_n + \frac{bx_n}{c + dx_n{}^p} = (1 - a)x_n + \frac{bx_n}{c + dx_n{}^p}, \tag{1.1.6}$$

where a, b, c, d, and p are control parameters whose values can be adjusted to accommodate different experimental data. Then F becomes

$$F(a,b,c,d,p,x) = (1 - a)x + \frac{bx}{c + dx^p}, \quad \text{or} \quad F(\mathbf{a},x) = (1 - a)x + \frac{bx}{c + dx^p},$$

with $\mathbf{a}=(a,b,c,d,p)\in \mathbf{R}^5$ (since $\mathbf{a}$ has five components). Hence, (1.1.6) can be written in the form

$$x_{n+1} = F(\mathbf{a},x_n). \tag{1.1.7}$$

However, since the evolution of the colony is usually studied with a, b, c, d, and p fixed, we may simply write $F(x)=(1-a)x+bx/(c+dx^p)$. Ann is very impressed by our analysis and is curious to see how our proposed model works. We go together to the computer lab, and open Mathematica. We use model (1.1.4). We ask her to suggest an initial value for the number of bacteria and she says that $x_0=1.5(\times10^6)$ is a good estimate of the number with which the experiment started. Using a suitable Mathematica program (see Appendix 1, Section 2) and (1.1.4), we construct the first 60 states of the population of bacteria; namely, we compute $x_1=F(1.5)$, $x_2=F(x_1)$,..., $x_{60}=F(x_{59})$ and we plot the points $(0,x_0)$, $(1,x_1)$,..., $(60,x_{60})$. To make more evident how the population of bacteria is evolving, we connect all pairs $(i-1,x_{i-1})$, (i,x_i), $i=1,2,...,60$ with segments (see Fig. 1.1.1). The graph shows clearly that after an initial period of adjustment, the number of bacteria in the colony oscillates between one state above and one below 2.5×10^6, which, according to Ann, represents how many bacteria per square inch can be sustained by the laboratory environment.

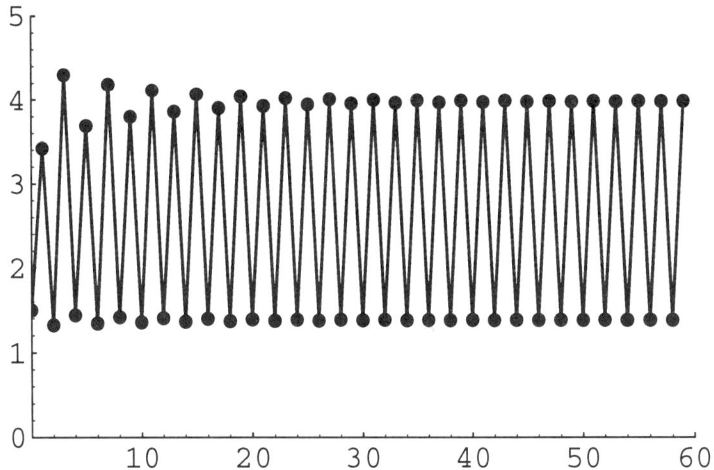

Fig. 1.1.1 It is evident that after an initial period of adjustment ("transient") the population of bacteria is oscillating periodically (with period 2).

We observe that the period of adjustment does not reflect the growth pattern observed by Ann at the beginning of the experiment. Different values of the parameters are needed to model this feature. For example, we could use a=.4, b=c=4, and p=2. We mention to Ann that the parameters cannot be expected to remain constant during an experiment. We also point out that a different choice of a, b, c, d, and p in (1.1.6) can model a behavior of the colony more complex than the steady growth or the periodic oscillations. For example, leaving a, b, c, and d as in (1.1.4) and selecting p=5 gives the system

$$x_{n+1} = .3x_n + \frac{12x_n}{1+x_n^5} \cdot \tag{1.1.8}$$

With the same initial condition x_0=1.5, we obtain a graph which suggests that the number of bacteria in the colony is evolving in an erratic manner (see Fig. 1.1.2).

Ann now believes that mathematics can be very useful in biology! She is curious to know if there is a model in which the three patterns of steady growth, periodic oscillations, and erratic evolution can be produced by changing **only** the death rate a. We tell her that the question is very interesting and we promise to investigate it.

Example 1.1.3 After a few days, Ann returns with another problem. Producing some newspaper clippings she points out that the ash whitefly (*Siphoninus phillyreae*) was introduced in Southern California around August 18, 1988 and multiplied so quickly that 48 of the 56 California counties were literally invaded by these pests.

About three years later the entomologists from UC Riverside imported, mass reared and released the ash whitefly's natural enemy, a tiny black wasp (*Encarsia inaron*). The summer infestation density of the ash whitefly before the mass release of *E. inaron* averaged 8 to 21 whiteflies/cm^2 of infested leaf. Within two years of *E. inaron* release the infestation density dropped to .32 to 2.18 whiteflies/cm^2 leaf (see [Pickett et al., 1996]). Ann would like to know if we can design a mathematical model of this situation.

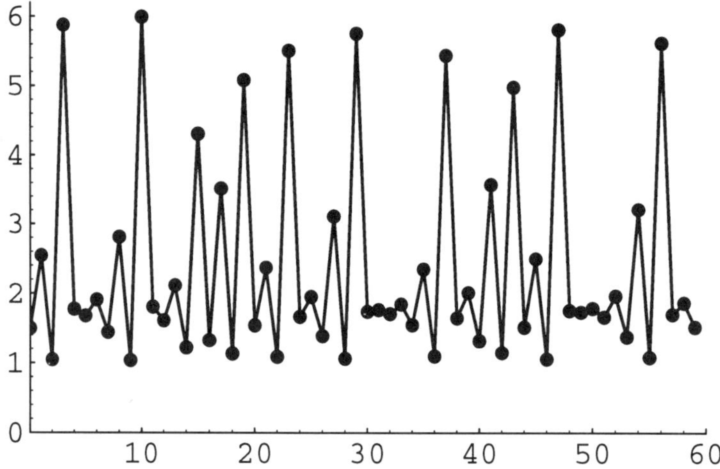

Fig. 1.1.2 It appears that when the evolution of the colony of bacteria is governed by (1.1.8) the number of bacteria follows an erratic pattern.

After giving the problem some serious consideration, we feel that the populations of whiteflies and tiny wasps should undergo periodic one-year oscillations, with possible different levels of density for both populations during the different seasons. At the outset we are inclined to adopt a model for the whiteflies similar to the one proposed before for the colony of bacteria, possibly with different constants and with an extra term to account for the whiteflies destroyed by the *E. inaron*. However, after some research in the library, we realize that this dynamical process belongs to the class of predator-prey problems, which have been studied extensively since 1920 (see Section 4 of this chapter for details). Guided by the literature in this area, and after some computer experiments we choose the form

$$\begin{cases} x_{n+1} = 1.01x_n - .008x_n y_n \\ y_{n+1} = .9y_n + .04y_n(1.01x_n - .065x_n y_n). \end{cases} \tag{1.1.9}$$

We show (1.1.9) to Ann and we explain that the state variable x represents the whiteflies (x=2 means an average of 2 whiteflies/cm^2 of infested leaf) and the state variable y the wasps (y=3 means an average of 3 black wasps/cm^2 of whiteflies infested leaf). We point out that in the absence of wasps, the population of whiteflies grows exponentially (see Fig. 1.1.3), since

$$x_1 = 1.01x_0,\ x_2 = 1.01x_1 = (1.01)^2 x_0, \ldots,\ x_n = (1.01)^n x_0, \ldots$$

and $(1.01)^n \to \infty$ as $n \to \infty$. This is not very realistic in the long run, but it was certainly true for the first few months after the infestation began. In the absence of whiteflies the wasps become extinct since

$$y_1 = .9y_0,\ y_2 = .9y_1 = (.9)^2 y_0, \ldots,\ y_n = (.9)^n y_0, \ldots$$

and $(.9)^n \to 0$ as $n \to \infty$. The function governing the system is

$$F(\mathbf{x}) = F(x,y) = (1.01x - .008xy,\ .9y + .04xy(1.01 - .065y)). \qquad (1.1.10)$$

F is a function from $\mathbf{R}^2$ into $\mathbf{R}^2$, although for obvious reasons we are interested only in the first quadrant of $\mathbf{R}^2$, namely $x \geq 0$ and $y \geq 0$. The component functions of F are

$$F_1(\mathbf{x}) = F_1(x,y) = 1.01x - .008xy,\ F_2(\mathbf{x}) = F_2(x,y) = .9y + .04xy(1.01 - .065y).$$

System (1.1.9) can be rewritten in the form

$$\mathbf{x}_{n+1} = F(\mathbf{x}_n). \qquad (1.1.11)$$

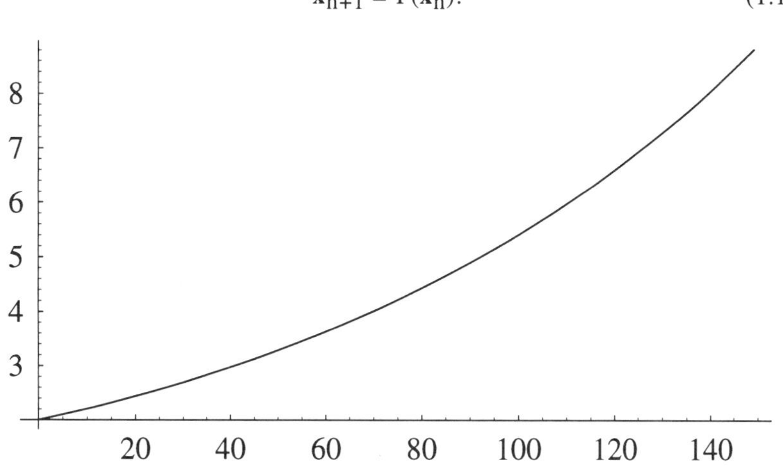

Fig. 1.1.3 The number of whiteflies/cm^2 of infested leaf grows exponentially in the absence of a natural enemy.

We assume that in agreement with the available data, the average number of whiteflies/cm^2 of infested leaf was $x_0=14$ at the mass release of black wasps, and let $y_0=2$. Hence, the initial state is $\mathbf{x}_0=(14,2)$. We compute the first 400 states of the populations of both species and plot in one graph (i,x_i), and in another (i,y_i) for $i=0,1,\ldots,400$. We superimpose the two graphs (Mathematica uses the Show command: see Section 2 of Appendix 1) to obtain Fig. 1.1.4, in which the larger dots represent the whiteflies. We see that their population decreases dramatically. We do not see in the graph the convergence of both populations to the steady state (2.15,1.25), which appears to be a reasonable long-term outcome and is in good agreement with the data provided.

Ann is truly amazed by the proposed model, but we feel that it is not perfect and that further adjustments are needed. For example, we can make the model more general by introducing some parameters, as we did in the case of the colony of bacteria. A starting point could be

$$\mathrm{F}(\mathbf{x}) = \mathrm{F}(x,y) = (\mathrm{a}x - .008xy,\ \mathrm{c}y + .04xy(\mathrm{a} - .065y)). \tag{1.1.12}$$

In (1.1.12) a and c are two control parameters, a>1 and c$\in$(0,1), which can be adjusted to fit experimental data and to better represent the behavior of the two populations. Then, to be more accurate, we should write

$$\mathrm{F}(\mathbf{a},\mathbf{x}) = (\mathrm{a}x - .008xy,\ \mathrm{c}y + .04xy(\mathrm{a} - .065y)) \tag{1.1.13}$$

where $\mathbf{x}=(x,y)$ is the vector in $\mathbf{R}^2$ which represents the two state variables of the system (number of whiteflies and number of black wasps) and $\mathbf{a}=(\mathrm{a,c})$ is a vector in $\mathbf{R}^2$ which represents the (positive) control parameters of the system. The dynamical system becomes

$$\mathbf{x}_{n+1} = \mathrm{F}(\mathbf{a},\mathbf{x}_n). \tag{1.1.14}$$

However, when parameters a and c are regarded as fixed numbers, we may prefer the symbol F($\mathbf{x}$) to F($\mathbf{a},\mathbf{x}$) and write (1.1.14) in the simpler form

$$\mathbf{x}_{n+1} = \mathrm{F}(\mathbf{x}_n). \tag{1.1.15}$$

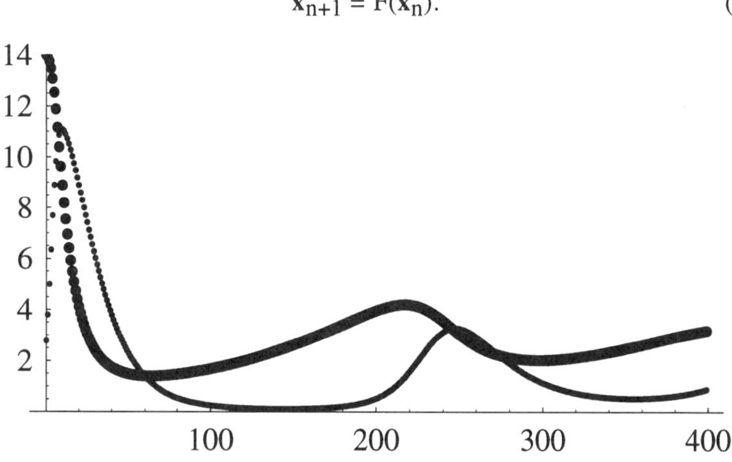

Fig. 1.1.4 The population of whiteflies (larger dots) decreases dramatically after the introduction of black wasps. The two populations converge to the state (2.15, 1.25), in agreement with the information provided.

Problems

Throughout the book, problems with a * have questions whose numerical and/or graphical solution may require a computer or a calculator. The Mathematica programs that can be used in the numerical investigation of these problems are contained in Appendix 1. When the problem requires you to find the time or find the fixed point or similar quantities, you should find the exact value whenever possible.

1. * An investor makes a deposit of $50,000 at 7% compounded monthly. Find the time needed for the principal to double, namely to reach $100,000.

2. * An investor makes a deposit of $5,000 at 6% compounded monthly. After 30 months the investor adds $3,000. At that time the bank changes the compounding period to a trimester. Find the balance after 45 months.

3. * A colony of bacteria starts with x_0=2.4($\times 10^5$) bacteria/inch^2. It is assumed that the growth of the colony is governed by $F(x)=.6x+4x^6e^{-1.5x}$ where $x(\times 10^5)$ represents the number of bacteria/inch^2. The time interval between the nth and the (n+1)th generation is one week. Write the dynamical system that governs the evolution of the colony and find out how many bacteria/inch^2 there are after 10 weeks (for a suitable Mathematica program to find x_{10}, see Section 2 of Appendix 1).

4. * Examine the same situation of Problem 3 assuming that x_0 is the same but that $F(x)=.4x+4x^6e^{-1.5x}$.

5. * Find those values of x such that $x=.3x+12x/(1+x^4)$ (see Example 1.1.2). These are the stationary states (or fixed points) of the system (you can use the programs of Appendix 1, Section 4).

6. * Assume that the growth of a colony of bacteria is governed by $F(x)=.6x+4x^6e^{-1.5x}$. Find those values of x such that $x=F(x)$ (see Appendix 1, Section 4). Generalize F with the introduction of control parameters.

7. In Example 1.1.3 [see (1.1.10)] there are two different state vectors **x** in the first quadrant such that **x**=F(**x**). One of them is **x**=**0**. Find the other one.

8. * Explore the consequences of changing the values of parameters a and c in the system of Example 1.1.3 [see equality (1.1.12)]. Introduce other parameters and investigate how the evolution of the system is affected by changes in these new parameters.

9. * Use the Orbit program (Appendix 1, Section 2) to illustrate the growth of the investment in Problem 1. Compare the outcomes with Everest Savings and Mercury Savings. Make some comments.

10. * Explore the possibility of solving the problem proposed by Ann in connection with Example 1.1.2.

2. Definition of Discrete Dynamical Systems

After looking at the previous examples, we see that a discrete dynamical system is a relation of the form

$$\mathbf{x}_{n+1} = F(\mathbf{x}_n). \tag{1.1.16}$$

The function F(**x**) of (1.1.16) may contain a number of control parameters **a**= (a,b,c,...). In this case the function should be written F(**a**,**x**). However, as mentioned before, when the parameters are considered fixed, we may prefer the notation F(**x**) to F(**a**,**x**). How to arrive at (1.1.16), its meaning, use, and limitations are described briefly below. We assume that a given real dynamical process has to be studied, as in the three examples presented before.

• We identify the state variables of the system. For example, in the biological process involving the two competing species of whiteflies and black wasps, we are interested in knowing how each species affects the evolution of the other. Hence, two state variables are needed. We used the coordinates of the two-dimensional state vector **x**=(x,y) to represent the average number of whiteflies and black wasps per infested leaf. In general, the state vector **x** is q-dimensional, i.e., $\mathbf{x}\in \mathbf{R}^q$.

• We identify possible control parameters of the system. In this step we look for those parameters that affect the evolution of the state variables. In the first example we found two: the interest rate i and the number m of compounding periods per year. Hence the parameter vector **a** had two components, **a**=(i,m), $\mathbf{a}\in \mathbf{R}^2$. In the second example we proposed five parameters: a,b,c,d, and p. Hence, **a** had five components, **a**=(a,b,c,d,p), $\mathbf{a}\in \mathbf{R}^5$. In general, **a** has m components, i.e., $\mathbf{a}\in \mathbf{R}^m$, with m and q usually different.

• The next step is to determine the mathematical relations that translate the laws governing the evolution of the process. As suggested by (1.1.16) and as we have seen in the three examples, these relations are embodied by a function F which depends on the state vector **x** and on the parameter vector **a**, although we have agreed to consider **a** constant. The range of F is in the same space where the state vector **x** lives. In fact, in the first two examples the state variable was simply x, and F was a function from **R** into **R**, although in both cases we were interested only in $x\geq 0$. In the second example the state variables were two: x and y. Thus we considered the vector **x**=(x,y) and F was a function from $\mathbf{R}^2$ into $\mathbf{R}^2$, although once more we were interested only in $x\geq 0$ and $y\geq 0$.

In Examples 1.1.2 and 1.1.3 we have seen that determining the exact form of F might be a nontrivial task. The situation analyzed in Example 1.1.3 was particularly challenging and we may still have doubts that our model is the best for the problem proposed. In general, F is defined on some subsets $U\subset \mathbf{R}^m$ and $V\subset \mathbf{R}^q$ with range in $\mathbf{R}^q$. We can write $F : U\times V\rightarrow \mathbf{R}^q$, where U×V is the standard Cartesian product of the two sets U and V. Notice that F is identified by its q component functions $F_i : U\times V\rightarrow \mathbf{R}$. For example, in (1.1.13) we have $F_1(\mathbf{a},\mathbf{x})=ax-.008xy$, $F_2(\mathbf{a},\mathbf{x}) =cy+.04xy(a-.065y)$.

Determining the proper form of the function F is a central part of the modeling process. It is not an easy task. As a rule of thumb, one can say that there is an optimal range for the level of complexity of any model. Below this range, the model is biased toward the theoretical side, and above this range the model loses its synthetic advantage. Large areas of uncertainty remain between these two extremes, and it may be difficult, or even impossible, to design an "optimal model," namely, a model in which all advantages and key features are incorporated.

• We are now ready to write the dynamical system (1.1.16), which tells us how the (n+1)th state of the vector $\mathbf{x}$, denoted by $\mathbf{x}_{n+1}$, is obtained once the nth state $\mathbf{x}_n$ is known. The 0-state, $\mathbf{x}_0$, is called the **initial state**. It can be any vector $\mathbf{x}$ in the domain of F. The evolution of the system starting from $\mathbf{x}_0$ is given by the sequence

$$\mathbf{x}_0,\ \mathbf{x}_1 = F(\mathbf{x}_0),\ \mathbf{x}_2 = F(\mathbf{x}_1) = F(F(\mathbf{x}_0)),\ \mathbf{x}_3 = F(\mathbf{x}_2) = F(F(F(\mathbf{x}_0))),\ \ldots .$$

We usually write $F^2(\mathbf{x})$, $F^3(\mathbf{x})$, ... in place of $F(F(\mathbf{x}))$, $F(F(F(\mathbf{x})))$, ... Hence, we have

$$\mathbf{x}_{n+1} = F(\mathbf{x}_n)=F^{n+1}(\mathbf{x}_0). \tag{1.1.17}$$

Equality (1.1.17) will be used repeatedly throughout the book.

Definition 1.1.1
The sequence $\{\mathbf{x}_0, \mathbf{x}_1,..., \mathbf{x}_n,...\}$ is denoted by $O(\mathbf{x}_0)$ and is called the **orbit** or **trajectory** of the system starting from $\mathbf{x}_0$.

As remarked previously, we shall always assume that as an orbit $O(\mathbf{x}_0)$ evolves, the control parameters are kept constant, although we realize that this limitation may be unrealistic in real processes (see Example 1.1.2).

• The time interval between two successive states of an orbit is usually suggested by the real process itself. For example, $\mathbf{x}_{n+1}$ could be separated from $\mathbf{x}_n$ by one hour, one day, one week, one month, etc. In the examples presented before, the time interval was the compounding period (in Example 1.1.1) and one week (in Example 1.1.2). No time interval was provided for Example 1.1.3.

The reader is probably thinking that in real processes the time variable evolves in a continuous rather than a discrete manner. Our approach does not follow this continuous evolution. We **sample** the state of the system at fixed time intervals. This strategy can be better understood if the reader thinks about the standard practice of measuring body temperature every 6 to 8 hours, even though the temperature undergoes continuous changes. With proper choice of the time step, the technique provides very useful information on the behavior of the real process. The time intervals are always assumed to be equally spaced and are denoted by 0, 1, ..., n, ...

The following additional examples should make the reader more familiar and comfortable with the topics discussed so far.

Example 1.1.4 Assume that the value x_{n+1} of the state variable x at time n+1 depends on the value x_n at time n according to the relation

$$x_{n+1} = ax_n(1 - x_n).$$

The function that governs the system is $F(x)=ax(1-x)$, and q=m=1. Once more, we should really consider F as depending on the variable x and on the parameter a, namely $F(a,x)=ax(1-x)$. This dynamical system is called **logistic**. For modeling reasons it is normally required that $x \in [0,1]$ and $a \in [0,4]$. Hence U=[0,4] and V=[0,1]. The time step is left open.

Given an initial state x_0, we have

$$x_1 = F(x_0) = ax_0(1 - x_0),\ x_2 = F(x_1) = ax_1(1 - x_1).$$

Since $x_1 = ax_0(1-x_0)$, we obtain

$$x_2 = a(ax_0(1 - x_0))(1 - ax_0(1 - x_0)) = a^2x_0(1 - x_0) - a^3x_0^2(1 - x_0)^2 = F^2(x_0).$$

Hence, x_2 is computed by replacing x_0 with $F(x_0)$ in the definition of F. Similarly, x_3, x_4, ... are obtained by replacing x_0 with $F^2(x_0)$, $F^3(x_0)$,... in the definition of F (for the numerical aspect of this process, see Appendix 1, Section 2, Subsection 1).

Example 1.1.5 Assume that in a two-species population the size of each species at time n+1 depends on the size at time n according to the relations

$$\begin{cases} x_{n+1} = ax_n(1 - x_n - y_n) \\ y_{n+1} = bx_ny_n. \end{cases}$$

Then $\mathbf{x}=(x,y) \in \mathbf{R}^2$ and $\mathbf{a}=(a,b) \in \mathbf{R}^2$. The function governing the system is

$$F(\mathbf{x}) = F(x,y) = (ax(1 - x - y),\ bxy), \tag{1.1.18}$$

with component functions $F_1(\mathbf{x})=ax(1-x-y)$ and $F_2(\mathbf{x})=bxy$.

Technically, we should regard F as a function of $\mathbf{x}=(x,y)$ and $\mathbf{a}=(a,b)$ and write

$$F(\mathbf{a},\mathbf{x}) = F((a,b),\ (x,y)) = (F_1(\mathbf{a},\mathbf{x}),\ F_2(\mathbf{a},\mathbf{x})),$$

where $F_1(\mathbf{a},\mathbf{x})=ax(1-x-y)$ and $F_2(\mathbf{a},\mathbf{x})=bxy$. However, we can still use form (1.1.18) when we study the evolution of the two species. In fact, in this case, the parameters a and b are considered constant, although their exact value may not be specified.

The state variable and the parameters are assumed to be positive. Hence, $U=\{(a,b): a\geq 0, b\geq 0\}$, $V=\{(x,y): x\geq 0, y\geq 0\}$. The time interval between two successive states is left open.

Goals of this book

In this book we **do not study** the strategies and methods used to derive (1.1.16) from a real dynamical process. This topic belongs to a book on mathematical modeling and is not addressed here. Also, we do not normally investigate and discuss the length of the time step between successive states. The starting point of our study are relations of the form (1.1.16). We plan to achieve the following distinct but strictly related goals:

1. Analyze the behavior of (1.1.16) for different values of $\mathbf{x}_0$, considering the parameters fixed.

2. Study the changes in this behavior as the parameters are changed.

We provide now a more detailed description of the two goals and illustrate, with a simple example, how they can be achieved.

• The first task is to determine the behavior of the orbits of (1.1.16) (we also say "the orbits of F") for different choices of the initial condition $\mathbf{x}_0$. We focus on the long-term behavior of the orbits. In other words, we look at what an orbit does for n very large, or more rigorously, as $n \to \infty$. We do not intend to neglect the "transient behavior," namely, what is happening at the beginning of an orbit, i.e., for small values of n. In fact, frequently we will be looking at an entire orbit $O(\mathbf{x}_0)$. Our main interest, however, is to determine the fate of $\mathbf{x}_n$ as $n \to \infty$. We call this part state space analysis of the dynamical system.

• The second task is to study the changes in the characteristics and behavior of (1.1.16) brought to bear by changes in the parameters. We call this part parameter space analysis of the dynamical system. During this stage of our investigation we should use, for obvious reasons, $F(\mathbf{a},\mathbf{x})$ rather than $F(\mathbf{x})$.

When **a** is a scalar (denoted by a, $a \in I$, $I \subset \mathbf{R}$, I an interval) the set of functions $\{F(a,\mathbf{x}) : a \in I\}$ is called a **one-parameter** family of maps. We shall pay particular attention to the properties of such families.

The following simple example illustrates how both tasks can be accomplished.

Example 1.1.6 Let $F(x)=ax+2$. This is a one-parameter family of maps and we can write $\{F(a,x)=ax+2, a \in \mathbf{R}\}$. To illustrate the two types of investigations mentioned above, first consider the parameter a fixed, for example, a=3/4. Hence, $F(x)=(3/4)x+2$. Notice that if we select $x_0=8$ we have $x_1=(3/4)8+2=8$. Thus, $x_1=x_0$. We call this value of x a **fixed point** of F since 8=F(8), and we denote it by x_s.

Let us study the orbits of the system $x_{n+1}=(3/4)x_n+2$. Using the fixed point 8 we can write

$$x_{n+1} = (3/4)(x_n - 8) + 8. \tag{1.1.19}$$

Given x_0, we have

$$x_1 = (3/4)(x_0 - 8) + 8,\ x_2 = (3/4)(x_1 - 8) + 8 = (3/4)^2(x_0 - 8) + 8, \dots .$$

In the expression for x_2 we have replaced x_1 with $(3/4)(x_0-8)+8$. After n+1 steps we arrive at

$$x_{n+1} = (3/4)^{n+1}(x_0 - 8) + 8. \tag{1.1.20}$$

Since $(3/4)^{n+1} \to 0$ as $n \to \infty$ we obtain that $x_{n+1} \to 8$ as $n \to \infty$. In other words, every orbit converges to the fixed point $x_s=8$ no matter what the initial condition $x_0 \in \mathbf{R}$ might be.

Let us investigate what changes the system goes through as the parameter a is changed. First, the fixed point is now the solution of $x=ax+2$. We see that there is no fixed point when a=1, and there is only one, $x_s(a)=2/(1-a)$, when a≠1. We have written $x_s(a)$ since the fixed point "changes" with a. When a≠1 we can use (1.1.19) with 3/4 replaced by a and 8 replaced by $x_s(a)$. By writing x_s instead of $x_s(a)$, the relation (1.1.19) assumes the form

$$x_{n+1} = a(x_n - x_s) + x_s, \tag{1.1.21}$$

and (1.1.20) becomes

$$x_{n+1} = a^{n+1}(x_0 - x_s) + x_s. \tag{1.1.22}$$

With $|a|<1$ every orbit converges to $x_s=2/(1-a)$ since $a^{n+1}\to 0$ as $n\to\infty$. With $|a|>1$ the orbit of every initial state $x_0\neq x_s$ goes to infinity. For a=–1 the fixed point is $x_s=1$, and given $x_0\neq 1$, we have $x_1=-x_0+2$, $x_2=-x_1+2=-(-x_0+2)+2=x_0$, $x_3=-x_2+2=-x_0+2=x_1$. Hence the orbit is x_0, $x_1=-x_0+2$, x_0, $-x_0+2$, x_0, ... We say that $O(x_0)$ is periodic of period 2. For a=1 we have $x_1=x_0+2$, $x_2=x_1+2=x_0+4$, $x_3=x_2+2=x_0+8$,... . Every orbit goes to $+\infty$.

We have completed the parameter space analysis of our system by specifying how the system $x_{n+1}=ax_n+2$ behaves for all possible values of the parameter $a\in\mathbf{R}$.

Example 1.1.6 gives a little taste of how the two goals can be achieved. It also shows that the two tasks, although different, are strictly related. In many cases we will not make an effort to identify which task we are pursuing.

Sometimes (1.1.16) is replaced by slightly more general forms, like

$$\mathbf{x}_{n+1} = F(\mathbf{x}_n, \mathbf{x}_{n-1}). \tag{1.1.23}$$

Equation (1.1.23) tells us that the state of the orbit at time n+1 depends directly from its states at time n and n–1. We call (1.1.23) a discrete dynamical system with a delay of a one-time unit. We can eliminate the delay by increasing the dimension of the system. We set $\mathbf{y}_{n+1}=\mathbf{x}_n$ and rewrite (1.1.23) as follows:

$$\begin{cases} \mathbf{x}_{n+1} = F(\mathbf{x}_n, \mathbf{y}_n) \\ \mathbf{y}_{n+1} = \mathbf{x}_n. \end{cases} \tag{1.1.24}$$

By setting $G(\mathbf{x}, \mathbf{y})=(F(\mathbf{x},\mathbf{y}),\mathbf{x})$ and $\mathbf{z}=(\mathbf{x},\mathbf{y})$ we arrive at

$$\mathbf{z}_{n+1} = G(\mathbf{z}_n), \tag{1.1.25}$$

which is of the same form as (1.1.16). More complicated cases are also possible, and they are solved analogously, as the following examples show.

Example 1.1.7 Let $x_{n+1}=ax_{n-1}(1-x_n)$. This one-dimensional dynamical system, containing a delay of a one-time unit, can be replaced by a two-dimensional system with no delay by setting $y_{n+1}=x_n$. We obtain the system

$$\begin{cases} x_{n+1} = ay_n(1-x_n) \\ y_{n+1} = x_n. \end{cases}$$

In vector form we get $\mathbf{x}_{n+1}=G(\mathbf{x}_n)$ where $\mathbf{x}=(x,y)$ and $G(\mathbf{x})=(ay(1-x),x)$.

Example 1.1.8 Let $x_{n+1}=ax_n(1-x_{n-2})$. The dynamical system contains a delay of two-time units. We can replace it with a three-dimensional system with no delay. Set $y_{n+1}=x_n$ and $z_{n+1}=y_n$. We obtain

$$\begin{cases} x_{n+1} = ax_n(1-z_n) \\ y_{n+1} = x_n \\ z_{n+1} = y_n. \end{cases}$$

In vector form we have $\mathbf{x}_{n+1}=G(\mathbf{x}_n)$, $\mathbf{x}=(x,y,z)$, $G(\mathbf{x})=(ax(1-z),x,y)$.

Example 1.1.9 Assume that $\mathbf{x}_{n+1}=F(\mathbf{x}_n,\mathbf{x}_{n-1},\mathbf{x}_{n-2})$. Let $\mathbf{y}_{n+1}=\mathbf{x}_n$ and $\mathbf{w}_{n+1}=\mathbf{y}_n$. We obtain the system

$$\begin{cases} \mathbf{x}_{n+1} = F(\mathbf{x}_n,\mathbf{y}_n,\mathbf{w}_n) \\ \mathbf{y}_{n+1} = \mathbf{x}_n \\ \mathbf{w}_{n+1} = \mathbf{y}_n. \end{cases}$$

Setting $\mathbf{z}=(\mathbf{x},\mathbf{y},\mathbf{w})$, $G(\mathbf{x},\mathbf{y},\mathbf{w})=(F(\mathbf{x},\mathbf{y},\mathbf{w}),\mathbf{x},\mathbf{y})$, we can write $\mathbf{z}_{n+1}=G(\mathbf{z}_n)$ which is of the form (1.1.16).

In this book we always assume that $\mathbf{x}_{n+1}$ depends directly **only on** $\mathbf{x}_n$. Systems with delay, i.e., when $\mathbf{x}_{n+1}$ depends directly on one or more states of the form $\mathbf{x}_{n-k}$, $k\geq 1$, are replaced by higher-dimensional systems with no delay.

Problems

1. Let $F(\mathbf{a},x)=x(ax-b)$. The dynamical system governed by F is $x_{n+1}=x_n(ax_n-b)$. Identify the state variables and the control parameters.

2. Let $\mathbf{x}_{n+1}=F(\mathbf{a},\mathbf{x}_n)$, where $F(\mathbf{a},\mathbf{x})=(ax(1-x-y),bxy)$. Write the dynamical system explicitly (namely, $x_{n+1}=\ldots,y_{n+1}=\ldots$), pointing out its state variables, its parameters, and its component functions.

3. Let $F(\mathbf{a},\mathbf{x})=(a(x-y),bx-y-xz,xy-cz)$. Write explicitly the discrete dynamical system governed by F. Identify the state variables, the control parameters, and the component functions.

4. Let $F(a,x)=-ax^2+1$. Write explicitly $F(a,F(a,x))$, usually denoted by $F^2(a,x)$ (you can use the programs of Appendix 1, Section 2).

5. Let $F(\mathbf{a},x)=x(a-bx)$. Write explicitly $F(\mathbf{a},F(\mathbf{a},x))=F^2(\mathbf{a},x)$.

6. Let $F(a,x) = -ax^2+1$. What is the degree in x of the polynomial $F^3(a,x)$? What about the degree in a? Generalize to F^n, the nth iterate of F.

7. Let $x_{n+1}=x_n+x_{n-1}$. This is the Fibonacci sequence and it can be regarded as a one-dimensional discrete dynamical system with a delay of one-time unit. Replace it with a two-dimensional system with no delay.

8. Let $x_{n+1}=ax_{n-1}(1-x_{n-2})$. Replace this one-dimensional system with a three-dimensional system with no delay.

9. Let

$$\begin{cases} x_{n+1} &= 2x_n - .2x_{n-1}y_{n-1} \\ y_{n+1} &= y_n + .1x_{n-1}y_{n-1}. \end{cases}$$

Replace the two-dimensional system having a delay of a one-time unit with a four-dimensional system with no delay.

10. Let $\mathbf{x}_{n+1}=F(\mathbf{x}_n)$ with F given by (1.1.12). Assume that at certain point in time the y species becomes extinct. Denote by x_0 the size of x at that time. Verify that x will grow exponentially from that point on (recall that $a>1$). Is this unrestricted growth reasonable?

11. Let $\mathbf{x}_{n+1}=F(\mathbf{x}_n)$ with F given by (1.1.12). Assume that at a certain point in time the x species becomes extinct. Denote by y_0 the size of y at that time. Verify that y will become extinct (recall that $0<c<1$). Is this outcome reasonable?

Section 2. STATIONARY STATES AND PERIODIC ORBITS

1. Stationary States

Definition 1.2.1
A point $\mathbf{x}_0$ is called a **stationary state** of (1.1.16) if

$$\mathbf{x}_1 = F(\mathbf{x}_0) = \mathbf{x}_0. \tag{1.2.1}$$

The symbol $\mathbf{x}_s$ will be used to denote a stationary state. Each $\mathbf{x}_s$ can be regarded either as a state of the dynamical system $\mathbf{x}_{n+1}=F(\mathbf{x}_n)$ satisfying (1.2.1) or as a vector $\mathbf{x}$ satisfying the system of equations $\mathbf{x}=F(\mathbf{x})$. For this reason we shall also call $\mathbf{x}_s$ a **fixed point** of F.

Example 1.2.1 Every stationary state of the system $x_{n+1}=ax_n(1-x_n)$ must satisfy the equation $x=ax(1-x)$. We see that $x_s=0$ is a stationary state regardless of the value of a. Assuming that $x\neq0$, we obtain

$$1 = \mathrm{a}(1 - x) \quad \text{or} \quad \mathrm{a}x = \mathrm{a} - 1. \tag{1.2.2}$$

Therefore, a second stationary state is $x_s(\mathrm{a})=(\mathrm{a}-1)/\mathrm{a}$. We have written $x_s(\mathrm{a})$ since the point changes with a. For every a we can visualize the fixed points of $F(x)=\mathrm{a}x(1-x)$ since they are given by the intersection of the graph of F with the line $y=x$ (see Fig. 1.2.1 for a=3).

Example 1.2.2 The stationary states of $(x_{n+1},y_{n+1})=(\mathrm{a}x_n-\mathrm{b}x_ny_n,\mathrm{c}y_n+\mathrm{d}x_ny_n)$ are the solutions of the system

$$\begin{cases} x = \mathrm{a}x - \mathrm{b}xy \\ y = \mathrm{c}y + \mathrm{d}xy. \end{cases}$$

The point (0,0) is a solution regardless of the values of a,b,c, and d. Assume that $0<\mathrm{c}<1$ and $\mathrm{a}>1$. It follows that $x=0$ if and only if $y=0$. Thus, additional stationary states satisfy the inequality $xy\neq0$. Dividing the first equation by x and the second by y, we obtain the system

$$\begin{cases} 1 = \mathrm{a} - \mathrm{b}y \\ 1 = \mathrm{c} + \mathrm{d}x. \end{cases} \tag{1.2.3}$$

Hence, $x = (1-\mathrm{c})/\mathrm{d}$, $y = (\mathrm{a}-1)/\mathrm{b}$ or $\mathbf{x}_s(\mathbf{a})=((1-\mathrm{c})/\mathrm{d},(\mathrm{a}-1)/\mathrm{b})$.

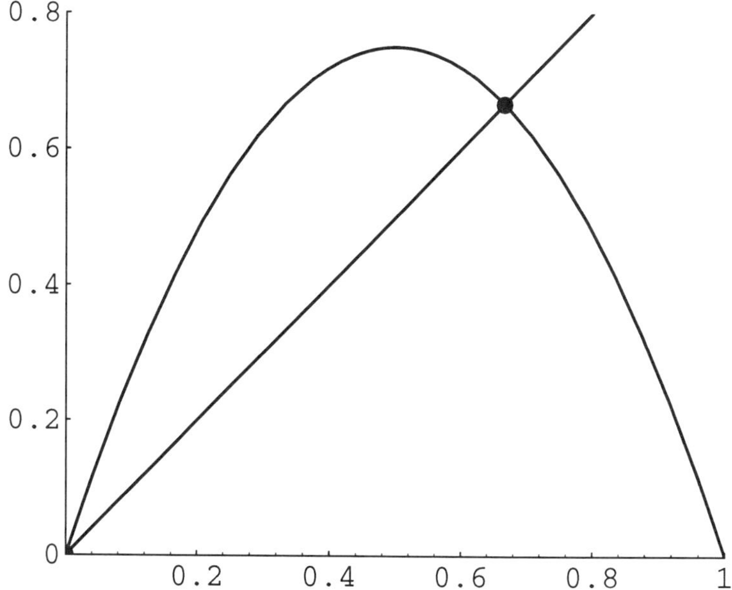

Fig. 1.2.1 The fixed points of the dynamical system governed by the function $F(x)=3x(1-x)$ are the intersections of the graph of F with the line $y=x$.

Notice that any orbit starting from a stationary state $\mathbf{x}_s$ will not move away from it. This situation is possible in theory, but not in practice, since every process undergoes small fluctuations which are normally not accounted for in a model. These fluctuations will force the system out of $\mathbf{x}_s$. This observation, together with others which will be discussed later, motivates the introduction of the idea of **stability**. Some fundamental definitions are presented in this chapter. Other concepts related to stability will be discussed in later chapters.

Stable stationary states

Definition 1.2.2
A stationary state $\mathbf{x}_s$ is **stable** if for every r>0 there exists δ>0 such that

$$\|\mathbf{x}_0 - \mathbf{x}_s\| \le \delta \text{ implies that } \|\mathbf{x}_n - \mathbf{x}_s\| \le r \text{ for all } n \ge 1. \tag{1.2.4}$$

In other words, once we have chosen how close we want to remain to $\mathbf{x}_s$ in the future (choice of r), we can find how close we must start at the beginning (existence of δ).

The symbol $\|\mathbf{x}\|$ denotes the "Euclidean norm" of $\mathbf{x}$, defined by

$$\|\mathbf{x}\| = \|(x_1, ..., x_q)\| = (x_1{}^2 + ... + x_q{}^2)^{.5}. \tag{1.2.5}$$

In **R** the Euclidean norm is the usual absolute value.

Example 1.2.3 Let $F(x)=1-x$. The point $x_s=.5$ is the only fixed point of F. For every other initial state x_0 we have $x_1=F(x_0)=1-x_0$, $F(F(x_0))=F(1-x_0)=x_0$. Thus, $|x_n-.5|=|x_0-.5|$ for all n=1,2,... . The fixed point $x_s=.5$ is stable (see Fig. 1.2.2). In the definition of stability the number δ can be selected equal to r.

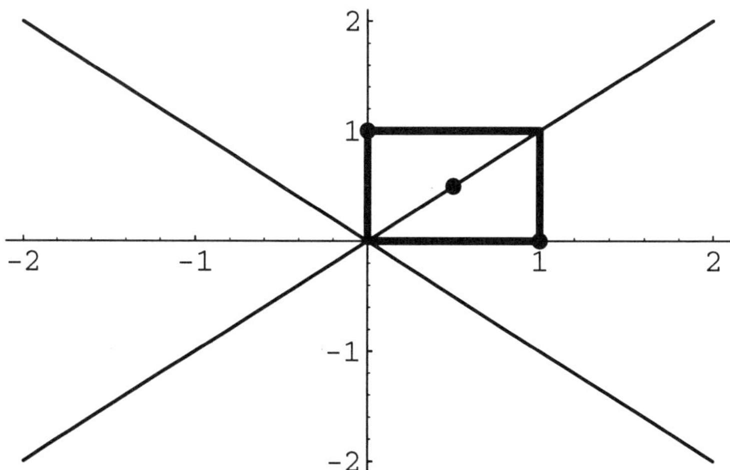

Fig. 1.2.2 This graph shows the fixed point .5 and graphically illustrates its stability (see Example 1.2.3).

Example 1.2.4 Let $F(\mathbf{x})=F(x,y)=(x\cos\theta-y\sin\theta,\ x\sin\theta+y\cos\theta)$, $\theta\in(0,2\pi)$. This is a counterclockwise rotation of the plane of an angle θ. Since $\theta\in(0,2\pi)$, every point $\mathbf{x}_0$ of the plane is moved from its initial position except the origin $\mathbf{0}$, which is the only fixed point of the rotation. Setting $\mathbf{x}_0=(x_0,y_0)\neq\mathbf{0}$ we obtain that $\mathbf{x}_1=F(\mathbf{x}_0)$ is given by

$$\begin{cases} x_1 = x_0\cos\theta - y_0\sin\theta \\ y_1 = x_0\sin\theta + y_0\cos\theta. \end{cases} \tag{1.2.6}$$

The new state $\mathbf{x}_2=F(\mathbf{x}_1)$ can be obtained from $\mathbf{x}_1$ with a rotation of θ, or from $\mathbf{x}_0$ with a rotation of 2θ. Hence (see Fig. 1.2.3),

$$\begin{cases} x_2 = x_1\cos\theta - y_1\sin\theta = x_0\cos2\theta - y_0\sin2\theta \\ y_2 = x_1\sin\theta + y_1\cos\theta = x_0\sin2\theta + y_0\cos2\theta. \end{cases}$$

In general, $\mathbf{x}_n$ is obtained from $\mathbf{x}_0$ with a rotation of $n\theta$, i.e., the nth iterate of F is

$$F^n(x,y) = (x\cos n\theta - y\sin n\theta,\ x\sin n\theta + y\cos n\theta). \tag{1.2.7}$$

Consequently,

$$\begin{cases} x_n = x_0\cos n\theta - y_0\sin n\theta \\ y_n = x_0\sin n\theta + y_0\cos n\theta. \end{cases} \tag{1.2.8}$$

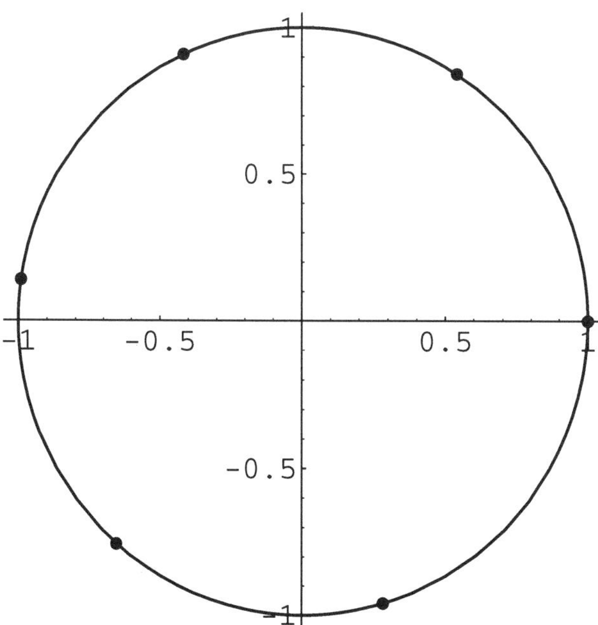

Fig. 1.2.3 First six states of the orbit O(1,0) for the map of Example 1.2.4 with $\theta=1$.

Using the Euclidean norm we have

$$\|F^n(\mathbf{x}_0)\|^2 = (x_0\cos n\theta - y_0\sin n\theta)^2 + (x_0\sin n\theta + y_0\cos n\theta)^2 = (x_0)^2 + (y_0)^2 = \|\mathbf{x}_0\|^2.$$

Hence, all states $\mathbf{x}_n$, n=0,1,... of the orbit $O(\mathbf{x}_0)$ are at the same distance from the fixed point $\mathbf{x}_s=\mathbf{0}$, since $\|\mathbf{x}_n\|=\|\mathbf{x}_0\|$. The origin is stable with δ=r.

Definition 1.2.3
A stationary state $\mathbf{x}_s$ that is not **stable** is said to be **unstable**.

Hence $\mathbf{x}_s$ is unstable if there exists r>0 such that for every $\delta>0$ we can find $\mathbf{x}_0$ with the following two properties:

- $\|\mathbf{x}_0 - \mathbf{x}_s\| \leq \delta$.
- $\|\mathbf{x}_n - \mathbf{x}_s\| > r$ for some $n \geq 1$.

Example 1.2.5 Let F : **R**→**R** be defined by

$$F(x) = \begin{cases} .5x & \text{for} \quad x \leq 0 \\ 2x & \text{for} \quad x > 0. \end{cases}$$

Every orbit starting to the left of the unique stationary state x_s=0 will converge to 0, while every orbit starting to the right of 0 will go to infinity. Thus, 0 is an unstable stationary state. We could choose r=1 in the definition of instability. Then, for every $\delta>0$ we can select any point $x_0\in(0,\delta)$. Since $x_n=2^n x_0$ it is clear that $x_n>1$ for some $n\geq1$. Some authors call 0 **semistable**.

Problems

1. Let $F(x)=2x/3+1$. Show that x=3 is a stationary state of $x_{n+1}= F(x_n)$.

2. Find the stationary states of the system governed by $F(x)=-1.5x^2+1$.

3. Find the fixed points of the system governed by $F(a,x)=-ax^2+1$. For what values of a does the function F have none, one, or two fixed points? (Hint: Consider the case a=0 separately.)

4. Let $F(x,y)=(-y,x)$. Verify that the origin is a stable stationary state.

5. Let $F(\mathbf{x})=\mathbf{x}+(1,2)$. How many components does the vector $\mathbf{x}$ have? Let $\mathbf{x}_{n+1}=F(\mathbf{x}_n)$. Does the system have any stationary states?

6. Let $F(\mathbf{x})=(x\cos1-y\sin1+1, x\sin1+y\cos1+2)$. Show that the discrete dynamical system $\mathbf{x}_{n+1}=F(\mathbf{x}_n)$ has one and only one fixed point. (Hint: Notice that the determinant of the matrix of the coefficients is different from 0.)

7. * Let $F(x)=x^3-2x+1$. Show that the dynamical system $x_{n+1}=F(x_n)$ has three stationary states (see Appendix 1, Section 4).

8. Let

$$\begin{cases} x_{n+1} = ax_n(1 - x_n - y_n) \\ y_{n+1} = bx_n y_n. \end{cases}$$

Find the stationary states of this system.

9. * Let $F(x)=\cos x$. Show that F has one and only one fixed point. Find the point with five correct decimal places (see Appendix 1, Section 4).

10. Let $F(x)=x^2$. Show that $x_s=0$ is stable and $x_s=1$ is unstable.

11. Let $\mathbf{x}=(x,y)$ and $F(\mathbf{x})=0.5\mathbf{x}+(1,2)$. Find the sfixed point of the system $\mathbf{x}_{n+1}=F(\mathbf{x}_n)$ and show that it is stable.

12. Let $\mathbf{x}=(x,y)$ and $F(\mathbf{x})=2\mathbf{x}+(1,2)$. Find the fixed point of the system $\mathbf{x}_{n+1}=F(\mathbf{x}_n)$ and show that it is unstable.

2. **Periodic Orbits**

Definition 1.2.4
An orbit $O(\mathbf{x}_0)$ of (1.1.16) is said to be periodic of period $p \geq 1$ if

$$\mathbf{x}_p = \mathbf{x}_0. \tag{1.2.9}$$

The smallest integer p such that (1.2.9) holds is called the **minimum period** (or simply the **period**) of the orbit.

The choice of the attribute "minimum" is motivated by the fact that the orbit cannot be decomposed into smaller loops. We also say that $\mathbf{x}_0$ is a periodic point of period p. A fixed point can be regarded as a periodic point of period 1. Unless otherwise stated, when we say that $O(\mathbf{x}_0)$ is a periodic orbit of period p or $\mathbf{x}_0$ is a periodic point of period p, we mean that p is the **period**.

Since $\mathbf{x}_p=F^p(\mathbf{x}_0)$ we can write (1.2.9) in the form $F^p(\mathbf{x}_0)=\mathbf{x}_0$. Whenever p is the period of $O(\mathbf{x}_0)$ we also have $F^m(\mathbf{x}_0)\neq\mathbf{x}_0$ for every $m<p$. Every point $\mathbf{x}_i$, $i=0,1,\ldots,p-1$, of the periodic orbit $O(\mathbf{x}_0)$ of period p is periodic of the same period. Thus, $O(\mathbf{x}_0)$ contains exactly p distinct periodic points of period p.

In this book a periodic orbit of period p will usually be identified in two ways:

- We may simply say "let $O(\mathbf{x}_0)$ be a periodic orbit of period p"; or
- "Let $\{\mathbf{x}_0,\ldots,\mathbf{x}_{p-1},\ldots\}$ be a periodic orbit of period p."

Sometimes, for the sake of simplicity, we use the symbol $\{\mathbf{x}_0,\ldots,\mathbf{x}_{p-1}\}$ to denote a periodic orbit of period p.

Example 1.2.6 Consider the system $x_{n+1}=ax_n(1-x_n)$, with $x\geq 0$, $a\in[0,4]$. The dynamical system is governed by the map $F(x)=ax(1-x)$. A point x_0 is periodic of period 2 if

$$x_2 = x_0 \quad \text{and} \quad x_1 \neq x_0.$$

Using the function F we can write the two conditions in the form $x_0=F^2(x_0)$ (F^2 is the second iterate of F) and $x_0\neq F(x_0)$. The inequality ensures that x_0 is not a fixed point of F. To find those values of x that satisfy both conditions, we can proceed as follows. Let x_0=p and x_1=q. From $x_1=F(x_0)$ and $x_2=F(x_1)=x_0$ we derive

$$q = ap - ap^2 \quad \text{and} \quad p = aq - aq^2. \tag{1.2.10}$$

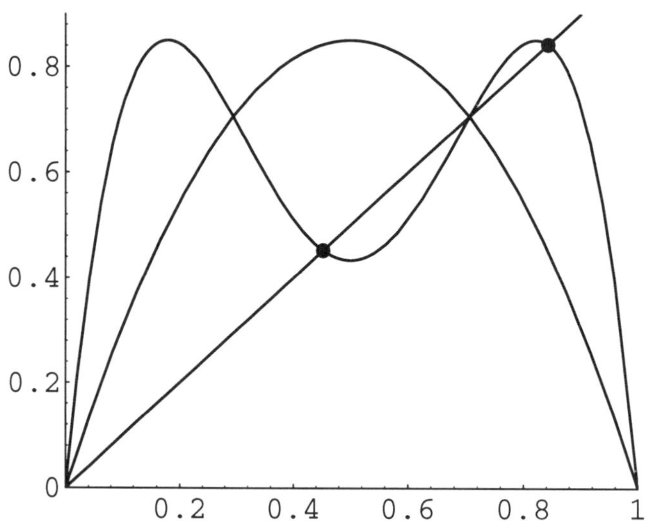

Fig. 1.2.4 The periodic points of period 2 of $F(x)=3.4x(1-x)$ are the intersections of the graph of $F^2(x)=F(F(x))$ with the line $y=x$ at those points where $F(x)\neq x$.

Subtracting the second equality of (1.2.10) from the first gives

$$q - p = -a(q - p) + a(q - p)(q + p).$$

Since $x_1\neq x_0$ we have $q-p\neq 0$, and we can divide the equality above by q–p to obtain $q+p=(1+a)/a$. Adding the two equalities (1.2.10) we arrive at

$$q + p = a(q + p) - a(q + p)^2 + 2apq.$$

Substituting $q+p=(1+a)/a$ and simplifying gives $pq=(1+a)/a^2$. Since we know the sum p+q and the product pq, we can determine p and q by solving the quadratic equation

$$m^2 - (1 + a)m/a + (1 + a)/a^2 = 0.$$

In fact, the roots of this equation have the properties $m_1+m_2= (1+a)/a$ and $m_1m_2 = (1+a)/a^2$. Hence,

$$p = m_1 = \frac{a + 1 + \sqrt{a^2 - 2a - 3}}{2a}, \qquad q = m_2 = \frac{a + 1 - \sqrt{a^2 - 2a - 3}}{2a}.$$

We see that no periodic orbit of period 2 exists for $a \leq 3$, and exactly one exists for $a > 3$ (recall that $a \in [0,4]$). Notice that both p and q are fixed points of F^2.

In general, every state $\mathbf{x}_i$ of a periodic orbit $O(\mathbf{x}_0)$ of period p is a stationary state of the pth iterate F^p of F, i.e.,

$$\mathbf{x}_i = F^p(\mathbf{x}_i), \quad i = 0, 1, 2, ..., p - 1. \tag{1.2.11}$$

Example 1.2.7 Let a=4 in the map of Example 1.2.6. Then

$$x_0 = p = (5 + \sqrt{5})/8 \quad \text{and} \quad x_1 = q = (5 - \sqrt{5})/8.$$

We can easily verify that

$$F^2((5 + \sqrt{5})/8) = F((5 - \sqrt{5})/8) = (5 + \sqrt{5})/8,$$

$$F^2((5 - \sqrt{5})/8) = F((5 + \sqrt{5})/8) = (5 - \sqrt{5})/8.$$

Stable periodic orbits

Definition 1.2.5
A periodic orbit of period p $\{\mathbf{x}_0,...,\mathbf{x}_{p-1},...\}$ is **stable** if each point $\mathbf{x}_i$, $i=0,1,...,p-1$ is a stable stationary state of the dynamical system governed by F^p.

The stability of $\mathbf{x}_0$ as a fixed point of F^p guarantees (it is actually equivalent to) the stability of all remaining points $\mathbf{x}_1,...,\mathbf{x}_{p-1}$ whenever F is continuous. This result will be established in Chapters 2 and 6 for the smaller but important class of differentiable maps.

Example 1.2.8 Let $F(x)=1-x$. Every point $x \neq .5$ is periodic of period 2. For example, F(0)=1 and F(1)=0. The periodic orbit {0,1} is stable. In fact, F^2 is the identity map, for which each point is a stable stationary state.

Example 1.2.9 Let $F(\mathbf{x})=F(x,y)=(-y,x)$. Every point $\mathbf{x} \neq \mathbf{0}$ is periodic of period 4. For example, from $\mathbf{x}_0=(1,0)$ we obtain $\mathbf{x}_1=(0,1)$, $\mathbf{x}_2=(-1,0)$, $\mathbf{x}_3=(0,-1)$, $\mathbf{x}_4=\mathbf{x}_0$, ... The orbit is stable, since F^4 is the identity map for which each point is a stable stationary state.

Definition 1.2.6
A periodic orbit of period p $\{\mathbf{x}_0,...,\mathbf{x}_{p-1},...\}$ which is not stable is said to be **unstable**.

Hence, the orbit is unstable if one of its states $\mathbf{x}_i$ is an unstable fixed point of F^p. The instability of $\mathbf{x}_0$ as a fixed point of F^p guarantees (it is actually equivalent to) the instability of all remaining points $\mathbf{x}_1,...,\mathbf{x}_{p-1}$ whenever F is continuous. This result will be established in Chapters 2 and 6 for the smaller but important class of differentiable maps.

The following is an example of an unstable periodic orbit.

Example 1.2.10 Let

$$F(x)=\begin{cases} -x & \text{for } -1\le x\le 1 \\ -2x+1 & \text{for } x>1 \\ -2x-1 & \text{for } x<-1. \end{cases}$$

Notice that F(1)=–1 and F(–1)=1. The periodic orbit {1,–1} is unstable. In fact, for $x>1$ we have $F(F(x))=F(-2x+1)=-2(-2x+1)-1=4(x-1)+1$. Hence, the orbit of a point $x_0>1$ according to the second iterate of F is given by

$$x_1 = 4(x_0-1)+1,\ x_2 = 4(x_1-1)+1 = 4^2(x_0-1)+1,\dots .$$

In general, we obtain $x_n=4^n(x_0-1)+1$. Hence, the orbit is going to $+\infty$ as $n\to\infty$. Thus, the state 1 is an unstable fixed point for the second iterate of F, and according to Definition 1.2.6, the periodic orbit {1,–1} is unstable.

Similarly, the periodic orbit {–1,1} is unstable. There is no need to verify this since we have already seen that the state 1 is an unstable fixed point of the second iterate of F. The reader should notice that the orbits starting at points $|x_0|<1$ do not move away from {1,–1}.

Problems

1. Find a periodic orbit of period 2 of the map $F(x)=-2x^2+1$.

2. Let $F(x)=-ax^2+1$. For what values of the parameter a does the function F have none or one periodic orbit of period 2? Are there any values of a for which there is more than one periodic orbit of period 2?

3. Assume that a dynamical system governed by a map $F(\mathbf{x})$ has exactly two periodic points of period 2, $\mathbf{x}_1$ and $\mathbf{x}_2$. Prove that $F(\mathbf{x}_1)=\mathbf{x}_2$ and $F(\mathbf{x}_2)=\mathbf{x}_1$.

4. Let $F(x,y)=(-y,x)$. Verify that every nonstationary orbit is periodic of period 4 and stable.

5. Verify that each periodic point of period p is a stationary state of F^{np} for every integer n=1, 2, ...

6. Assume that **x** is a periodic point of period 2 and of period 3 for a dynamical system governed by a function F. Prove that **x** is a stationary state.

7. Let $F(\mathbf{x})=F(x,y,z)=(y,z,-x)$. Find the periodic orbit $O(\mathbf{x}_0)$, where $\mathbf{x}_0=(1,1,2)$ and show that it is stable.

8. Let $F(x)=-ax^2+1$ and let $x_0=1$. Find x_1 and x_2. Using the chain rule determine $(d/dx)F^3(x)$ at the point $x=1$. Recall that by $F^3(x)$ we mean the third iterate of $F(x)$.

9. Assume that x_0 is a periodic point of period 2 for a dynamical system governed by a differentiable function F : **R**→**R**. Prove that the derivatives of the second iterate of F at x_0 and at x_1=F(x_0) are the same.

10. Assume that x_0 is a periodic point of period 3 for a dynamical system governed by a differentiable function F : **R**→**R**. Prove that the derivatives of the third iterate of F at x_0, x_1=F(x_0), and x_2=F(x_1) are the same.

11. Let F(x)=2x–[2x], where [x] is the greatest integer contained in x. Find a periodic point of period 3 in the interval [0,1].

12. Let F(x)=2x–[2x]. Prove that F has in [0,1] only two periodic points of period 2.

13. Let F(**x**)=F(x,y)=($x$cos θ–$y$sin θ,$x$sin θ+$y$cos θ), with θ=π/4. F is a rotation of the plane around the origin. Every nonstationary orbit is periodic of the same period. Find the period.

14. Let F be as in Problem 13, but with θ=3π/5. Every nonstationary orbit is periodic of the same period. Find the period.

Section 3. CHAOTIC DYNAMICAL SYSTEMS

1. Limit Points, Limit Sets, and Aperiodic Orbits

The orbit O($\mathbf{x}_0$) of the dynamical system

$$\mathbf{x}_{n+1} = F(\mathbf{x}_n) \tag{1.3.1}$$

has infinitely many states, namely, $\mathbf{x}_0$, $\mathbf{x}_1$, $\mathbf{x}_2$,..., $\mathbf{x}_n$,... . These states may not be all distinct. They are all equal when $\mathbf{x}_0$=$\mathbf{x}_s$ is a fixed point and there are only p distinct when the orbit O($\mathbf{x}_0$) is periodic of period p.

Definition 1.3.1
A point **z** is said to be a **limit point** of O($\mathbf{x}_0$) if there exists a subsequence $\{\mathbf{x}_{n_k}$: k=0,1, ...} of O($\mathbf{x}_0$) such that $\|\mathbf{x}_{n_k} - \mathbf{z}\| \to 0$ as k→∞.

A stationary orbit has only one limit point: $\mathbf{x}_s$. A periodic orbit O($\mathbf{x}_0$) of period p has exactly p limit points: $\mathbf{x}_0$,...,$\mathbf{x}_{p-1}$.

Example 1.3.1 In the system governed by the map F(x)=x(1–x), the point 0 is the only limit point of the stationary orbit O(0). It is also a limit point of the orbit x_0 = 1/2, x_1 = F(1/2) = 1/4, x_2 = F(1/4) = 3/16, ...

Example 1.3.2 In the system governed by $F(x)=1-x$ the point 1/4 is a limit point of the periodic orbit $x_0=1/4$, $x_1=3/4$, $x_2=1/4=x_0,\dots$.

Example 1.3.3 In the system governed by the map $F(x,y)=(y,-x)$ the limit points of the orbit $O((1,0))$ are 4: $\mathbf{x}_0=(1,0)$, $\mathbf{x}_1=(0,-1)$, $\mathbf{x}_2=(-1,0)$, $\mathbf{x}_3=(0,1)$.

Definition 1.3.2
The **limit set** $L(\mathbf{x}_0)$ of the orbit $O(\mathbf{x}_0)$ is the set of all limit points of the orbit.

Some properties of the limit set of an orbit will be studied in later chapters. In the meantime, the following is a fundamental equality between $L(\mathbf{x}_0)$ and its image under the continuous function F which governs the dynamical system:

$$F(L(\mathbf{x}_0)) = L(\mathbf{x}_0)\,, \quad \text{with } F(L(\mathbf{x}_0)) = \{F(\mathbf{z}) : \mathbf{z} \in L(\mathbf{x}_0)\}. \tag{1.3.2}$$

In other words, for every $\mathbf{z}\in L(\mathbf{x}_0)$ we have:

- $F(\mathbf{z}) \in L(\mathbf{x}_0)$, i.e. $F(L(\mathbf{x}_0)) \subset L(\mathbf{x}_0)$.
- There exists $\mathbf{w} \in L(\mathbf{x}_0)$ such that $F(\mathbf{w}) = \mathbf{z}$, i.e., $L(\mathbf{x}_0) \subset F(L(\mathbf{x}_0))$.

It is not hard to show that whenever $L(\mathbf{x}_0)$ has only finitely many points p, they actually constitute a periodic orbit of period p of the system (see Chapter 6, Section 1, Problems 5 and 6). This observation motivates the following definition.

Definition 1.3.3
An orbit $O(\mathbf{x}_0)$ is said to be **asymptotically stationary** if its limit set is a stationary state, and **asymptotically periodic** if its limit set is a periodic orbit. An orbit $O(\mathbf{x}_0)$ such that $\mathbf{x}_{n+p}=\mathbf{x}_n$ for some $n\geq 1$ and some $p\geq 1$ is said to be **eventually stationary** if $p=1$ and **eventually periodic** if $p>1$.

Hence, every eventually stationary or eventually periodic orbit is, respectively, asymptotically stationary or asymptotically periodic. The converse is not true. The orbit

$$x_0 = 1/2,\ x_1 = 1/4,\ x_2 = 3/16,\ \dots$$

of Example 1.3.1 is asymptotically stationary, but not eventually stationary.

Example 1.3.4 Let $F(x)=4x(1-x)$ and let $x_0=1/2$. Then $x_1=1$ and $x_2=x_3=\dots=0$. Hence $O(1/2)$ is eventually stationary ($n=2$, $p=1$).

Example 1.3.5 Let $F : [0,1]\to[0,1]$ be defined by $F(x)=2x-[2x]$, where $[x]$ is the greatest integer less than or equal to x. Notice that 0 is a stationary state of F, and

$$F(3/8) = 3/4,\ \ F(3/4) = 1/2,\ F(1/2) = 0.$$

Hence, the orbit

$$x_0 = 3/8,\ x_1 = 3/4,\ x_2 = 1/2,\ x_3 = x_4 = \dots = x_n = \dots = 0$$

is eventually stationary (n=3, p=1). The orbit

$x_0 = 1/24,\ x_1 = 1/12,\ x_2 = 1/6,\ x_3 = 1/3,\ x_4 = 2/3,\ x_5 = 1/3,\ x_6 = 2/3,\ x_7 = 1/3,$ etc.

is eventually periodic (n=3, p=2).

Definition 1.3.4
The orbit $O(\mathbf{x}_0)$ is **aperiodic** if its limit set $L(\mathbf{x}_0)$ is not finite.

From the previous discussion it is clear that either $O(\mathbf{x}_0)$ is **asymptotically periodic**, i.e., $L(\mathbf{x}_0)$ is finite, or **aperiodic**, i.e., $L(\mathbf{x}_0)$ is infinite.

Example 1.3.6 Figure 1.3.1 provides numerical evidence that the orbit of the dynamical system $F(x)=4x(1-x)$ starting from $x_0=.3$ is aperiodic. To visualize the behavior of the orbit we plot the points $(x_n, x_{n+1})=(x_n, F(x_n))$, for n=0, ...,1000. They appear to fill up the entire graph of F in the interval [0,1]. This numerical evidence supports the conclusion that the orbit visits every open subinterval of [0,1] no matter how small. The result will be established theoretically in Chapter 6, and it is clearly impossible if the orbit is asymptotically periodic.

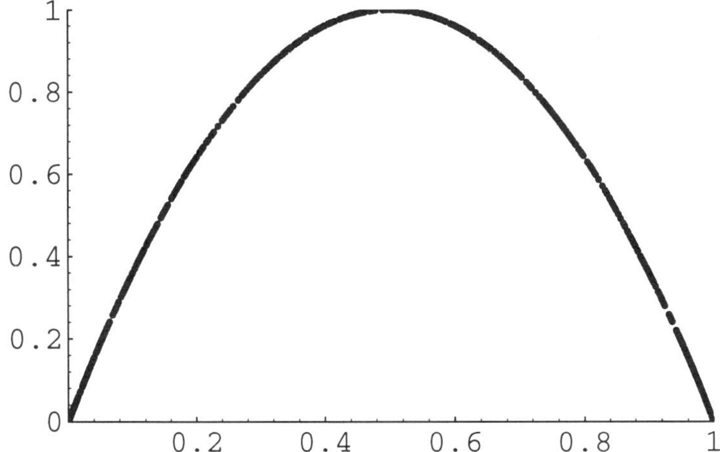

Fig. 1.3.1 The orbit O(.3) of $F(x)=4x(1-x)$ is "represented" on the graph of the function $F(x)$, by plotting $(x_n,x_{n+1})=(x_n,F(x_n))$ for n=0,...,1000.

Example 1.3.7 Let $F(x,y)=(x\cos 1-y\sin 1, x\sin 1+y\cos 1)$. F is a counterclockwise rotation of the plane of an angle $\theta=1$ radian. We offer here some numerical evidence that $O(\mathbf{x}_0)$ is aperiodic, where $\mathbf{x}_0=(1,0)$.

In Fig. 1.3.2 we have plotted 1000 iterates of the system $\mathbf{x}_{n+1}=F(\mathbf{x}_n)$ starting with $\mathbf{x}_0=(1,0)$. It appears that the orbit visits every arc of the circle, no matter how small. We express this fact by saying that the orbit is "dense" in the circle. This feature cannot be accomplished by an asymptotically periodic orbit.

It is not easy to establish theoretically the aperiodic character of an orbit, since it depends on its asymptotic behavior. Later we shall see some criteria for making this task easier. The following result ensures the existence of aperiodic orbits for discrete dynamical systems in the real line.

Theorem 1.3.1 [Li-Yorke, 1975] *Let* I *be an interval and* F : I→I *be continuous. Assume that* F *has a periodic orbit of period* 3. *Then* F *has a periodic orbit of every period and there is an infinite set* S *contained in* I *such that every orbit starting from a point of* S *is aperiodic.*

Theorem 1.3.1 looks simple, but its application to specific cases may be challenging.

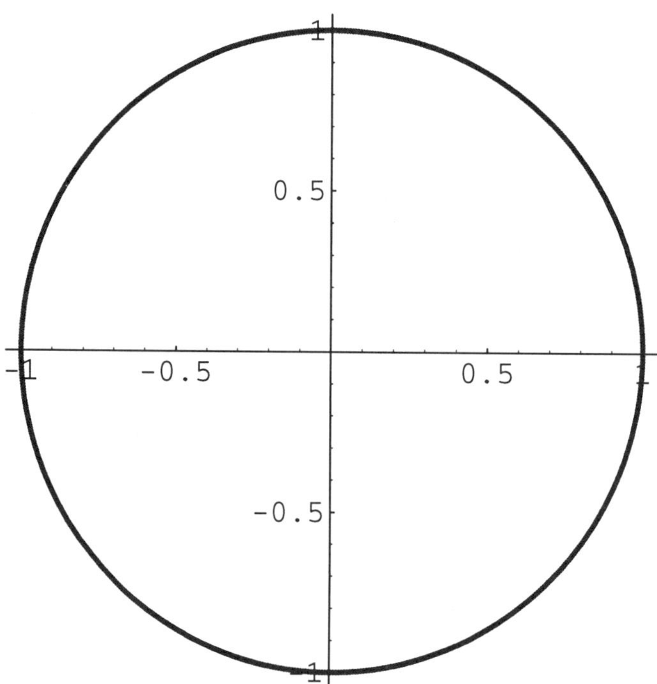

Fig. 1.3.2 For the dynamical system governed by the map $F(x,y)=(x\cos 1-y\sin 1, x\sin 1+y\cos 1)$ we have plotted the first 1000 iterates of O(1,0) (see Appendix 1, Section 2, Subsection 4). It appears that the orbit visits every open arc, no matter how small, of the circumference.

Example 1.3.8 Consider F : [0,1]→[0,1] defined by

$$F(x)=\begin{cases} 3x & \text{for } 0\le x\le 1/3 \\ 17/9 - 8x/3 & \text{for } 1/3<x\le 2/3 \\ 1/9 & \text{for } 2/3<x\le 1 \end{cases}$$

(see Fig. 1.3.3). The map is continuous and F(1/9)=1/3, F(1/3)=1, and F(1)=1/9. Hence, x=1/9 is periodic of period 3. According to Theorem 1.3.1 there is an infinite set S such that the orbit of every point of S is aperiodic. Determining S is not a trivial task.

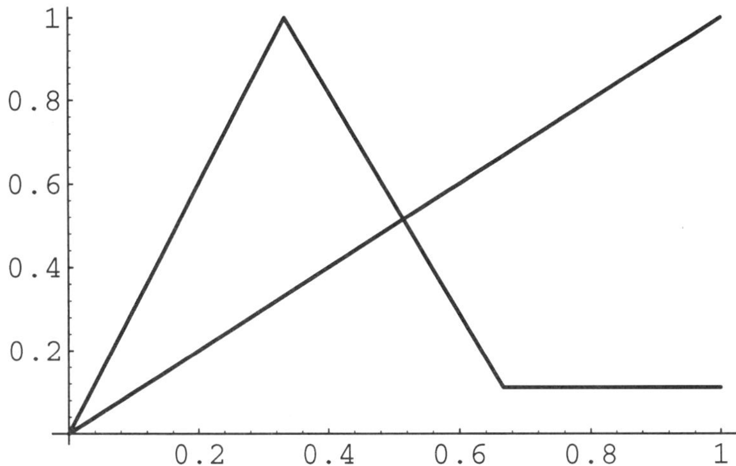

Fig. 1.3.3 Graph of the map of Example 1.3.8 together with the line $y=x$.

Problems

1. Consider the sequence 1/2, 2/3, 1/4, 3/4, 1/5, 4/5,... . Find its limit points.

2. Consider the sequence $x_n=\sin(n\pi/2)$. Find the limit points of the sequence.

3. Let $F(x)=.5x(1-x)$. Prove that for every $x\in[0,1]$, we have $L(x)=\{0\}$.

4. Let $F(x)=2x/3+1$. Show that $L(x_0)=\{3\}$ for every $x_0\in\mathbf{R}$. (Hint: See Example 1.1.6.)

5. * Let $F(x)=x+\pi-[x+\pi]$, where $[x+\pi]$ denotes the greatest integer contained in $x+\pi$. Using the programs of Appendix 1, Section 2, produce numerical evidence that F has aperiodic orbits in [0,1].

6. Let $F(x)=2x-[2x]$, where $[2x]$ denotes the greatest integer contained in $2x$. When $x_0\in[0,1]$ is written in base 2, namely $x_0=.a_1a_2a_3...$ with a_i either 0 or 1, we obtain that $F(x_0)=F(.a_1a_2a_3...)=.a_2a_3...$. Use this fact to show that F has aperiodic orbits.

7. Consider the map of Example 1.3.8. Show that the limit set of every orbit starting from a point x, $2/3\leq x\leq 1$, is $L(x)=\{1/9,1/3,1\}$.

8. Let $F(x)=2x-[2x]$. F has a periodic orbit of period 3 (see Problem 11 of Section 2, Subsection 2). Can the result of Li-Yorke be applied to F?

9. Let

$$F(x)=\begin{cases} 2x & 0\le x\le .5 \\ 2-2x & .5<x\le 1. \end{cases}$$

Find a periodic orbit of period 3 for F. Can the result of Li-Yorke be applied to F?

10. Let $F(x)=2|x|-1$. Show that F has aperiodic orbits in $[-1,1]$.

2. Unstable Orbits and Chaotic Systems

Every dynamical system with aperiodic orbits could be considered chaotic. This approach is, however, not appropriate. It qualifies as chaotic certain dynamical systems which do not belong to this category. For example, the system of Example 1.3.7 is a rotation (a rigid motion) of the plane. Hence, it cannot be considered chaotic. However, as established in the following theorem, all its nonstationary orbits are aperiodic.

Theorem 1.3.2 *Let* $F(x,y)=(x\cos 1-y\sin 1, x\sin 1+y\cos 1)$. *Let* $\mathbf{x}_0=(x_0,y_0)$ *be such that* $\|\mathbf{x}_0\|=1$. *Then every arc of the unit circle* $x^2+y^2=1$ *contains points of* $O(\mathbf{x}_0)$.

Proof. Recall that F is a counterclockwise rotation of the plane by 1 radian around the origin. Assume that there is an arc A with positive length which does not contain points of $O(\mathbf{x}_0)$. Then there exists a largest possible open arc A_0, containing A, such that A_0 does not contain any points of $O(\mathbf{x}_0)$, but its end points belong to $O(\mathbf{x}_0)$. Counterclockwise rotation of A_0 by 1 radian provides another open arc, A_1 and we have $F(A_0)=A_1$. Hence, A_0 and A_1 have to be disjoint. In fact, if $A_0=A_1$ then $1=2\pi$, which is impossible, and if $A_0\neq A_1$ but $A_0\cap A_1\neq\varnothing$ then A_0 contains one of the end points of A_1, which is impossible as well because both end points of A_1, just as the end points of A_0, belong to $O(\mathbf{x}_0)$.

Rotating once more, we obtain a second arc, A_2. We cannot have $A_2=A_1$ or $A_2\cap A_1\neq\varnothing$ for the same reasons mentioned above. Similarly, we cannot have $A_2=A_0$, since this would imply that $2=2\pi$, or $A_2\cap A_0\neq\varnothing$ since the end points of A_2 belong to $O(\mathbf{x}_0)$. Continuing in this manner we obtain seven arcs $A_0, A_1, \ldots, A_6$ such that for every $i\neq j$ we have $A_i\cap A_j=\varnothing$. This is obviously absurd since the total length of the seven arcs exceeds the length of the circumference. QED

From the previous theorem it follows that every point $\mathbf{z}$ of the circle is a limit point of the orbit $O(\mathbf{x}_0)$, since every open arc, however small, centered at $\mathbf{z}$ contains infinitely many states of the orbit. In a similar manner we can analyze the orbit of every point $\mathbf{x}_0\neq\mathbf{0}$ by adjusting the argument above to the circle centered at $\mathbf{0}$ and with radius $r=\|\mathbf{x}_0\|$.

On intuitive grounds, the dynamical system governed by the rotation $F(x,y)=(x\cos 1-y\sin 1, x\sin 1+y\cos 1)$ cannot be considered chaotic, even though all nonstationary orbits of the system are aperiodic. In particular, we see that orbits whose initial states are close to each other will remain close forever. The evolution of the system from any initial condition $\mathbf{x}_0$ is determined quite accurately even if the state $\mathbf{x}_0$ is not known exactly. Any error in the measurement of the initial coordinates is not magnified by successive iterations. Similar conclusions can be obtained for every rotation $P(\theta)$ of $\mathbf{R}^2$ of an angle θ such that θ/π is irrational.

Hence, chaotic systems should have something else besides aperiodic orbits. One ingredient that has been considered by experimental scientists is the so-called "unpredictability" property, namely the property that orbits starting at points very close to each other can be quite far apart at some later time, even though the orbits remain confined in a bounded region. Since, on experimental grounds, the initial state of an orbit is never known accurately, we cannot "predict" where the system will be at some later time.

To provide a scientific frame for the intuitive idea just mentioned we introduce the definition of **stable** and **unstable** orbits. As expected, the presence of unstable orbits plays an important role in dynamical systems in general, and in chaotic systems in particular.

Definition 1.3.5
An orbit $O(\mathbf{x}_0)$ is **stable** if for every $r>0$ there exists $\delta>0$ such that $\|\mathbf{x}_0-\mathbf{y}_0\|\leq\delta$ implies that $\|\mathbf{x}_n-\mathbf{y}_n\|\leq r$ for every $n\geq1$.

Here is an example of a stable orbit that is not periodic.

Example 1.3.9 Let $F(\mathbf{x})=(x\cos1-y\sin1, x\sin1+y\cos1)$. The dynamical system $\mathbf{x}_{n+1}= F(\mathbf{x}_n)$ is a rotation of an angle $\theta=1$ radian. For example, the point $(1,0)$ is moved to the point $(\cos1,\sin1)$, while the point $(0,1)$ is moved to $(-\sin1,\cos1)$. Consider the initial condition $\mathbf{x}_0=(1,0)$. According to Theorem 1.3.2 the orbit $O(\mathbf{x}_0)$ is aperiodic. We now show that $O(\mathbf{x}_0)$ is stable. In fact, let $\mathbf{y}_0\in\mathbf{R}^2$. Then (see also Example 1.2.4) $\|\mathbf{x}_n-\mathbf{y}_n\|=\|F^n(\mathbf{x}_0)-F^n(\mathbf{y}_0)\|=\|\mathbf{x}_0-\mathbf{y}_0\|$, an equality that is expected since F is a rotation (a rigid motion).

An orbit $O(\mathbf{x}_0)$ of a dynamical system $F : \mathbf{R}^q\to\mathbf{R}^q$ which is not stable is said to be **unstable.** Hence, $O(\mathbf{x}_0)$ is unstable if there exists $r>0$ such that for every $\delta>0$ we can find $\mathbf{y}_0$ with the following two properties:

- $\|\mathbf{y}_0 - \mathbf{y}_0\| \leq \delta$.
- $\|\mathbf{y}_n - \mathbf{x}_n\| > r$ for some $n \geq 1$.

We have seen examples of unstable stationary states (see Example 1.2.5) and unstable periodic orbits (see Example 1.2.10). An example of unstable non-stationary and nonperiodic orbit is provided below.

Example 1.3.10 Consider the dynamical system $x_{n+1}=4x_n(1-x_n)$ in $[0,1]$. We have seen graphical evidence that the orbit $O(0.3)$ is aperiodic (see Example 1.3.6). In Chapter 6 we shall establish theoretically that the system has an orbit $O(x_0)$ [not necessarily $O(.3)$] such that every point of $[0,1]$ is a limit point of the orbit.

Figure 1.3.4 gives graphical evidence of "instability" and "unpredictability". We consider two orbits $O(.3)$ and $O(y_0)$ with $y_0=.3000001$. The two initial states are closer that 10^{-6}! On the horizontal axis we plot the iteration number n, and on the vertical axis we plot the absolute value of the distance between x_n and y_n. We clearly see that at the beginning the two orbits are close to each other.

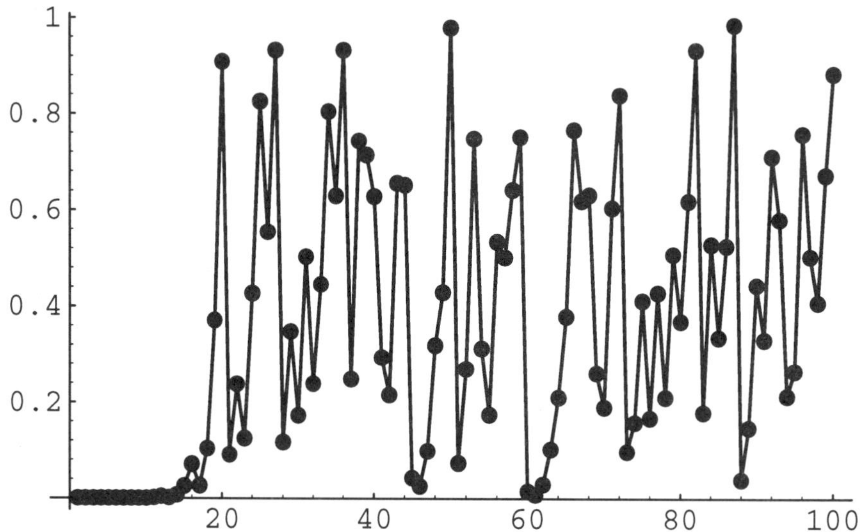

Fig. 1.3.4 The distances $|x_n - y_n|$, n=1,2,...,100 of O(.3) and O(.3000001) clearly show that the system governed by $F(x) = 4x(1-x)$ is unpredictable.

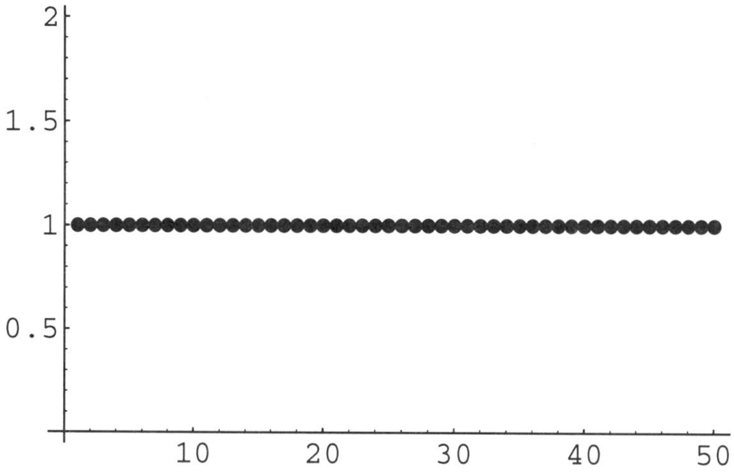

Fig. 1.3.5 The first 50 distances between corresponding states of O((1,0)) and O((1,1)) are plotted. In the horizontal axis we have the iteration number and in the vertical axis the corresponding distance.

Starting with iteration 20 the two orbits have a very different evolution. Sometimes they are quite close, but other times they are almost as far as 1, which is the largest distance we could find between x_n and y_n. Not knowing exactly what our initial condition is makes it impossible to predict what our position will be at some later time.

No orbit of the system of Example 1.3.7 satisfies the requirement of instability. In fact, no matter what initial conditions $\mathbf{x}_0$ and $\mathbf{y}_0$ we select, the sequence $\{\|\mathbf{x}_n-\mathbf{y}_n\|: n=0,1, ...\}$ is constant since F, being a rigid motion of the plane, preserves the distances. We provide graphical evidence (see Fig. 1.3.5) of this property by plotting $(n,\|\mathbf{x}_n-\mathbf{y}_n\|)$ for the two orbits O((1,0)) and O((1,1)) and n=1,...,50. We see that the distance does not change as time progresses. A comparison of Fig. 1.3.4 with Fig. 1.3.5 clearly shows how dramatic is the difference between the two dynamical systems.

Chaotic behavior

We postpone to Chapter 6 a formal definition of chaotic behavior. For the time being we shall consider a discrete dynamical system $\mathbf{x}_{n+1}=F(\mathbf{x}_n)$ chaotic in a set X if X is F-invariant [i.e., $F(X)\subset X$], and F has in X aperiodic and unstable orbits. In Chapter 6 these two properties will be combined with additional requirements to arrive at our definition of chaotic behavior.

At this point we do not have any theoretical tool, except Theorem 1.3.1, for proving that a system $\mathbf{x}_{n+1}=F(\mathbf{x}_n)$ has aperiodic and unstable orbits in an invariant set X. Hence, we shall simply use numerical evidence. Accordingly, $x_{n+1}=F(x_n)=4x_n(1-x_n)$ will be considered chaotic in the invariant interval [0,1] (see Figs.1.3.1 and 1.3.4). In Chapter 2 we will see that F is chaotic in [0,1] according to another definition of chaos, namely chaos in the Li-Yorke sense. Finally, in Chapter 6 we will prove that F is chaotic in [0,1] according to our definition of chaotic behavior.

We remark once more that the presence of aperiodicity and instability force some orbits with very close initial states to evolve in a completely different manner. Since our knowledge of any real process is never perfect due to measurement errors, noise, or a combination of both, we cannot make predictions on the evolution of a system having these two properties. Small differences at the outset can bring about dramatic changes in the future. This occurrence has sometimes been called the "butterfly effect." The name is motivated by the fact that weather patterns are frequently chaotic, and one could say, with a bit of exaggeration, that a butterfly can drastically change the course of the pattern with a simple movement of its wings!

We close this section by repeating that aperiodic and unstable orbits are needed but are not enough for characterizing chaos. Experimental scientists may sometimes label a system as chaotic only on the base of the presence of such orbits. This approach cannot be given general validity. Not all systems with aperiodic and unstable orbits in an invariant set X can be regarded as chaotic in X. The paper by Martelli-Dang-Seph [Martelli-Dang-Seph, 1998] has an example of a map $F : X\to X$, $X=\{\mathbf{x}\in \mathbf{R}^2: 0<r\leq\|\mathbf{x}\|\leq R\}$, which cannot be considered chaotic in X, although it has aperiodic and unstable orbits in X (see Problem 1, Section 2, Chapter 6).

Problems

1. * Let $F(x)=2x^2-1$. Verify numerically that the dynamical system governed by F has aperiodic and unstable orbits in [−1,1] (see Appendix 1, Section 2).

2. Let $F(x)= -2|x|+1$. Show that the interval $[-1,1]$ is mapped onto itself by the function F. Consider the dynamical system governed by F in $[-1,1]$. Can the presence of aperiodic orbits be derived from Theorem 1.3.1?

3. Let F be the map of the previous problem and let $I=[0,0.1]$. Find $n\geq 1$ such that $F^n(I)=[-1,1]$.

4. * Let $F(x)=4x^3-3x$. Prove that $[-1,1]$ is mapped onto itself by F. Produce numerical evidence that the dynamical system defined by F in $[-1,1]$ has aperiodic and unstable orbits (see Appendix 1, Section 2). Can the presence of aperiodic orbits be derived from the theorem of Li-Yorke? (Hint: Use Appendix 1, Section 1.)

5. Let $F(x,y)=(x\cos\theta - y\sin\theta, x\sin\theta + y\cos\theta)$. Assume that the ratio θ/π is irrational. Following the proof of Theorem 1.3.2, show that the orbit of the point (1,0) visits every arc, no matter how small, of the circle $x^2+y^2=1$.

6. * Consider the map $F(x,y)=(1.4+0.3y-x^2,x)$. Investigate numerically the orbits of the system $\mathbf{x}_{n+1}=F(\mathbf{x}_n)$ (see Appendix 1, Section 2).

Section 4. **EXAMPLES OF DISCRETE DYNAMICAL SYSTEMS**

In this section we present four examples of discrete dynamical systems taken from the current literature. Our goal is to study the behavior of these systems rather than show how they are derived or to test their appropriateness, objectivity, and usefulness. Thus, we shall keep to a minimum the analysis of the modeling steps used in converting each real dynamical process into its corresponding mathematical model, and the discussion concerning the interpretation of the conclusions derived from the model in terms of the real counterpart. A more detailed study of each system is provided in Chapter 7, using the theoretical results established in the previous chapters. In this section we simply introduce the function governing each system and determine its fixed points.

One-dimensional example: blood-cell biology

Under normal circumstances the population of red blood cells in a healthy human being oscillates within a certain tolerance interval. For men, this interval is (5.4±.8) million/μL, while for women it is (4.8±.6)million/μL. In the presence of a disease such as anemia, this behavior changes, sometimes dramatically. In modeling this process we are faced with conflicting constraints. On one hand, the time appears in the process as a continuous variable, while on the other, the number of cells changes only by integer values.

Therefore, in a model based on a continuous approach and which proposes a differential equation on a function $x(t)$ representing the number of cells per unit volume at time t, there is the problem of assessing how well $x(t)$ models its discontinuous real counterpart. Conversely, in a model where time is discrete, there is the problem of establishing what length the time step should have to preserve the qualitative features of the real process.

The model we present here uses this last approach. Time is divided into discrete generations, numbered 1,2,...,n,... and it is assumed that a quantity x_{n+1} representing cells per unit volume (in millions) at time n+1 depends on its state x_n according to the relation

$$x_{n+1} = x_n + p_n - d_n . \tag{1.4.1}$$

In (1.4.1) p_n and d_n account for the number of cells produced (p_n) and destroyed (d_n) during the nth generation. We assume (see [Lasota, 1977]) that $d_n = ax_n$, $a \in [0,1]$, and

$$p_n = b(x_n)^r e^{-sx_n}$$

with b, r, and s all positive. Therefore, our one-dimensional discrete dynamical system is

$$x_{n+1} = (1 - a)x_n + b(x_n)^r e^{-sx_n} \tag{1.4.2}$$

and $F(\mathbf{a},x) = F((a,b,r,s),x) = (1-a)x + bx^r e^{-sx}$. It is easily seen that $F(\mathbf{a},0)=0$ for every choice of **a**. Hence, 0 is always a stationary state.

In the case a=1, which means that during the time interval under consideration all cells that were alive at time n are destroyed, we obtain the simpler model

$$x_{n+1} = b(x_n)^r e^{-sx_n}. \tag{1.4.3}$$

We shall provide a more detailed analysis of (1.4.2) and (1.4.3) in Chapter 7. In the meantime, observe that for a=.8, b=10, r=6, and s=2.5 (see Fig. 1.4.1) the system has three fixed points and a bounded, closed, and invariant interval to the right of the second fixed point.

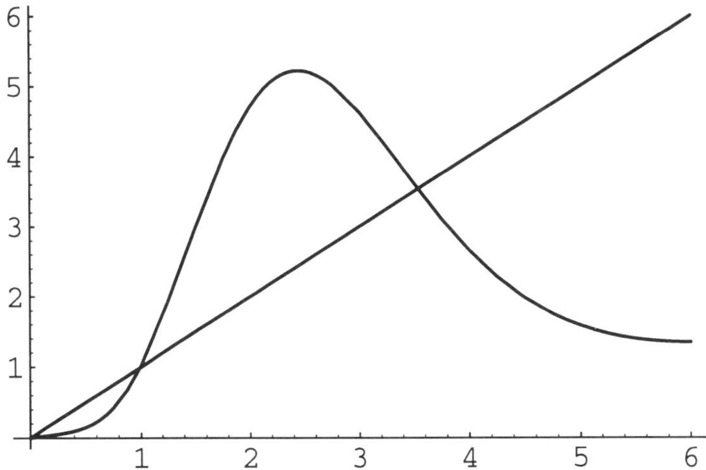

Fig. 1.4.1 Graph of the function F governing the blood-cell population for the following values of the parameters : a=.8, b=10, r=6, and s=2.5.

The two strictly positive fixed points are $x_{s2}\approx.989813$, and $x_{s3}\approx3.53665$. The invariant interval is $I_0\approx[1.478,5.221]$. Numerical evidence suggests that the system is chaotic in this interval (see Figs. 1.4.2 and 1.4.3).

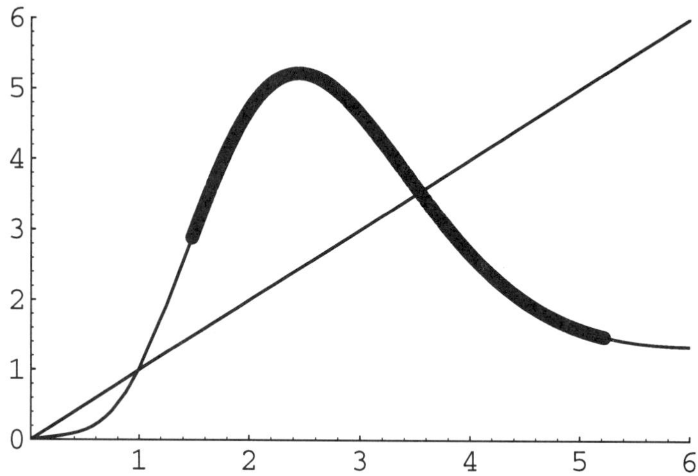

Fig. 1.4.2 O(5) of the system (1.4.2) with a=.8, b=10, r=6, and s=2.5 appears to be aperiodic since it fills up the graph of $F(x)=.2x+10x^6e^{-2.5x}$ for $x\in I_0$.

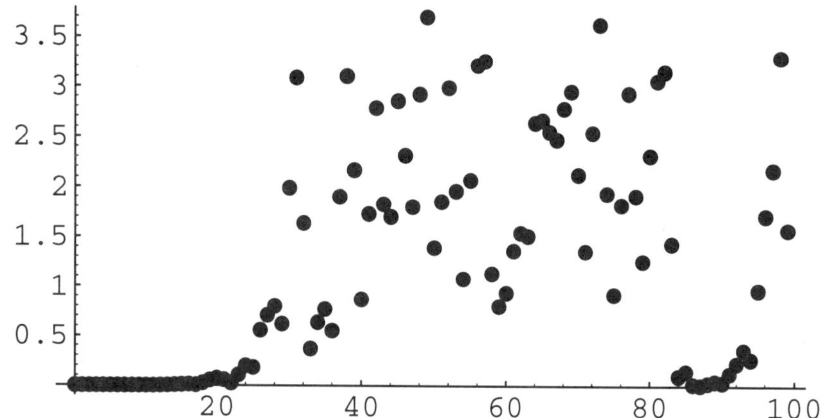

Fig. 1.4.3 The orbit of x_0=5 of (1.4.2) with a=.8, b=10, r=6, and s=2.5 appears to be unstable. Plotted here are the points $(n,|x_n-y_n|)$ for n=0,...,100, x_0=5, and y_0= 5.00001.

Two-dimensional examples: predator-prey models

During World War I the population of selachians in the north side of the Adriatic Sea rose dramatically, confronting the biologist Umberto D' Ancona with the seemingly unsolvable puzzle of finding the reason for the increase. He presented the problem to a mathematics expert, his father-in-law, Vito Volterra [Volterra, 1931], who provided a brilliant solution to the puzzle by building a simple, yet objective and useful model.

Volterra divided the fish population into two categories, predators (the selachians) and prey. Assuming that the number of encounters per unit time between pairs of the two groups is proportional to the product of their sizes and favors the predators' increase while decreasing the other group, Volterra proposed the following system of differential equations for studying the dynamics of the two populations. The variables $x(t)$ and $y(t)$ represent the prey's and predators' population, respectively, at time t; and a,b,c,d, and ε are nonnegative parameters:

$$\begin{cases} x'(t) = ax(t) - bx(t)y(t) - \varepsilon x(t) \\ y'(t) = -cy(t) + dx(t)y(t) - \varepsilon y(t). \end{cases} \tag{1.4.4}$$

The last terms of the two equations, $-\varepsilon x(t)$ and $-\varepsilon y(t)$, account for the fishing activity. Due to war hostilities this activity was significantly reduced from 1915 to 1919, the period characterized by the large increase in the proportion of selachians captured by the few fishing vessels operating in the area. The stationary states of (1.4.4) are those values of $x(t)$ and $y(t)$ which are time independent. Setting $x'(t)=y'(t)=0$ we obtain the two points (0,0) and $((c+\varepsilon)/d, (a-\varepsilon)/b)$. The first point represents the extinction of both predators and prey, a situation that may arise, but which we want to avoid. The second suggests that the parameter ε should be smaller than a. System (1.4.4) cannot be solved explicitly, in the sense that we cannot find the explicit form of the functions $x(t)$ and $y(t)$. However, it can be shown (see, for example, [Braun, 1982]) that given initial conditions $x(0)=x_0$ and $y(0)=y_0$, there are only two functions $x(t)$ and $y(t)$ which solve the system and satisfy the given conditions. The two functions are periodic of the same period T. We can obviously assume that T is one year. Moreover, their average values over the entire period are given by

$$(c+\varepsilon)/d = (1/T)\int_0^T x(t)dt, \quad (a-\varepsilon)/b = (1/T)\int_0^T y(t)dt. \tag{1.4.5a}$$

In the absence of any fishing activity ($\varepsilon=0$) we obtain

$$c/d = (1/T)\int_0^T x(t)dt, \quad a/b = (1/T)\int_0^T y(t)dt. \tag{1.4.5b}$$

Therefore, we see that a moderate fishing activity favors the prey, while its absence favors the predators, in agreement with the observation made by D'Ancona.

A numerical investigation of (1.4.4) can be performed by using the **one-step Euler method**. A time step $\Delta t=h$ is selected and the two derivatives are replaced by

$$\frac{x(t+h) - x(t)}{h} \quad \text{and} \quad \frac{y(t+h) - y(t)}{h} \tag{1.4.6}$$

respectively. We arrive at the system

$$\begin{cases} x(t+h) = (1+ah)x(t) - bhx(t)y(t) - \varepsilon hx(t) \\ y(t+h) = (1-ch)y(t) + dhx(t)y(t) - \varepsilon hy(t). \end{cases} \tag{1.4.7}$$

Choose t=0, and the same initial conditions $x(0)=x_0$, $y(0)=y_0$ mentioned before for the continuous system. Substitution into (1.4.7) gives

$$\begin{cases} x_1 = x(h) = (1+ah-\varepsilon h)x_0 - bhx_0y_0 \\ y_1 = y(h) = (1-ch-\varepsilon h)y_0 + dhx_0y_0. \end{cases} \tag{1.4.8}$$

Choose t=h. From (1.4.7) and (1.4.8) we obtain

$$\begin{cases} x_2 = x(2h) = (1+ah-\varepsilon h)x_1 - bhx_1y_1 \\ y_2 = y(2h) = (1-ch-\varepsilon h)y_1 + dhx_1y_1. \end{cases} \tag{1.4.9}$$

In general, after n+1 steps we arrive at

$$\begin{cases} x_{n+1} = x((n+1)h) = (1+ah-\varepsilon h)x_n - bhx_ny_n \\ y_{n+1} = y((n+1)h) = (1-ch-\varepsilon h)y_n + dhx_ny_n. \end{cases} \tag{1.4.10}$$

In the discrete dynamical system (1.4.10) we have $\mathbf{x}=(x,y)$, $\mathbf{a}=(a,b,c,d,\varepsilon,h)$,

$$F(\mathbf{a}, \mathbf{x}) = (\,(1+ha-\varepsilon h)x - bhxy,\ (1-ch-\varepsilon h)y + dhxy),$$

and $F_1(\mathbf{a},\mathbf{x})=(1+ha-\varepsilon h)x-bhxy$, $F_2(\mathbf{a},\mathbf{x})=(1-ch-\varepsilon h)y+dhxy$. (0,0) is always a stationary state of our system. Setting

$$x_1 = x_0, \quad y_1 = y_0, \quad x_1 \neq 0, \quad y_1 \neq 0, \tag{1.4.11}$$

in (1.4.10) we obtain the fixed point

$$x = \frac{c+\varepsilon}{d}, \quad y = \frac{a-\varepsilon}{b}. \tag{1.4.12}$$

We see that the stationary states of the original system and of its discrete form are the same. However, the long-term behavior of (x_n, y_n) is different from the one of $(x(t), y(t))$. In particular, (x_n, y_n) is not periodic (see Fig. 1.4.4).

Therefore, we should not consider system (1.4.10) a good model of the biological process studied by Umberto D'Ancona, particularly if we are interested in the long-range evolution of the process. The numerical technique provides only an approximate version of (1.4.4). Given an initial condition $\mathbf{x}_0=(x_0,y_0)$, a selected time interval $[0,t_0]$ and "an index of proximity" δ, we can choose h so that the orbit of (1.4.10) remains closer than δ to the corresponding orbit of (1.4.4) in $[0,t_0]$. However, after t_0 the accuracy falls off and the long-range behavior of the two systems is different. The orbits of the continuous version are closed, smooth curves around the nontrivial stationary state, while the orbits of the discrete system are increasing spirals which eventually exit the first quadrant.

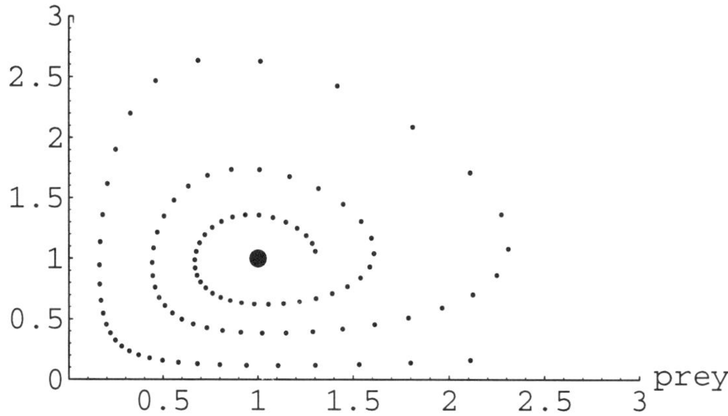

Fig. 1.4.4 Plot of predators versus prey for (1.4.10) with 1+ha=1.2, bh=.2, ε=0, 1–ch=.8, dh=.2. The initial condition is x_0 =1.3, y_0=1. Notice the "spiraling out" behavior as the number of iterations increases.

We now propose a discrete predator-prey model along the lines of Volterra's ideas. A similar model was proposed in Example 1.1.2 for understanding the predator-prey interaction between whiteflies and black wasps. With the variables x and y once more denoting prey and predators, we govern the evolution of the two populations with the system

$$\begin{cases} x_{n+1} = (r - \varepsilon)x_n - bx_ny_n \\ y_{n+1} = (s - \varepsilon)y_n + dx_ny_n\,(r - \varepsilon - by_n) \end{cases} \tag{1.4.13}$$

where r>1, s∈(0,1) and they are both close to 1, while b, d, ε are small (<<1) positive parameters. We see that (0,0) is again a fixed point of our system. Setting a=r–1 and c=1–s, we obtain that the non-zero fixed point is

$$x = \frac{c + \varepsilon}{d}, \quad y = \frac{a - \varepsilon}{b}. \tag{1.4.14}$$

We see, once more, that a moderate harvesting activity favors the prey. We shall study (1.4.13) in Chapter 7. In the meantime we present in Fig. 1.4.5 the first 300 iterations of the orbit starting at (3,5) with r=1.1, ε=.01, b=.04; s=.95, and d=.01. The nonzero fixed point is (6,2.25). The larger dots represent the prey. We notice the periodic behavior of the two populations.

Other discrete models of predator-prey interaction have been proposed. For example [Lauwerier, 1986b], the following dynamical system belongs to this class:

$$\begin{cases} x_{n+1} = ax_n(1 - x_n - y_n) \\ y_{n+1} = by_n x_n. \end{cases} \tag{1.4.15}$$

We will analyze (1.4.15) in Chapter 7.

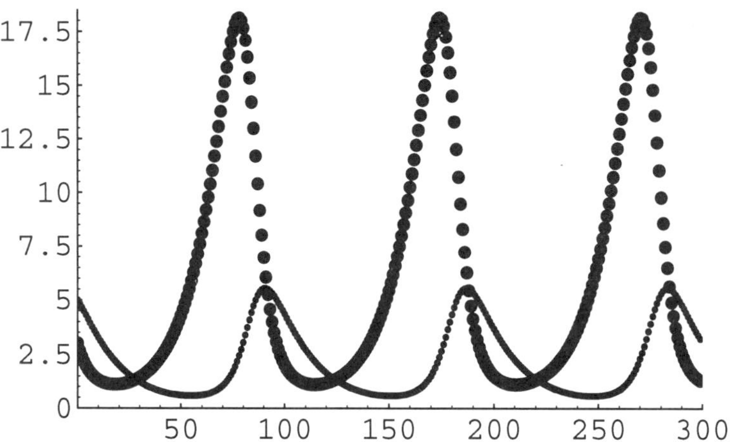

Fig. 1.4.5 First 300 iterations of the orbit starting at (3,5). Notice the periodic behavior of the two populations and their off-phase oscillations.

Three-dimensional example: meteorology

In 1963, the MIT meteorologist E. N. Lorenz [Lorenz, 1963] proposed a simplified model of atmospheric behavior which involved a system of differential equations in three variables $x(t)$, $y(t)$, and $z(t)$. To understand the meaning of these variables, let us imagine that in the upper atmosphere there is a layer of air of uniform depth H, where the temperature difference between the upper and lower surfaces is maintained at a constant value ΔT. In an ideal situation the temperature fall from the lower to the upper surface is linear, i.e., the graph of T versus h, from h_0 to h_0+H, is a straight line with negative slope. In general, there will be a deviation from this linear profile. The variable z is proportional to this distortion. The variable x is proportional to the intensity of the convective motion between the two layers, while the variable y is proportional to the temperature difference between the ascending and descending currents.

With these variables at hand, Lorenz obtained the system

$$\begin{cases} x' = -ax + ay \\ y' = -xz + rx - y \\ z' = xy - bz. \end{cases} \tag{1.4.16}$$

We now apply the **one-step Euler method** to (1.4.16). A time step Δt $=h$ is selected and the three derivatives are replaced by

$$\frac{x(t+h)-x(t)}{h}, \quad \frac{y(t+h)-y(t)}{h}, \quad \frac{z(t+h)-z(t)}{h}. \tag{1.4.17}$$

Substituting into (1.4.16), we obtain

$$\begin{cases} x(t+h) &= (1-ha)x(t) + hay(t) \\ y(t+h) &= -hx(t)z(t) + hrx(t) + (1-h)y(t) \\ z(t+h) &= hx(t)y(t) + (1-bh)z(t). \end{cases} \tag{1.4.18}$$

Choose t=0, set $x_0=x(0)$, $y_0=y(0)$, $z_0=z(0)$, $x_1=x(h)$, $y_1=y(h)$, $z_1=z(h)$, to obtain

$$\begin{cases} x_1 &= (1-ha)x_0 + hay_0 \\ y_1 &= -hx_0z_0 + hrx_0 + (1-h)y_0 \\ z_1 &= hx_0y_0 + (1-bh)z_0. \end{cases} \tag{1.4.19}$$

In general, after n+1 steps, we arrive at [see also (1.4.8), (1.4.9) and (1.4.10)]

$$\begin{cases} x_{n+1} &= (1-ha)x_n + hay_n \\ y_{n+1} &= -hx_nz_n + hrx_n + (1-h)y_n \\ z_{n+1} &= hx_ny_n + (1-bh)z_n. \end{cases} \tag{1.4.20}$$

Hence, $\mathbf{x}=(x,y,z)$, $\mathbf{a}=(a,b,r,h)$ and

$$F(\mathbf{a}, \mathbf{x}) = ((1-ha)x + hay, -hxz + hrx + (1-h)y, hxy + (1-bh)z).$$

We see that $\mathbf{x}_0=(0,0,0)$ is always a fixed point, regardless of the value of the parameter **a**. The other fixed points are found by setting

$$x_1 = x_0, \quad y_1 = y_0, \quad z_1 = z_0$$

in (1.4.19). From the first equation we obtain $x_0=y_0$. After substituting this into the second and third equations and dropping the subscripts, we arrive at

$$\begin{cases} xz &= rx - x \\ x^2 &= bz. \end{cases}$$

We divide the first equation by x, since $x=0$ implies that $y=z=0$. We obtain $z=r-1$. Thus, $x^2=b(r-1)$, which requires that $r>1$ ($r=1$ again gives $x=y=z=0$). We find two more stationary states (see Fig. 1.4.6):

$$\mathbf{x}_1 = (\sqrt{b(r-1)}, \sqrt{b(r-1)}, r-1), \ \mathbf{x}_2 = (-\sqrt{b(r-1)}, -\sqrt{b(r-1)}, r-1).$$

Just as in the case of the prey-predator model, we should not consider system (1.4.20) as equivalent to the process studied by Lorenz, particularly if we are interested in its long-range evolution. The numerical technique provides only an approximate version of (1.4.16). Given an initial condition $\mathbf{x}_0=(x_0,y_0,z_0)$, a time interval $[0,t_0]$ and an index of proximity δ, we can choose h so that the orbit of (1.4.20) remains closer than δ to the corresponding orbit of (1.4.16) in $[0,t_0]$. However, after t_0 the accuracy decreases and the long-range behavior of the two systems may be quite different, at least from a quantitative point of view.

In other words, two corresponding orbits, namely, two orbits starting with the same initial conditions, will remain close for a specific time interval, but may become significantly separated from each other in the long run. The stationary states of the two systems are the same, and some qualitative features remain in the transition from the continuous system to its discrete form. We shall analyze (1.4.20) in Chapter 7.

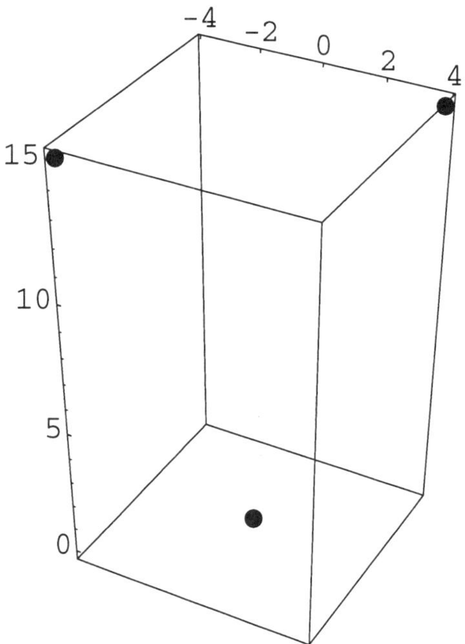

Fig. 1.4.6 The three stationary states of the Lorenz system for a=8, b=1, and r=16.

Multidimensional example: neural networks

A neural network is a crude but powerful model of the human brain, a kind of electronic device that attempts to mimic the way neurons of the brain work together. The potential of neural networks is still largely unknown, but expectations are very high based on the success of the applications in place so far.

Chase Manhattan Bank uses a neural net system to detect credit card frauds, while other banks use neural systems to analyze loan risks. Our fastest available computers take several years of CPU time to solve problems that neural nets can answer in a few seconds. There is a price to pay; the computer solution is exact, while the neural network solution has a 5-10% margin of error. However, we do not have many choices, since in many circumstances we need the solution within a few seconds after the problem has been posed.

Our brain has billions of neurons, but real neural nets are far from this level. If we imagine a net with N neurons, the connections among them can be arranged in a square matrix, called the **connectivity matrix** T, with N^2 entries. The entry t_{ij}, located in row i and column j of T, represents the action of neuron j on neuron i. This action can be **excitatory** ($t_{ij}>0$) or **inhibitory** ($t_{ij}<0$). When neuron j and i are not directly connected, we have $t_{ij}=0$. A function ϕ_j, j=1,2,..,N, called the "neuron response function," controls the action of neuron j and works like a switch. As long as the energy of neuron j remains below a threshold level, the state of the neuron is purely passive. The neuron can be influenced by other neurons but cannot have any effect on their state. When the threshold level is reached, the function ϕ_j triggers its response. A typical neuron response function is 0 for $x \le 0$, 1 for $x \ge 1$ and x for $x \in [0,1]$. The dynamical evolution of a neural net after receiving an initial input can be modeled by the system of equations [Hopfield, 1982]

$$x_{k,n+1} = x_{k,n} - c_k x_{k,n} + t_{k,1}\phi_1(x_{1,n}) + \dots + t_{k,N}\phi_N(x_{N,n}) + x_{k,I} \qquad (1.4.21)$$

for k=1,2,...,N. Equation (1.4.21) represents the energy level of neuron k at the n+1 stage of the evolution of the system. The quantity c_k, $0<c_k<1$, accounts for a certain leakage of neuron's k energy; the quantities $t_{k,j}$ for j=1,2,...,N are the connections between all other neurons and neuron k, and $t_{k,j}\ \phi_j(x_{j,n})$, j=1,2,...,N represents the action of neuron j on neuron k at the nth state of the system. For simplicity, it is frequently assumed that all neuron response functions are the same. Then $\phi_1=\dots=\phi_N=\phi$. The term $x_{k,I}$ accounts for the effect of the input $\mathbf{x}_I$ on neuron k. With a net of 100 neurons we have a system of 100 equations!

We can express (1.4.21) in a more compact form. Denote by I the N×N identity matrix; by C the N×N diagonal matrix whose diagonal entries are $c_1, c_2, \dots, c_N$; by $\mathbf{x}$ the neuron vector $\mathbf{x}=(x_1, \dots, x_N)$; and by $\mathbf{x}_I$ the input vector. Let $\Phi(\mathbf{x})=(\phi(x_1),\dots,\phi(x_N))$ (we assume that all neuron response functions are equal). System (1.4.21) can be rewritten in vector form:

$$\mathbf{x}_{n+1} = (I - C)\mathbf{x}_n + T\Phi(\mathbf{x}_n) + \mathbf{x}_I\,. \qquad (1.4.22)$$

We assume that when an input $\mathbf{x}_I$ is received and it finds the net in a certain initial state $\mathbf{x}_0$, a dynamical process is triggered according to (1.4.22). In other words, the states $\mathbf{x}_1, \mathbf{x}_2, \dots, \mathbf{x}_n, \dots$ are produced by the recursion scheme (1.4.22). When (and if) the process reaches a stationary state $\mathbf{x}_s$, the net, by means of this state, has provided an answer to the problem posed by the input $\mathbf{x}_I$. The entries of the connectivity matrix are a crucial element of the process since they play a determinant role in reaching the fixed point $\mathbf{x}_s$. Hence, much research has been done to find efficient methods to determine them.

The neurons of the net are usually arranged into three layers: **input, hidden,** and **output layer.** Correspondingly, the entries of $\mathbf{x}$ and $\mathbf{x}_I$ are separated into three sections, the first representing the input neurons and the last the output neurons. Since it is assumed that external inputs do not directly affect the neurons of the hidden and output layers, we obtain that only the first part of $\mathbf{x}_I$ has nonzero entries. The matrix T is now divided into nine blocks: T_{II}, T_{IH}, T_{IO},..., T_{OO} . It is normally assumed that the neurons of the input layer are not directly connected to those of the output layer. Thus, the entries of the blocks T_{IO} and T_{OI} are all zero. In this way the search for the optimal entries of T is simplified. In Chapter 7 we analyze (1.4.22) in more detail using the theory developed in the previous chapters.

Problems

1. * Let $x_{n+1}=(1-a)x_n+4(x_n)^6 \exp(-2x_n)$. Verify that for every $a\in(0,1)$ the system has exactly three fixed points (you can use the programs of Appendix 1, Section 4).

2. In Problem 1 assume that the x-coordinate of the second fixed point is smaller than 2.5. Prove that the derivative at that point is larger than 1.

3. The map $F(x,y)=(1+y-ax^2, bx)$ was studied by M. Hénon [Hénon, 1976]. Find the stationary states of the Hénon map.

4. Verify that the Hénon map is one-to-one, provided that $b\neq 0$.

5. Let $F(x,y)=(x+|x-y|-2, y-x+1)$. Find the fixed points of the dynamical system defined by F.

6. Write the complete set of equations (1.4.21) when N=3.

7. Assume that the response function ϕ_j of neuron j has the form $\phi_j(x)=a_jx+b_j$, j=1, 2, ..., N. Rewrite the system (1.4.21) for N=3 .

8. Let $x'(t)=1+x^2$. Find the general solution of this differential equation by separation of variables. Then find the unique solution that satisfies the initial condition $x(0)=1$. Apply the one-step Euler method to the differential equation using the strategy outlined in the two-dimensional example. Select a small time step and compare the orbit of the discrete system with initial condition $x_0=1$ with the corresponding solution of the differential equation.

9. The logistic differential equation is $x'(t)=ax(t)(1-x(t))$. Solve the differential equation (Bernoulli's type) and show that for every initial condition $x(0)=x_0>0$ we have that $x(t)\to 1$ as $t\to\infty$. Apply the one-step Euler method to the logistic equation using the strategy outlined in the two-dimensional example. Compare the discrete system so obtained with the discrete logistic system mentioned in Example 1.1.4.

10. Show that $x=0$ implies that $y=z=0$ in the search for fixed points of (1.4.20).

11. *Verify that by changing the initial conditions in system (1.4.2) with a=.8, b=10, r=6, and s=2.5, we obtain graphs which are qualitatively similar to the graphs of Figs. 1.4.2 and 1.4.3.

12. *Consider the system (1.4.13) with r=1.1, ε=.01, b=.04, s=.95, and d=.01. Select several different initial conditions (x_0,y_0), with $x_0>0$ and $y_0>0$. Plot the points (x_n,y_n) for n=0,1,...,1000. Can you conclude that the orbits are periodic?

CHAPTER 2

ONE-DIMENSIONAL DYNAMICAL SYSTEMS

SUMMARY

In this chapter we present an extensive analysis of discrete one-dimensional systems, starting in Section 1 with some ideas on conjugacy and on linear (or affine) systems. We then turn our attention to nonlinear systems and we provide, in Section 2, sufficient criteria for the stability of a stationary state and a periodic orbit. In Section 3 we present results on global attractivity of fixed points, and in Section 4 we study bifurcation points and bifurcation diagrams. Finally, in Section 5 we introduce the Li-Yorke definition of chaotic behavior.

Section 1. COBWEB AND CONJUGACY

A discrete dynamical system is said to be one-dimensional if it is of the form

$$x_{n+1} = F(x_n), \tag{2.1.1}$$

where the state variable x is a scalar, i.e., $x \in \mathbf{R}$. Frequently, F also depends on one or more parameters. In this case we replace $F(x)$ with $F(\mathbf{a},x)$ where $\mathbf{a}=(a_1,...,a_m)$ (see Example 2.1.1 below). Then (2.1.1) becomes

$$x_{n+1} = F(\mathbf{a},x_n) \tag{2.1.2}$$

where F is a function defined in the Cartesian product $\mathbf{R}^m \times \mathbf{R}$ with values in $\mathbf{R}$.

Unless otherwise stated, we assume that F and F_x are continuous. The parameter $\mathbf{a}$ is frequently a real number (a scalar). The dynamical process and/or the form of the function F may impose restrictions on the values of x and $\mathbf{a}$ allowed in (2.1.1) and (2.1.2).

Example 2.1.1 Let $F(a,x)=ax(1-x)$. Since $F_x(a,x)=a-2ax$, F, F_x are continuous. For every $a\in[0,4]$ the dynamical system

$$x_{n+1} = ax_n(1 - x_n) \tag{2.1.3}$$

represents the evolution of a process in the interval [0,1] (see Example 1.1.4). Hence the function F, although defined for every x and a, is frequently restricted to $x\in[0,1]$ and $a\in[0,4]$. We could identify (2.1.3) with a simplified model of the dynamics of a population that lives in a limited environment. The growth of the population is controlled by the parameter a, which may be considered as the vital coefficient of the population.

One could think to obtain a more realistic model by considering the function $F(\mathbf{a},x)=F(a,b,x)=x(a-bx)$, $\mathbf{a}=(a,b)$. The dynamical system is now

$$x_{n+1} = x_n(a - bx_n). \tag{2.1.4}$$

The improvement, however, is only apparent, since a change of variable reduces the study of (2.1.4) to the study of (2.1.3) (see Problem 3 of Subsection 2 below).

1. The Cobweb Method

Assume that $\mathbf{a}$ in (2.1.2) is fixed. We can now perform an analysis of the dynamical system governed by F using the graph of F, since x is the only variable left in the system.

Recall that the fixed points of F are the x-coordinates of the points of intersection between the graph of F and the line $y=x$ (see Fig. 1.2.1). The trajectory starting from x_0 can be visualized in the plane by drawing a vertical segment from $(x_0,0)$ to $(x_0,F(x_0))=(x_0,x_1)$ on the graph of F; then horizontally to the line $y=x$ to meet (x_1,x_1), and then again vertically to $(x_1,F(x_1))=(x_1,x_2)$ on the graph of F. Continue this process until sufficient information is obtained on the behavior of the orbit starting from x_0. The picture that provides this information is called a **cobweb diagram**.

For cxample, when the orbit $O(x_0)$ converges to a fixed point x_s, the cobweb diagram starting from $(x_0,0)$ will be a sequence of vertical and horizontal segments spiraling toward (x_s,x_s). In Fig. 2.1.1 the situation is illustrated by a cobweb diagram of the logistic map $F(x,a)=ax(1-x)$ for a=2.9. The positive fixed point of F is $x_s=19/29$ and the initial state is $x_0=.1$.

Similarly, when the orbit $O(x_0)$ converges to a periodic orbit of period p $\{z_0,z_1,..., z_{p-1},...\}$, the cobweb starting from $(x_0,0)$ will approach the closed cycle

$$(z_0,z_0),\ (z_0,z_1),\ ...,\ (z_{p-1},z_{p-1}),\ (z_{p-1},z_0),\ (z_0,z_0).$$

In Fig. 2.1.2 we illustrate this situation by selecting the logistic map $F(x,a)=ax(1-x)$ with a =3.4. In this case, as we shall see later, there is an attractive periodic orbit of period 2. The cobweb method illustrates how the orbit starting from $x_0=.1$ converges to the periodic orbit of period 2.

Finally, when the orbit starting from x_0 is aperiodic, the behavior will be graphically illustrated by the cobweb method. The sequence of vertical and horizontal segments will not display any recognizable pattern. In Fig. 2.1.3 we have selected a=4 in the logistic map. The initial condition is once again $x_0=.1$. To make sure that we are not in the presence of a periodic orbit with a very high period, we can continue the cobweb diagram and verify that new lines continue to appear until the graph becomes almost a black box.

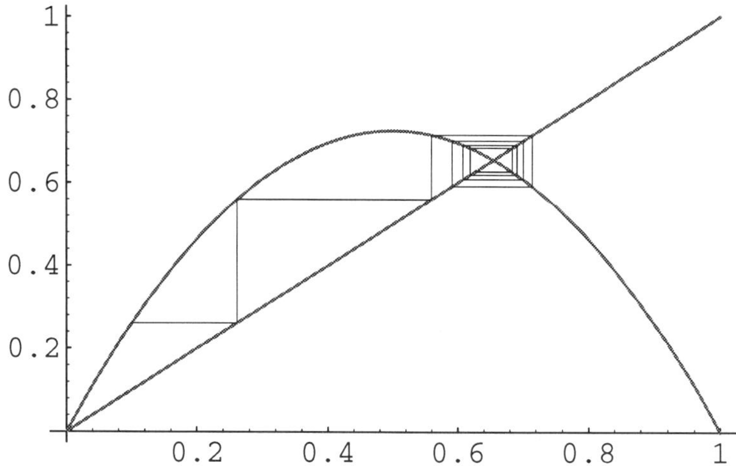

Fig. 2.1.1 The cobweb method shows how the orbit starting from $x_0=.1$ converges to the positive fixed point of the dynamical system governed by $F(x)=2.9x(1-x)$.

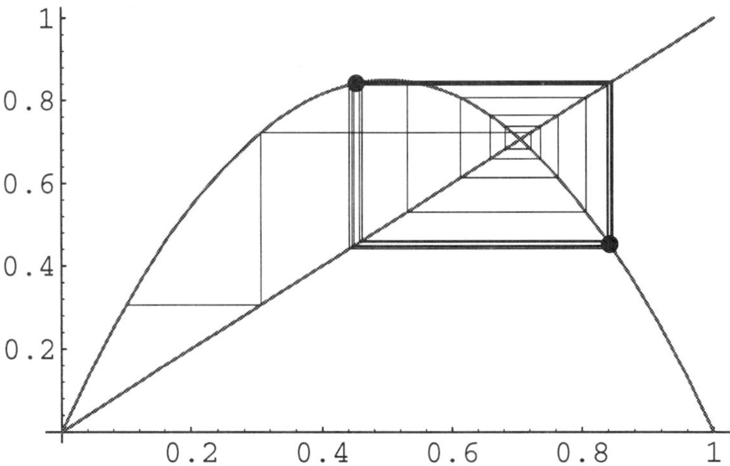

Fig. 2.1.2 The orbit of x_0=.1 passes very close to the fixed point 12/17 of the dynamical system governed by F(x)=3.4x(1–x), before moving toward a periodic orbit of period 2.

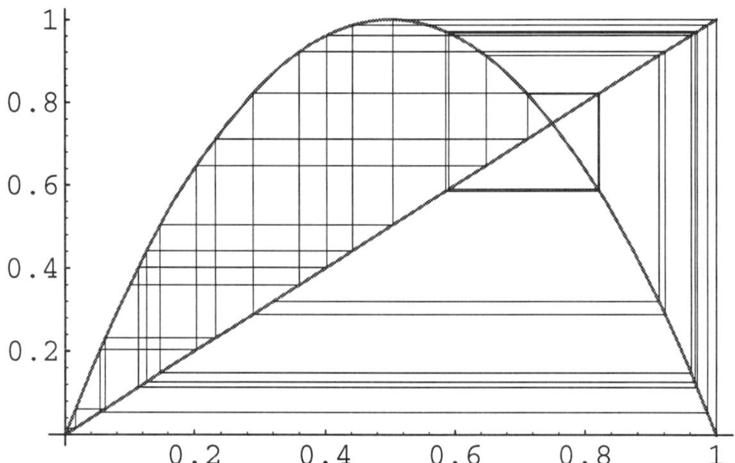

Fig. 2.1.3 Starting at x_0=.1 we follow the orbit of the dynamical system F(x)=4x(1–x). No pattern seems to emerge. The orbit is apparently aperiodic.

Problems

1. The starting point of the orbit O(.3) is marked on the horizontal axis of the graph for Problem 1. Find the next three states of the orbit.

2. Given the graph for Problem 2 and the state x_3 marked on the x-axis, find x_4 and x_2 on the x-axis. Assume that $x_2 < x_3$.

Graph for Problem 1

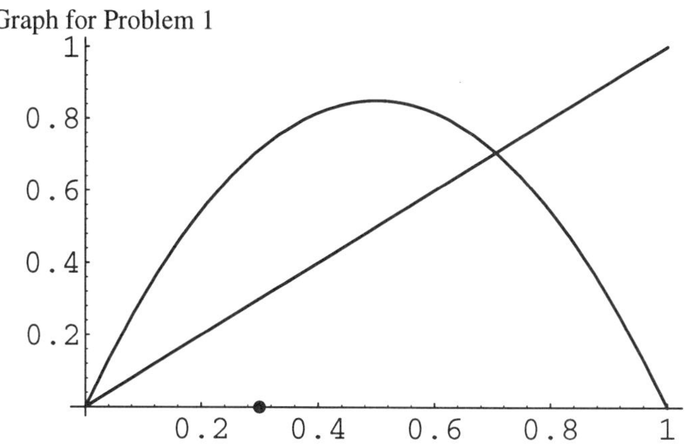

Graph for Problem 2

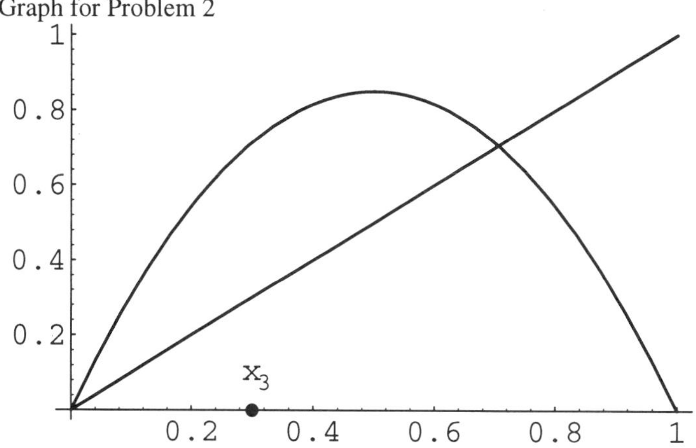

Graph for Problem 3

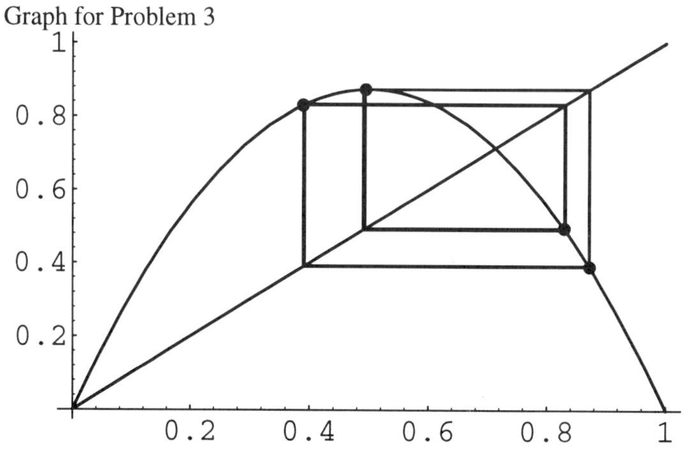

3. After computing 40 iterations and discarding the first 32 the cobweb diagram of the orbit O(.2) of the dynamical system $x_{n+1}=3.48949x_n(1-x_n)$ appears as shown in the graph for Problem 3. Justify the following two statements:

- O(.2) is not periodic.
- O(.2) is asymptotically periodic.

Find the period of the orbit approached by O(.2).

4. The point marked on the graph of F is $(x_1, F(x_1))$. Find x_3. Assuming that $x_0>x_1$ find x_0.

Graph for Problem 4

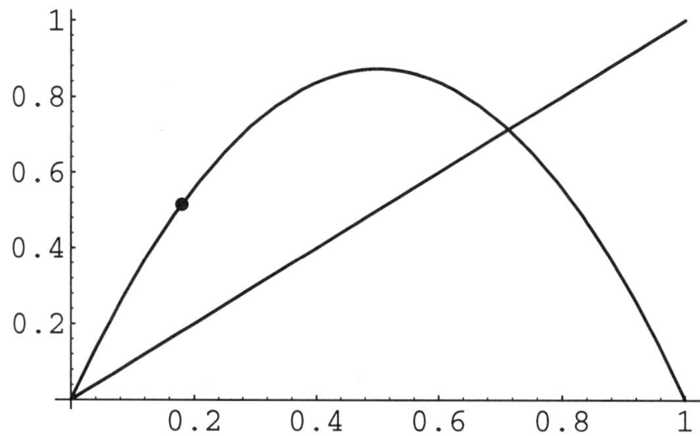

2. Conjugacy

We mentioned previously that a change of variable may reduce a system to a simpler one. This idea is made more precise with the concept of conjugate maps.

Definition 2.1.1
Let I and J be two intervals and $\phi : I \to J$ be continuous, one-to-one, and onto. We say that ϕ is a **conjugacy** between $G : I \to I$ and $F : J \to J$ if

$$\phi(G(x)) = F(\phi(x)) \tag{2.1.5}$$

for all $x \in I$. The maps F and G are said to be **conjugate** (by ϕ).

Example 2.1.2 Let $G(x)=4x(1-x)$, $x\in[0,1]$, $F(y)=2y^2-1$, $y\in[-1,1]$, and $y=\phi(x)=1-2x$. Notice that ϕ is continuous, one-to-one, and onto from [0,1] to [−1,1]. We have

$$\phi(G(x)) = \phi(4x(1-x)) = -2(4x(1-x))+1 = -8x(1-x)+1 = 8x^2-8x+1.$$

Similarly,

$$F(\phi(x)) = F(1-2x) = 2(1-2x)^2-1 = 2(4x^2-4x+1)-1 = 8x^2-8x+1.$$

Hence, the two maps F and G are conjugate in the given intervals by the map ϕ.

Since ϕ is invertible, (2.1.5) implies that

$$G = \phi^{-1} o F o \phi, \tag{2.1.6}$$

where the symbol o stands for the composition of functions.

Example 2.1.3 With the maps F, G, and ϕ of Example 2.1.2 we have $\phi^{-1}(y)=(1-y)/2$ and we obtain

$$\phi^{-1}(F(\phi(x))) = \phi^{-1}(F(1-2x)) = \phi^{-1}(8x^2 - 8x + 1) =$$

$$\frac{-8x^2 + 8x - 1 + 1}{2} = 4x(1-x) = G(x).$$

From (2.1.6) we derive $GoG=\phi^{-1}oFo\phi o\phi^{-1}oFo\phi=\phi^{-1}oFoFo\phi$. Denoting FoF and GoG by F^2 and G^2, respectively, we obtain $G^2=\phi^{-1}oF^2o\phi$. The reader should keep in mind that unless otherwise stated, the symbol $F^2(x)$ means $F(F(x))$ and not $F(x)\times F(x)$. In general, using an induction argument and omitting the composition symbol between the functions, we arrive at

$$G^n = \phi^{-1}F^n\phi \quad \text{or, equivalently,} \quad \phi G^n = F^n\phi. \tag{2.1.7}$$

Equality (2.1.7) implies that G^n and F^n are conjugate by the same map ϕ for every n=1,2,... . The study of the dynamical system generated by G, i.e., the analysis of the orbits

$$x_n = G(x_{n-1}) = G^n(x_0), \quad x_0 \in I, \tag{2.1.8}$$

can be performed by investigating the orbits of the system governed by F. In fact, given an initial state $x_0 \in I$, and using the equality $x_n=G^n(x_0)$ we derive from (2.1.7)

$$x_n = G^n(x_0) = \phi^{-1}(F^n(\phi(x_0))). \tag{2.1.9}$$

In particular, if y_s is a fixed point of F, then $x_s=\phi^{-1}(y_s)$ is a fixed point of G:

$$G(x_s) = \phi^{-1}(F(\phi(x_s))) = \phi^{-1}(F(y_s)) = \phi^{-1}(y_s) = x_s.$$

Similarly, if $\{y_0,\dots,y_{p-1},\dots\}$ is a periodic orbit of period p of F, then $\{x_0=\phi^{-1}(y_0),\dots, x_{p-1}=\phi^{-1}(y_{p-1}),\dots\}$ is a periodic orbit of period p of G:

$$x_p= G^p(x_0) = \phi^{-1}(F^p(\phi(x_0))) = \phi^{-1}(F^p(y_0)) = \phi^{-1}(y_0) = x_0.$$

Moreover, for every q<p,

$$x_q = G^q(x_0) = \phi^{-1}(F^q(\phi(x_0))) = \phi^{-1}(F^q(y_0)) \neq \phi^{-1}(y_0) = x_0.$$

At the end of this chapter we shall analyze the implications of conjugacy with respect to stability and instability of stationary states and periodic orbits. In the meantime we illustrate the usefulness of this concept with some examples.

Example 2.1.4 We want to study the dynamical system generated in the real line by the map $G(x)=ax+b$. The numbers a and b are fixed.

Assume that $a\neq 1$ and set $y=\phi(x)=x-b/(1-a)$. Then $G(x)=\phi^{-1}(F(\phi(x)))$, where $F(y)=ay$. The study of the trajectories of the dynamical system generated by F is very simple [see (2.1.10) and (2.1.11)].

When a=1 the map $G(x)=x+b$ is a translation. Every orbit diverges to $+\infty$ when b>0 and to $-\infty$ when b<0. For b=0 every point is a stationary state.

Example 2.1.5 Let $I=[-1,1]$ and F, G : $I\to I$ be defined by $F(x)=2|x|-1$, $G(x)=2x^2-1$. The maps F and G are conjugate by $\phi : I\to I$, $\phi(x)=(2/\pi)\arcsin x$. First observe that ϕ is strictly increasing in $[-1,1]$ since $\phi'(x)=2/(\pi(1-x^2)^{1/2})>0$. Thus, ϕ is one-to-one. Moreover, $\phi(-1)=-1$, $\phi(1)=1$. Hence, ϕ is onto. Let us verify that $\phi(G(x))= F(\phi(x))$ for every $x\in[-1,1]$. Both F and G are even functions and ϕ is an odd function. It follows that $F(\phi(x))$ and $\phi(G(x))$ are even. Consequently, it is enough to prove the equality when $x\in[0,1]$, i.e.,

$$(4/\pi)\arcsin(x-1) = F(\phi(x)) = \phi(G(x)) = (2/\pi)\arcsin(2x^2-1).$$

The equality holds if and only if $2\arcsin x-\arcsin(2x^2-1)=\pi/2$ for every $x\in[0,1]$. This is true when $x=0$. To make sure that it holds for every $x\in[0,1]$, we simply need to show that the derivative of the function $k(x)=2\arcsin x-\arcsin(2x^2-1)$ is 0 for $x\in(0,1)$. In fact, in this case, k is a constant function and the equality holds for every $x\in[0,1]$. We leave this exercise to the reader (see Problem 9). The plots of $\phi(G(x))$ and $F(\phi(x))$ (see Fig. 2.1.4) confirm the theoretical analysis.

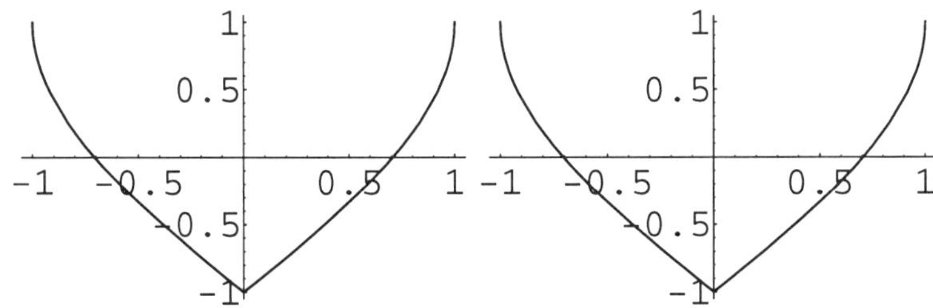

Fig. 2.1.4 On the left we have the graph of $\phi(G(x))$ and on the right the graph of $F(\phi(x))$ in $[-1,1]$. We see that the two graphs are identical, thus implying the equality $\phi(G(x))=F(\phi(x))$ for all $x\in[-1,1]$.

The only fixed points of F are 1 and $-1/3$. Since $\phi(1)=1$ and $\phi(-1/2)=-1/3$, we obtain that 1 and $-1/2$ are the only fixed points of G. They can be obtained directly from the quadratic equation $2x^2-1=x$. The periodic orbit of period 2 of F: $\{y_0=-3/5, y_1=1/5\}$ is mapped into the periodic orbit of period 2 of G: $\{x_0=-\sin(.3\pi),$

$x_1=\sin(.1\pi)\}$. A similar approach can be used to find periodic orbits of period 3,4, ... of G.

The study of the dynamical system governed by F is certainly easier than the analysis of the one generated by G. The conjugacy allows us to transfer the results found for F into corresponding results for G.

Linear and affine systems

Assume that in (2.1.2) the function F has the form $F(\mathbf{a},x)=ax+b$. Then the dynamical system is said to be **linear** if b=0 and **affine** if b≠0. In Example 2.1.2, we have seen that the system

$$x_{n+1} = ax_n + b, \qquad a \neq 1, \tag{2.1.10}$$

can be studied by analyzing the system

$$x_{n+1} = ax_n. \tag{2.1.11}$$

The orbits of (2.1.11) are of the form $\{a^n x_0 : n=1,2,..., x_0 \in \mathbf{R}\}$. Thus, if $|a|<1$, every orbit converges to the origin. When $|a|>1$, every nonstationary orbit diverges to infinity. For a=1, every orbit is stationary, while for a=–1, every nonstationary orbit is periodic of period 2.

These results for (2.1.11) provide corresponding results for (2.1.10). For a≠1 the stationary state of (2.1.10) is $x_s=b/(1-a)$. When $|a|<1$ every orbit of (2.1.10) converges to x_s, while when $|a|>1$, every nonstationary orbit diverges to infinity. For a=–1, every nonstationary orbit is periodic of period 2.

Another way to study (2.1.10) when $|a|\neq 1$ is to find the fixed point $x_s=F(a,x_s)= b/(1-a)$ and then write

$$ax + b = a(x-x_s) + ax_s + b = a(x-x_s) + x_s \tag{2.1.12}$$

(since $ax_s+b=x_s$). Given x_0 we obtain

$$x_1 = a(x_0-x_s)+x_s,$$
$$x_2 = a(x_1-x_s)+x_s = a(a(x_0-x_s)+ x_s-x_s)+x_s = a^2(x_0-x_s)+x_s$$

and, in general,

$$x_n = a^n(x_0-x_s)+x_s. \tag{2.1.13}$$

This result confirms our previous statement that $x_n \to x_s$ if $|a|<1$ and $x_n \to \infty$ if $|a|>1$.

Example 2.1.6 Let $F(x)=(2/3)(x-1)$. The fixed point of F is $x=-2$. According to (2.1.12), F can be rewritten in the form $F(x)=(2/3)(x+2)-2$. Therefore, using (2.1.13) we have $F^n(x)=(2/3)^n(x+2)-2$, which shows that every orbit of the dynamical system $x_{n+1}=(2/3)(x_n-1)$ converges to –2.

Problems

1. Prove by induction that $G=\phi^{-1}F\phi$ implies that $G^n=\phi^{-1}F^n\phi$ for every $n\geq 1$.

2. Show that the two systems $y_{n+1}=y_n(3-4y_n)$, $y\in[0,.75]$ and $x_{n+1}=3x_n(1-x_n)$, $x\in[0,1]$ are topologically conjugate. (Hint : Set $y=\phi(x)=.75x$.)

3. Show that the two systems

$$y_{n+1} = y_n(a-by_n)\ ,\ y\in[0,\ a/b] \text{ and } x_{n+1} = ax_n(1-x_n),\ x\in[0,1]$$

are topologically conjugate. (Hint : Set x=by/a.)

4. Verify that the two systems

$$x_{n+1} = 4x_n(1-x_n)\ ,\ x\in[0,1] \text{ and } y_{n+1}=-2y_n{}^2+1,\ y\in[-1,1]$$

are conjugate. (Hint : Use a map of the form $\phi(x)$=ax+b and select a,b so that ϕ maps the interval [0,1] onto the interval [–1,1].)

5. Use the idea of Problem 4 to verify that the two systems $y_{n+1}= -2|y_n|+1$, $y\in[-1,1]$ and $x_{n+1}=G(x_n)$, $x\in[0,1]$, with

$$G(x) = \begin{cases} 2x & 0 \le x \le 1/2 \\ 2-2x & 1/2 < x \le 1 \end{cases}$$

are conjugate.

6. Verify that $G(x)=2x^2-1$ and $T(x)=4x^3-3x$ map the interval [–1,1] onto itself. Explain why the two maps cannot be conjugate in this interval.

7. The dynamical system $x_{n+1}=\cos x_n$ has a fixed point $x_s\neq 0$. Find a dynamical system that is conjugate to the given system and whose fixed point is $x_s=0$.

8. Let $\phi(x)=(2/\pi)\arcsin x$. Find $\psi=\phi^{-1}$ and show that $\psi(F(y))=G(\psi(y))$, $y\in[0,1]$, where F and G are the maps in Example 2.1.5.

9. Show that $(d/dx)[2\arcsin x-\arcsin(2x^2-1)]=0$ for every $x\in(0,1)$.

10. Find two periodic orbits of period 3 of $F(x)=2|x|-1$. Using the conjugacy given in Example 2.1.5, find the corresponding orbits of $G(x)=2x^2-1$.

11. Show that the dynamical system governed by $F(\mathbf{a},x)=x(a-bx)$ changes the interval I=[0,a/b] into itself, provided that $0\le a\le 4$.

12. Find the fixed point of $F(x)=-3x/4+.5$ and write the dynamical system governed by F in the form (2.1.13).

13. Show that the dynamical system $x_{n+1}=-x_n+2$ has only one stationary state and that every other orbit of the system is periodic of period 2.

14. Find the fixed point of the system $x_{n+1}=3x_n+2$ and write the system in the form (2.1.13).

Section 2. **SINKS AND SOURCES**

In the next two sections we study stability and instability of fixed points and periodic orbits for one-dimensional dynamical systems. We start with properties of fixed points and periodic orbits that are peculiar to the one-dimensional case.

1. **Stationary States and Periodic Orbits**

Recall from Chapter 1 that the fixed points of the dynamical system (2.1.2) are the solutions of the equation

$$x = F(\mathbf{a},x). \tag{2.2.1}$$

Example 2.2.1 Let $F(a,b,x)=ax^3-bx$. The stationary states are obtained from the equation $x=ax^3-bx$, whose solutions are $x=0$ and $x=\pm((1+b)/a)^{1/2}$.

Similarly, the periodic points of period p are the solutions of the equation

$$x = F^p(\mathbf{a},x), \tag{2.2.2}$$

which are not solutions of the equation

$$x = F^m(\mathbf{a},x) \tag{2.2.3}$$

for all $m<p$. In other words, a point x_0 is **periodic of period** p if x_0 is a fixed point of F^p and it is not a fixed point of F^m for all $m<p$. The integer p is called the **period** of the orbit. Notice that the points $x_1=F(\mathbf{a},x_0),\ldots,x_{p-1}=F(\mathbf{a},x_{p-2})$ are also periodic of the same period p. As mentioned in Chapter 1, when we say that $O(x_0)$ is **periodic of period** p we mean that p is the **period** of $O(x_0)$.

Recall that a periodic orbit of period p will usually be identified in two ways:

- We may simply say "let $O(x_0)$ be a periodic orbit of period p"; or
- "Let $\{x_0,\ldots,x_{p-1},\ldots\}$ be a periodic orbit of period p."

Sometimes, for the sake of simplicity, we use the symbol $\{x_0,\ldots,x_{p-1}\}$ to denote a periodic orbit of period p.

Example 2.2.2 The periodic points of period 2 of the dynamical system governed by $F(x)=4x(1-x)$ are the solutions of the equation

$$x = 4\times 4x(1-x)\times[1-4x(1-x)] = 16x(1-x)[1-4x(1-x)],$$

which are not solutions of the equation $x=4x(1-x)$, i.e. $x\neq 0$ and $x\neq 3/4$. Assuming that $x\neq 0$, we obtain

$$1 = 16(1-x)(1-4x+4x^2) \quad \text{or} \quad 64x^3-128x^2+80x-15=0.$$

Dividing by $x-3/4$, we obtain $16x^2-20x+5=0$, whose solutions are

$$x_0 = (5 - \sqrt{5})/8, \quad x_1 = (5 + \sqrt{5})/8.$$

These coincide with the solutions of Example 1.2.6 for a=4 (see also Example 1.2.7). Since the points x_0, x_1 are the only periodic points of period 2 of F we have $F(x_0)=x_1$ and $F(x_1)=x_0$. Thus, $F^2(x_0)=x_0$ and $F^2(x_1)=x_1$. Recall that, unless otherwise stated, $F^2(x)$ means $F(F(x))$.

As mentioned in Chapter 2, Section 1 the fixed points of $F(a,x)$ are the x-coordinates of the intersections of its graph with the line $y=x$ (see Fig. 2.2.1). Similarly, the periodic points of period 2 are the x-coordinates of the intersections of the graph of $F^2(a,x)$ with the line $y=x$, that do not belong to the graph of $F(a,x)$.

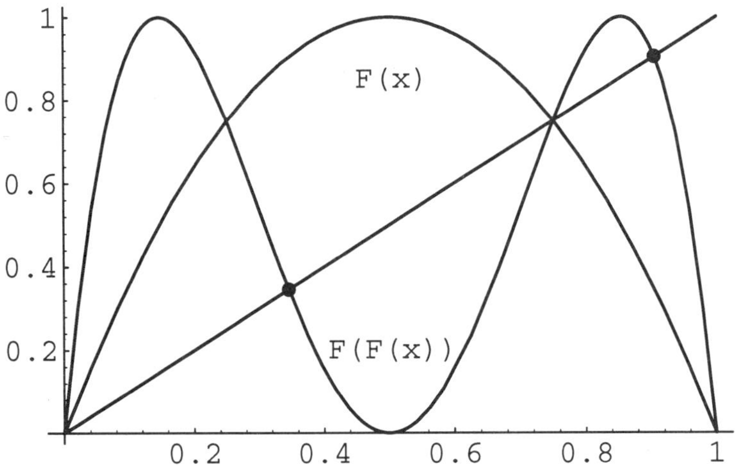

Fig. 2.2.1 Graph of $F(x)=4x(1-x)$ and of its second iterate, $F(F(x))$. The x-coordinates of the points where the graph of the second iterate intersects the line $y=x$ are fixed points of $F(F(x))$. The x-coordinates of the two points marked in the graph are not fixed points of F. Hence, they are periodic points of period 2 of F.

One question that remains to be answered is the following. Aside from solving equation (2.2.1) or (2.2.2), which may pose a nontrivial challenge if they involve transcendental functions or polynomials of high order, how can one be sure that there are fixed points and periodic orbits? The following discussion helps in answering this question. We shall use a well-known theorem from calculus.

Theorem 2.2.1 *Assume that* $h : [a,b] \to \mathbf{R}$ *is continuous. Then the image of* h *is a bounded and closed interval.*

The above theorem implies the following result, which is useful for proving the existence of fixed points.

Corollary 2.2.1 (Intermediate Value Theorem) *Let* $h : [a,b] \to \mathbf{R}$ *be continuous. Assume that* h *changes sign in* [a,b]. *Then there exists* $c \in [a,b]$ *such that* $h(c)=0$.

Proof. By Theorem 2.2.1, the image of h is a bounded and closed interval. Since h changes sign in [a,b], the interval must contain 0. QED

From Corollary 2.2.1 we derive two useful conditions for the existence of fixed points.

Corollary 2.2.2 *Let* $F : [a,b] \to \mathbf{R}$ *be continuous and such that*

- *either* $F(a)$ *and* $F(b) \in [a,b]$;
- *or* a *and* $b \in F([a,b])$.

Then F *has a fixed point in* $[a,b]$.

Proof. When $a=F(a)$ or $b=F(b)$ there is nothing to prove, since either a or b are fixed points. Assume that both equalities are false. Define $h(x)=x-F(x)$ and notice that in both cases h changes sign in the interval $[a,b]$. In fact, in the first case we have $h(a)<0<h(b)$. In the second case there are $c,d \in [a,b]$ such that $F(c) \leq a$ and $F(d) \geq b$. Hence, $h(c)>0>h(d)$. In both cases there is a point $x_s \in (a,b)$ such that $h(x_s)=0$, i.e., $x_s=F(x_s)$. QED

Example 2.2.3 Let $F(x)=x-e^{-x}+2$. Then $h(x)=x-F(x)=e^{-x}-2$. Since $h(-1)>0$ and $h(0)<0$, the function F has a fixed point in the interval $[-1,0]$.

Example 2.2.4 Let $F(x)=-e^{x}+\cos(x+1)$. Then $h(x)=x-F(x)=x-\cos(x+1)+e^{x}$. Since $h(-1)=-2+e^{-1}<0$ and $h(0)=-\cos 1+1>0$, F has a fixed point in $[-1,0]$.

The existence of periodic points of period p can be established using the strategy of Corollary 2.2.2 with F replaced by F^p. However, we have to make sure that the fixed points of F^p are not fixed for F^m with $m<p$.

The existence of a periodic orbit of period 2 implies the existence of a fixed point. In fact, let $p=F(q)$ and $q=F(p)$, with $p \neq q$. We can obviously assume that $p<q$. Then, $F(p)>p$ and $F(q)<q$ and, by Corollary 2.2.2, F has a fixed point in $[p,q]$.

There is a more general and remarkable result, due to the Russian mathematician A.N. Sarkovskii [Sarkovskii, 1964], which provides a hierarchical arrangement of the fixed points of a continuous function F which maps an interval $I \subset \mathbf{R}$ into itself.

The statement of the result requires that we first present the natural numbers $\mathbf{N}=\{1,2,3,...\}$ in a different order. We shall call it Sarkovskii's **ordering** of **N**. The numbers are organized in a two-dimensional array with infinitely many rows and columns. The first row contains all odd numbers, starting from 3, in increasing order. The second row contains, in increasing order, all integers that are obtained by multiplying the elements of the first row by 2. The third row contains all integers that are obtained by multiplying the elements of the first row by $4=2^2$, etc. . The last row contains all powers of 2 in decreasing order, and ends with 2, 1.

3	5	7	9	11	13	15.....................	
6	10	14	18	22	26	30.....................	
12	20	28	36	44	52	60.....................	
24	40	56	72	88...			
....							
....							
....			32	16	8	4	2 1.

The reader can verify that every integer is included in the list above, and appears in it exactly once.

Theorem 2.2.2 (Sarkovskii) *Let* I *be an interval and let* $F : I \to I$ *be continuous. Assume that* F *has a periodic orbit of period* p. *Then* F *has a periodic orbit of every period, which follows* p *in the Sarkovskii ordering of* **N**.

From Theorem 2.2.2 we derive that a one-dimensional system with a periodic orbit of period 3 has a periodic orbit of every period. Moreover, the presence of a periodic orbit of period p>1 implies the presence of a fixed point. These results are certainly remarkable given the modest assumption from which they are derived.

Example 2.2.5 Consider the map $F(x)=2|x|-1$. F has the periodic orbit of period 3: $x_0= -1/7$, $x_1=-5/7$, and $x_2=3/7$ (see Problem 10 of Section 1, Subsection 1). Hence, we can apply Sarkovskii's theorem and obtain the existence of periodic points of every period.

It is possible to construct a map with a periodic point of period 5, but not a periodic point of period 3 (see Fig. 2.2.2), as the following example shows.

Example 2.2.6 Let $F:[1,5]\to[1,5]$ be such that F(1)=3, F(2)=5, F(3)=4, F(4)=2, and F(5)=1. In each segment [i,i+1], i=1,2,3,4, define F so that its graph is a straight line segment joining (i, F(i)) with (i+1, F(i+1)). The reader should verify that F does not have any periodic orbit of period 3 (see Problem 12).

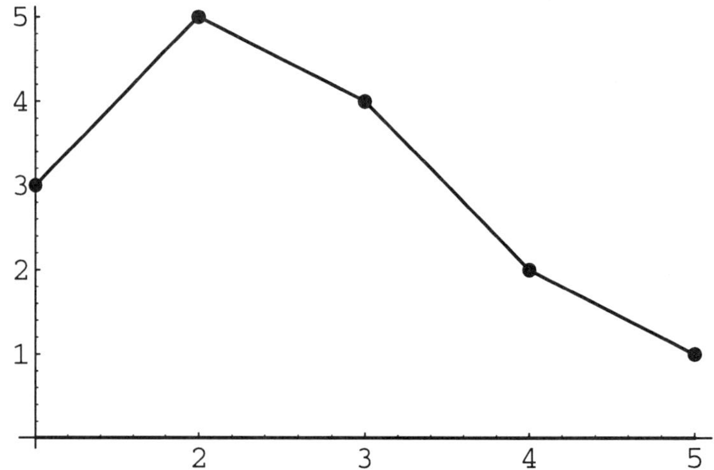

Fig. 2.2.2 Example of a map with a periodic orbit of period 5 but not of period 3.

Problems

1. Discuss the stationary states of the system $F(a,b,x)=ax^3-bx$ as functions of the parameters a and b.

2. Assume that $x\in I$ has the property

$$x = F^2(x) \quad \text{and} \quad x = F^3(x).$$

Prove that x=F(x). In general, prove that from

$$x = F^n(x) \quad \text{and} \quad x = F^m(x),\ n > m,$$

we derive $x=F^{n-m}(x)$.

3. Let F : [0,1]→[0,1] be defined by

$$F(x) = \begin{cases} 2x & 0 \le x \le .5 \\ 2(1-x) & .5 < x \le 1. \end{cases}$$

Find a periodic point of period 4 of F.

4. Let F be as in Problem 3. Find a periodic point of period 5.

5. Let F(a,x)=–a|x|+1. Find the values of a for which the function maps [–1,1] into itself. Find those values of a∈ [0,2] for which the map has a periodic orbit of period 3.

6. * Let F(x)=x+4cos$x$. Show that F has infinitely many fixed points. Verify graphically that the map has periodic points of period 2.

7. * Let F(x)=–2|x|+1. Plot the graph of F and of F^2. Derive from the two graphs the number and location of the periodic points of period 2 of F.

8. * Let F(x)=2x–[2x]. Plot the graph of F and the graph of its third iterate. Derive the number and location of the periodic points of period 3 of F.

9. Let F(x)=x–arctan$x$+2. Show that the discrete dynamical system governed by F does not have any fixed point or periodic orbit of any period.

10. On the Graph for Problem 10 mark the fixed points of the map.

Graph for Problem 10

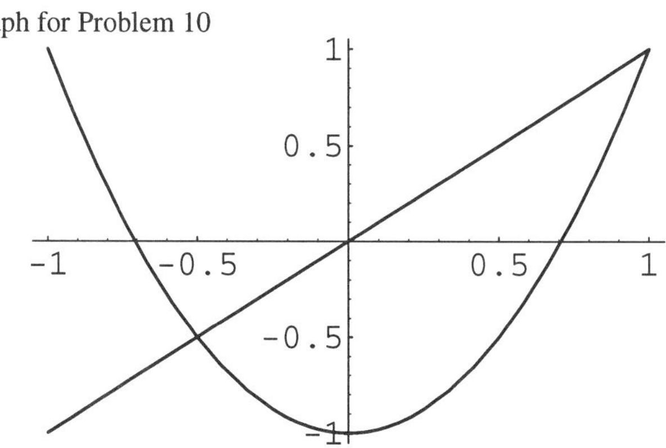

Graph for Problem 11

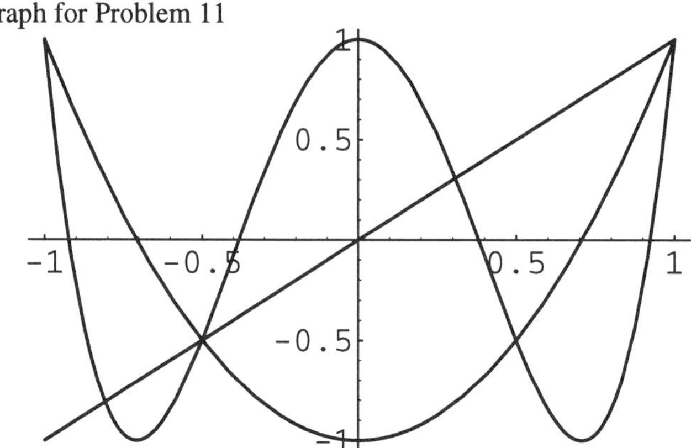

11. On the Graph for Problem 11, mark the periodic points of period 2 that are not fixed points. The map is the one from Problem 10.

12. * Verify that the map of Example 2.2.6 does not have any periodic point of period 3. Find a periodic point of period 2.

13. Prove that every natural number appears once and exactly once in the Sarkovskii ordering of **N**. (Hint: Recall that every integer can be uniquely factored as a product of prime numbers.)

2. Sinks

In Chapter 1 we introduced the idea of stability of stationary states and periodic orbits. We present here a condition that ensures stability, together with some other notions and results strictly related to stability and instability of fixed points and periodic orbits. Here, and in the sequel, we frequently use Lemma 2.2.1 and Theorem 2.2.3. Lemma 2.2.1 is sometimes called the permanence of sign theorem since it implies that a continuous function which is different from 0 at a point x_0 of its domain does not change sign at points x close enough to x_0.

Lemma 2.2.1 *Let* I *be an interval and* $g : I \to \mathbf{R}$ *be continuous at a point* $x_0 \in I$. *Assume that* $|g(x_0)| < 1$. *Let* $k \in (|g(x_0)|, 1)$. *Then there is* $r > 0$ *such that*

$$|g(x)| \leq k \quad \textit{for all } x \in [x_0 - r, x_0 + r] \cap I. \tag{2.2.4}$$

Remark 2.2.1 Lemma 2.2.1 remains true if < and ≤ are replaced by > and ≥, respectively. Moreover, for both versions to be true it is enough that

$$\lim_{x \to x_0} g(x) = L, \qquad |L| < 1 \text{ or } |L| > 1.$$

Theorem 2.2.3 (Mean Value Theorem, MVT) *Let* $f : [a, b] \to \mathbf{R}$ *be continuous on* $[a,b]$ *and differentiable on* (a,b). *Then there exists* $c \in (a,b)$ *such that*

$$f(b) - f(a) = f'(c)(b - a). \tag{2.2.5}$$

We can now establish a condition for x_s to be stable.

Theorem 2.2.4 *Let* I *be an open interval and* x_s *be a fixed point of a continuous function* F : I→I. *Assume that there is* r>0 *such that* F *is differentiable on* (x_s-r, x_s+r), *except possibly at* x_s *and* $|F'(x)|\leq 1$. *Then* x_s *is stable.*

Proof. Let $x_0 \in J$. By the MVT, $F(x_0)-F(x_s)=F'(c)(x_0-x_s)$. Hence,

$$|x_1 - x_s| = |F(x_0) - F(x_s)| \leq |x_0 - x_s|,$$

and in general,

$$|x_{n+1} - x_s| = |F(x_n) - F(x_s)| \leq |x_n - x_s| \leq \ldots \leq |x_0 - x_s|.$$

Thus, the stationary state x_s is stable (choose δ=r in the definition of stability). QED

Example 2.2.7 Let

$$F(x) = \begin{cases} x & x \leq 0 \\ .9x & x > 0. \end{cases}$$

The origin is a stable fixed point since $|F'(x)|\leq 1$ for every $x\neq 0$.

Theorem 2.2.4 can easily be extended to periodic orbits of period p by rephrasing it for the pth iterate of F. Recall first that each point of the periodic orbit of period p $\{x_0, x_1, \ldots, x_{p-1}, \ldots\}$ is a fixed point of F^p.

Theorem 2.2.5 *Let* I *be an open interval and* $\{x_0,\ldots, x_{p-1},\ldots\}$ *be a periodic orbit of period* p *of a continuous function* F : I→I. *Assume that there is* r>0 *such that for every* j=0,1,...,p−1, F *is differentiable on* (x_j-r, x_j+r) *except possibly at* x_j *and*

$$\left|\frac{d}{dx} F^p(x)\right| \leq 1. \tag{2.2.6}$$

Then the periodic orbit is stable.

Proof. Since each point x_j, j=0,1,...,p−1 is a fixed point of F^p, inequality (2.2.6) and Theorem 2.2.4 imply that x_j is a stable fixed point for F^p. QED

Example 2.2.8 Let $F(x)=1-x$. The periodic orbit of period 2 {0,1} is stable since $(d/dx)F^2(x)=1$ for every $x\in \mathbf{R}$.

Conditions more demanding than stability or instability can be imposed, leading to the definitions of **sink** and **source** presented below.

Definition 2.2.1
Let I be an open interval and F : I→I be continuous. A fixed point x_s of F is said to be a **sink** (or an **attractor**) if there exists r>0 such that for all $x_0\in I$ which are at distance less than or equal to r from x_s, $|x_0-x_s|\leq r$, we have $|x_n-x_s|\to 0$ as $n\to\infty$.

We say that a periodic orbit of period p $\{x_0,...,x_{p-1},...\}$ of F is a **sink** (or an **attractor**) if each state x_i is a sink for the dynamical system governed by F^p.

We now establish a condition for x_s to be a sink.

Theorem 2.2.6 *Let* I *be an open interval and* x_s *be a fixed point of a continuous function* $F : I \to I$. *Let* $d>0$ *be such that* F *is differentiable in* (x_s-d, x_s+d), *with continuous derivative at* x_s. *Then* $|F'(x_s)|<1$ *implies that* x_s *is a sink.*

Proof. Choose $k\in(|F'(x_s)|,1)$. By Lemma 2.2.1, there is a positive number $r<d$ such that $|F'(x)|\le k<1$ for all $x\in[x_s-r, x_s+r]=J$. Select an initial state $x_0\in J$. Then, by the mean value theorem, we have

$$x_1 - x_s = F(x_0) - F(x_s) = F'(c)(x_0 - x_s)$$

for some c between x_0 and x_s. Consequently,

$$|x_1 - x_s| \le k|x_0 - x_s|.$$

This inequality shows that x_1 is again inside J and we can apply to x_1 the same argument used for x_0. We obtain

$$|x_2 - x_s| \le k|x_1 - x_s| \le k^2|x_0 - x_s|.$$

Continuing in this manner we arrive, after n steps, at the inequality

$$|x_n - x_s| \le k^n|x_0 - x_s|.$$

Since $k\in[0,1)$ we derive that $|x_n-x_s|\to 0$ as $n\to\infty$. QED

Remark 2.2.2 In the case of Theorem 2.2.6 we can estimate how fast a sequence of iterates starting from a point $x_0\in J$ converges to x_s, without explicitly using the fixed point. In fact, from

$$|x_0 - x_s| \le |x_0 - x_1| + |x_1 - x_s| \quad \text{and} \quad |x_1 - x_s|\le k|x_0-x_s|$$

we obtain $|x_0-x_s|(1-k)\le|x_1-x_0|$. Substitution into $|x_n-x_s|\le k^n |x_0-x_s|$ gives

$$|x_n - x_s| \le \frac{k^n}{1-k}|x_1 - x_0|. \tag{2.2.7}$$

This gives an estimate of how close is x_n to x_s without the knowledge of x_s.

Example 2.2.9 Let $F(a,b,x)=ax^3-bx$. Regardless of the choice of the parameters a and b, the point $x_s=0$ is a fixed point of F. Since $F_x(a,b,x)=3ax^2-b$, we have $F_x(a,b,0)=-b$. Thus, for $|b|<1$, $x_s=0$ is a sink.

For **periodic orbits** of period p the condition for being a sink is established using F^p, the pth iterate of F, in place of F. First notice that using the chain rule and the periodicity of the orbit, we have

$$\frac{d}{dx} F^p(x_i) = F'(x_{i-1})F'(x_{i-2})...F'(x_0)F'(x_{p-1})...F'(x_i), \quad 0 \le i \le p-1,$$

i.e., the derivative of the pth iterate F^p of F at x_i is the product of the derivatives of F at the points $x_0, x_1,...,x_{p-1}$. Consequently, the fundamental equality

$$\frac{d}{dx} F^p(x_i) = \frac{d}{dx} F^p(x_j), \qquad 0 \le i, j \le p-1 \tag{2.2.8}$$

holds. For example, for p=2,

$$\frac{d}{dx} F^2(x_0) = F'(F(x_0))F'(x_0) = F'(x_1)F'(x_0) = F'(x_0)F'(x_1) = \frac{d}{dx} F^2(x_1).$$

As a consequence of these remarks and of Theorem 2.2.6, we obtain the following result.

Theorem 2.2.7 *Let* I *be an open interval and* $\{x_0,..., x_{p-1},...\}$ *be a periodic orbit of period* p *of a continuous function* $F : I \to I$. *Let* d>0 *be such that* F *is differentiable in* (x_i-d, x_i+d) *with a continuous derivative at* x_i *for all* i=0,1,...,p–1. *Then the periodic orbit is a sink if*

$$\left|\frac{d}{dx} F^p(x_0)\right| < 1. \tag{2.2.9}$$

Proof. Recall that each point x_i, i=0,1,...,p–1 is a fixed point of F^p. By (2.2.9) and Theorem 2.2.6, the point x_0 is a sink for F^p. Equality (2.2.8) together with inequality (2.2.9) imply that each point x_i is a sink for the dynamical system defined by $F^p : I \to I$. Hence, the periodic orbit is a sink. QED

Example 2.2.10 Let $F(x)=3.2x(1-x)$. Then

$$x_0 = \frac{2.1 + \sqrt{.21}}{3.2}, \qquad x_1 = \frac{2.1 - \sqrt{.21}}{3.2},$$

is a periodic orbit of period 2 (see Example 1.2.6). From $F'(x)=3.2-6.4x$, we derive

$$F'(x_0) = -1 - 2\sqrt{.21} \quad \text{and} \quad F'(x_1) = -1 + 2\sqrt{.21}.$$

Hence,

$$(F^2)'(x_0) = (-1 - 2\sqrt{.21})(-1 + 2\sqrt{.21}) = 1 - .84 = .16.$$

The periodic orbit $\{x_0, x_1\}$ is a sink.

Inequality (2.2.9) is certainly verified if $|F'(x_i)|<1$ for all i=0,1,...,p–1. However, it may happen that $|F'(x_i| \ge 1$ for some $i \in \{0,1,...,p-1\}$, but (2.2.9) still holds. In particular, we may have $|F'(x_i)|>1$ for some $i\in\{0,1,...,p-1\}$. The following example illustrates this situation and it shows the peculiarity of the dynamical evolution of orbits close to a periodic sink with this property.

Example 2.2.11 Let

$$F(x)=\begin{cases}-x/4+1 & 0\leq x<3/4\\ -(13/4)(x-1) & 3/4\leq x\leq 1.\end{cases}$$

Notice that $F(0)=1$ and $F(1)=0$. Hence $\{0,1\}$ is a periodic orbit (of period 2). Moreover, $F'(0)=-1/4$ and $F'(1)=-13/4$. Hence, at $x_1=1$ the absolute value of the derivative is larger than 1. However, the derivative of the second iterate at 0 or at 1 is 13/16. Thus, the orbit is a sink.

Analyzing the behavior of an orbit whose initial state is very close to 0 we observe that in going from a neighborhood of 0 to a neighborhood of 1 the orbit gets closer to the periodic orbit, while in the successive iteration it gets farther. However, at the end of each two-step cycle, the state of the orbit is closer to the corresponding periodic state than it was before.

Sometimes the derivative of the function governing a dynamical system satisfies an inequality weaker than the one required by Theorem 2.2.6. In fact, it may happen that there is $d>0$ such that $|F'(x)|<1$ for all $x\in(x_0-d,\ x_0+d)$, except at $x=x_s$, where the derivative may not exist or $|F'(x_s)|=1$. Under these weaker assumptions, we can prove that x_s is a sink. To obtain the result, and other results presented later in this chapter, we need a definition and a lemma from calculus.

Definition 2.2.2
A sequence $\{x_n,\ n=1,2,...\}$ is **increasing (strictly increasing)** if

$$x_n\leq x_{n+1}\qquad (x_n<x_{n+1})\qquad n=1,2,\ldots.$$

By reversing the inequalities we obtain a **decreasing (strictly decreasing)** sequence.

In some occasions the entire sequence does not satisfy the definition, but if finitely many terms are deleted, the sequence that is left does. In such cases we say that the original sequence is **eventually increasing** or **decreasing**.

Lemma 2.2.2 *Let* $\{a_n : n=1,2,3,...\}$ *be an eventually decreasing sequence. Assume that* $\{a_n : n=1,2,3,...\}$ *is bounded below. Then the sequence is convergent, i.e., there exists a real number* a *such that*

$$|a_n-a|\to 0\ \textit{as}\ n\to+\infty.$$

Lemma 2.2.2 holds true if **decreasing** is replaced by **increasing** and **below** is replaced by **above**. We can now prove the following theorem.

Theorem 2.2.8 *Let* I *be an open interval and* x_s *be a fixed point of a continuous function* $F: I\to I$. *Let* $r>0$ *be such that* F *is differentiable in* $(x_s-r,\ x_s+r)$, *except possibly at* x_s *and* $|F'(x)|<1$. *Then* x_s *is a sink.*

Proof. First, notice that given $x_0 \in [x_s - r, x_s + r]$ and using the MVT, we have

$$|x_1 - x_s| = |F(x_0) - F(x_s)| < |x_0 - x_s|.$$

This shows that the sequence $\{|x_n - x_s|, n=0,1,...\}$ is strictly decreasing. According to Lemma 2.2.2, it must be convergent to some $c \in [0,r)$. We show that c=0. The set of limit points $L(x_0)$ of $O(x_0)$ is contained in (and may coincide with) the set $\{x_s - c, x_s + c\}$. According to (1.3.2), we have $F(L(x_0)) = L(x_0)$. However, again by the MVT, c>0 implies that

$$|F(x_s + c) - F(x_s)| < |x_s + c - x_s| = c \quad \text{and} \quad |F(x_s - c) - F(x_s)| < |x_s - c - x_s| = c.$$

Therefore, both $F(x_s+c)$ and $F(x_s-c)$ are closer to x_s than c. Since this contradicts the equality $F(L(x_0))=L(x_0)$, we must have c=0. QED

Example 2.2.12 Let $F : \mathbf{R} \to \mathbf{R}$ be defined by $F(x) = -x^3 + 3x^2 - 2x + 1$. Then $x_s = 1$ is a fixed point of F. Since $F'(1)=1$ we cannot use Theorem 2.2.6. Let us see if Theorem 2.2.8 can be applied to this case. We have $F'(x) = -3x^2 + 6x - 2$. To see how the derivative behaves for x close to 1 we can use the substitution $x = 1 + z$. We obtain $F'(1+z) = 1 - 3z^2$. We see that for z close to 0 (x close to 1) the derivative is positive and smaller than 1. Hence Theorem 2.2.8 can be applied and the fixed point is a sink.

Theorem 2.2.8 can easily be extended to periodic orbits of period p by rephrasing it for the pth iterate of F. We leave this as an exercise for the reader (see Problem 13).

Remark 2.2.3 Assume that a certain dynamical system has a periodic orbit (of period 2) $\{x_0, x_1\}$. The periodic orbit $\{x_1, x_0\}$ is different from $\{x_0, x_1\}$, since their initial states are not the same. However, at least in certain cases, it may be convenient to identify $O(x_1)$ with $O(x_0)$. For example, let $\{x_0, x_1\}$ be a sink. Select an initial condition y_0 sufficiently close to x_1. The orbit $O(y_0)$ converges to the periodic orbit $\{x_1, x_0\}$ and we must say so if the manner in which $O(y_0)$ approaches the periodic orbit is important. However, assume that the message we want to convey is simply that $L(y_0)=L(x_0)$. Then we may say that $O(y_0)$ converges to $O(x_0)$ or to $\{x_0, x_1\}$. This is particularly convenient when the period p of $O(x_0)$ is large, and there are a lot of points z such that $L(z)=L(x_0)$.

Problems

1. * Let $F(a,x) = ax - \arctan(x)$. Find the values of a for which $x=0$ is a sink.

2. * Let $F(a,x) = x - \ln(ax+1)$. Verify that 0 is a fixed point. Find the values of a for which 0 is a sink.

3. * Determine the values of the parameter a for which the stationary states of the system governed by $F(a,x) = ax^3 - (a-1)x$ are sinks.

4. * Determine the values of a for which the fixed points and the periodic orbit of period 2 of the system governed by $F(a,x) = -ax^2 + 1$ are sinks.

5. * Are there any values of a for which the periodic orbit of period 2 of the system governed by $F(a,x)=-a|x|+1$ is a sink?

6. * Let $F(a,x)=ax(1-x)$. Determine the values of the parameter a for which the fixed points and periodic orbit(s) of period 2 of F are sinks.

7. * Let $x_n=3^{1/n}$, n=1,2,... . Show that the sequence is decreasing and bounded below. Find its limit.

8. * Let $x_n=(n-1)/n$, n=1,2,... . Show that the sequence is increasing and bounded above. Find its limit.

9. * Consider the sequence $x_0=2$, $x_1=4+x_0r$, $x_2=4+x_1r$,..., where $r\in(0,1)$. Prove that the sequence is increasing and that it is convergent.

10. Let $k\in(0,1)$ and let a, and x_0 be real numbers. Define $x_1=a+kx_0$, $x_2=a+kx_1$... . Study the sequence $\{x_n$: n=0,1,...$\}$ for different choices of a and x_0. Prove that the sequence is always convergent.

11. * Let $F(x)=x^3-\arctan x$. Verify that $x=0$ is a sink (Hint: Apply Theorem 2.2.8.) Explore the conclusion numerically.

12. * Let $F(a,x)=ax^3-\arctan x$. Prove that for every a>0 the origin is a sink.

13. Rephrase Theorem 2.2.8 for periodic orbits.

14. Assume that x_s is a fixed point of F and there exists r>0 such that $|(d/dx)F^2(x)|<1$ whenever $0<|x_s-x|<r$. Prove that x_s is a sink.

15. * Use the result of Problem 14 to prove that $x_s=2/3$ is a sink for the map $F(x)=3x(1-x)$. Notice that Theorems 2.2.6 and 2.2.8 cannot be applied in this case. Verify numerically that for x_0 very close to 2/3, the sequence $|x_n-2/3|\to 0$, but it is not decreasing.

3. Sources

Definition 2.2.3
Let I be an open interval, $F: I\to I$ be continuous. A fixed point x_s of F is a **source** (or a **repeller**) if there exists r>0 such that for every $x_0\in I$, $0<|x_0-x_s|\le r$, we can find a positive integer m with the property

$$|x_m - x_s| > r.$$

We say that a periodic orbit $O(x_0)$ of period p of F is a **source** (or a **repeller**) if each state of the orbit is a source for the dynamical system generated by F^p.

Theorem 2.2.9 *Let* I *be an open interval and* x_s *be a fixed point of a continuous function* F : I→I. *Assume that there exists* d>0 *such that* F *is differentiable in* $(x_s-d,\ x_s+d)$, *and* F′ *is continuous at* x_s. *Then,* $|F'(x_s)|>1$ *implies that* x_s *is a source.*

Proof. Choose a number $k\in(1,|F'(x_s)|)$. By Remark 2.2.1 there is 0<r<d such that $|F'(x)|\geq k$ for all $x\in[x_s-r,\ x_s+r]=J$. Let $x_0\in J$, $x_0\neq x_s$. By the MVT we have

$$x_1 - x_s = F(x_0) - F(x_s) = F'(c)\ (x_0 - x_s)$$

for some c between x_0 and x_s. Consequently,

$$|x_1 - x_s| \geq k|x_0 - x_s|.$$

If $|x_1-x_s|>r$ we are done. Otherwise, the reasoning above can be repeated for x_1 and gives

$$|x_2 - x_s| \geq k|x_1 - x_s| \geq k^2|x_0 - x_s|.$$

As long as $x_1, x_2, ..., x_{j-1}$ remain in the interval $[x_s-r,\ x_s+r]$ we obtain

$$|x_j - x_s| \geq k^j|x_0 - x_s|.$$

Since k>1, the orbit will eventually exit the interval $[x_s-r,\ x_s+r]$, i.e., there exists an integer m≥1 such that $|x_m-x_s|\geq k^m\,|x_0-x_s|>r$. Hence, x_s is a repeller. QED

Example 2.2.13 Let $F(a,b,x)=ax^3-bx$. Regardless of the choice of the parameters a and b, the point $x_s=0$ is a stationary state of the dynamical system governed by F. Since $F_x(a,b,0)=-b$, the stationary state $x_s=0$ is a source if $|b|>1$.

For periodic orbits of period p, the condition for being a source is established using F^p in place of F.

Theorem 2.2.10 *Let* I *be an open interval and* $\{x_0,..., x_{p-1},...\}$ *be a periodic orbit of period* p *of a continuous function* F : I→I. *Assume that there is* d>0 *such that* F *is differentiable in* $(x_i-d,\ x_i+d)$ *with a continuous derivative at* x_i *for all* i=0,1,...,p−1. *Then the periodic orbit is a source if*

$$\left|\frac{d}{dx} F^p(x_0)\right| > 1. \tag{2.2.10}$$

Proof. By Theorem 2.2.9 and equality (2.2.8) each point x_i is a repeller for the dynamical system defined by F^p. Hence, the periodic orbit is a source. QED

Example 2.2.14 Let $F(x)=3.6x(1-x)$. The orbit $\{x_0,x_1\}$:

$$x_0 = \frac{2.3 + \sqrt{.69}}{3.6}, \qquad x_1 = \frac{2.3 - \sqrt{.69}}{3.6},$$

is periodic of period 2. From $F'(x)=3.6-7.2x$, we derive

$$F'(x_0) = -1 - 2\sqrt{.69} \quad \text{and} \quad F'(x_1) = -1 + 2\sqrt{.69}.$$

Hence,

$$(F^2)'(x_0) = (-1 - 2\sqrt{.69})(-1 + 2\sqrt{.69}) = 1 - 2.76 = -1.76.$$

The periodic orbit $\{x_0, x_1\}$ is a source.

Inequality (2.2.10) is certainly satisfied if $|F'(x_i)|>1$ for all $i=0,1,2,...,p-1$. However, it can happen that for some x_i we have $|F'(x_i)|\leq 1$, but (2.2.10) is still true. The following example illustrates this situation.

Example 2.2.15 Let $F : [0,1]\to[0,1]$ be defined by

$$F(x) = \begin{cases} -2x + 1 & 0\leq x<1/4 \\ -(2/3)(x - 1) & 1/4\leq x\leq 1. \end{cases}$$

Then $\{0,1\}$ is a periodic orbit (of period 2) such that $|F'(0)|=2$ and $|F'(1)|=2/3$. Hence, at $x_1=1$ the absolute value of the derivative of F is smaller than 1. However, the derivative of the second iterate of F at 0 or at 1 is $4/3>1$. Hence, the orbit is a source.

Carefully analyzing the behavior of an orbit whose initial state is very close to 0 we see that in going from a neighborhood of 0 to a neighborhood of 1 the orbit gets farther away from the periodic orbit, while in the successive iteration it gets closer. However, at the end of each two-step cycle, the state of the orbit is farther away from the corresponding periodic state than it was before.

A result analogous to Theorem 2.2.8 ensures that a fixed point x_s is a source under conditions less demanding than those of Theorem 2.2.9.

Theorem 2.2.11 *Let* I *be an open interval and* x_s *be a fixed point of a continuous function* $F : I\to I$. *Let* $r>0$ *be such that* F *is differentiable in* (x_s-r, x_s+r) *except possibly at* x_s *and* $|F'(x)|>1$. *Then* x_s *is a source.*

Proof. Let x_0 be such that $0<|x_0-x_s|\leq r$. From the MVT we derive that $\{|x_n-x_s|, n=0,1, ...\}$ is strictly increasing as long as the states of the orbit $O(x_0)$ remain at a distance not larger than r from x_s. In fact ,

$$|x_{i+1} - x_s| = |F(x_i)- F(x_s)| = |F'(c_i)|\,|x_i-x_s| > |x_i - x_s|.$$

Assume that $\{|x_n-x_s|, n=0,1,...\}$ never exceeds r. By Lemma 2.2.2 the sequence is convergent, i.e., $|x_n-x_s|\to c\leq r$ as $n\to\infty$. Consequently, the orbit $O(x_0)$ either converges to x_s-c or to x_s+c or to the periodic orbit of period 2 $\{x_s-c,x_s+c\}$. The situation is analogous to the case, examined before, when x_s was a sink. This time from the inequalities $|F(x_s-c)-x_s|>c$ and $|F(x_s+c)-x_s|>c$ we obtain a contradiction with $F(L(x_0)= L(x_0)$. Hence, there is $m\geq 1$ such that $|x_m-x_s|>r$ and x_s is a source. QED

Example 2.2.16 Let $F : \mathbf{R}\to\mathbf{R}$ be defined by $F(x)=x^3-3x^2+4x-1$. The only fixed point of the dynamical system governed by F is $x_s=1$. Since $F'(x)=3x^2-6x+4$, and $F'(1) =1$ we cannot apply Theorem 2.2.9. To see if Theorem 2.2.11 can be used let us write $x=1+h$. Then $F'(1+h)=1+3h^2$. Hence, by Theorem 2.2.11, the fixed point $x_s=1$ is a source.

Theorem 2.2.11 can be extended to periodic orbits of period p by rephrasing it for F^p.

Remark 2.2.4 According to Definitions 2.2.1 and 2.2.3, a periodic orbit $\{x_0,..., x_{p-1},...\}$ is a sink or a source if each state of the orbit is a sink or a source for F^p. However, as announced in Chapter 1 (see Definition 1.2.5) the condition for being a sink or a source is given using only x_0 (see Theorems 2.2.7 and 2.2.10).

Instability

In Chapter 1 we introduced the notion of instability for a stationary state $\mathbf{x}_s$ or a periodic orbit $O(\mathbf{x}_0)$ of period p. Repelling stationary states or periodic orbits are obviously unstable. The converse is not true, as the following examples show.

Example 2.2.17 Let $F(x)=x-x^2$. The only fixed point is $x_s=0$ (see Fig. 2.2.3). For $0<x<1$ we have $|F'(x)|<1$ and $0<F(x)<1$. Thus every orbit $O(x_0)$ with $x_0\in[0,1]$ converges to 0 (use the same reasoning as in Theorem 2.2.8). For every $x<0$ we have $F'(x)>1$ and $F(x)<0$. The trajectory of every point $x_0<0$ goes to $-\infty$. The origin is an unstable point, but not a repeller. Some authors call 0 a semistable fixed point.

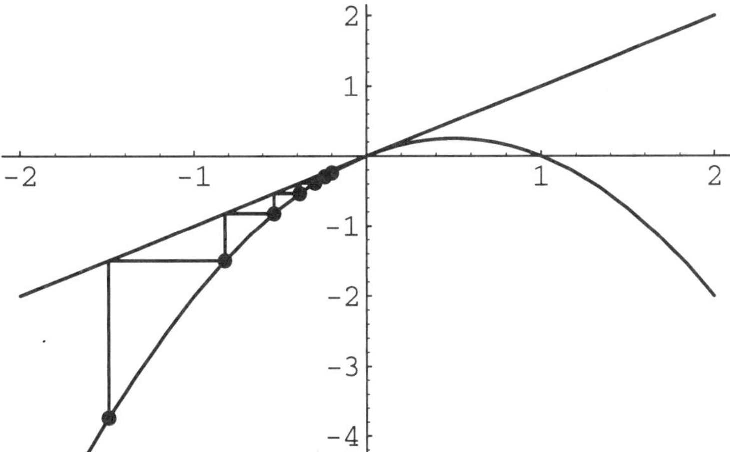

Fig. 2.2.3 Orbit O(−.15) of the dynamical system defined by the map of Example 2.2.17. It is evident that the orbit is going to $-\infty$.

Example 2.2.18 Let

$$F(x)=\begin{cases} -2x-1 & x\le -1 \\ -.5x+.5 & -1<x\le -.25 \\ -2.5x & -.25<x\le .25 \\ -.5x-.5 & .25<x\le 1 \\ -2x+1 & 1<x. \end{cases}$$

The interval [–1,1] is mapped onto itself by F, F(–1)=1, and F(1)=–1 (see Fig. 2.2.4). For every initial condition $x_0 \in [-1,1]$ except $x_s=0$, the orbit $O(x_0)$ converges to the periodic orbit {–1, 1}. For every $|x_0|>1$ the orbit goes to infinity.

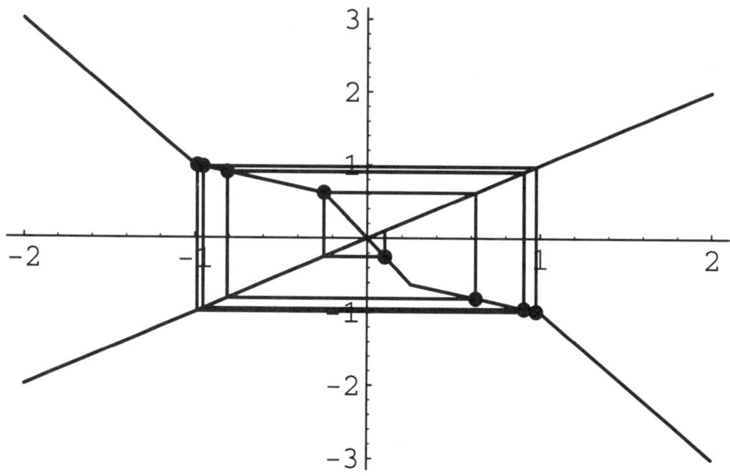

Fig. 2.2.4 Orbit O(.1) of the dynamical system defined by the map of Example 2.2.18. We see that the orbit converges to the periodic orbit of period 2: {–1,1}.

Problems

1. * Let $F(a,x)=ax-\arctan x$. Find the values of a for which $x=0$ is a source.

2. * Let $F(a,x)=a^2x-\ln(ax+1)$. Verify that $x=0$ is a fixed point. Find all values of a for which the point is a source.

3. * Determine the values of the parameter a for which the stationary states of the system governed by $F(a,x)=ax^3-(a-1)x$ are sources.

4. * Determine the values of a for which the fixed points and the periodic orbit of period 2 of the system governed by $F(a,x)=-ax^2+1$ are sources.

5. * Let $F(a,x)=ax(1-x)$. Determine for which values of a the periodic orbit of period 2 of F is a source.

6. * Let $F(x)=2x-\arctan(x)$. Verify that $x=0$ is a source.

7. * Let $F(a,x)=ax^3+\arctan x$. Are there any values of a>0 for which the origin is a source?

8. Let x_s be a fixed point of F. Assume that there exists r>0 such that $|(d/dx)F^2(x)|>1$ for all $x\in(x_s-r,x_s+r)$, $x\neq x_s$. Prove that x_s is a source.

9. Let

$$F(x)=\begin{cases}-.5x & x\le 0\\ -4x & 0<x.\end{cases}$$

Notice that 0 is a fixed point. Verify that Theorems 2.2.9 and 2.2.11 cannot be applied to F. Use the result of Problem 8 to prove that 0 is a source.

Section 3. **GLOBAL SINKS**

Until now we have confined our attention to local behavior of one-dimensional systems, in the sense that our analysis of stationary states and periodic orbits has been limited to the evolution of the system close to these points. We now want to gain some information on the **global behavior** of a dynamical system, at least in some simple cases. Similar situations on multidimensional systems are studied in Chapter 5.

Frequently, a numerical investigation of a system seems to suggest that all its orbits converge to a stationary state or to a periodic orbit. However, this first impression may be misleading. A system may have more than one stationary state, and it may also have several, or even infinitely many periodic orbits. In the dynamical evolution of the system, these stationary states and periodic orbits will neither converge to the fixed point nor to the periodic orbit where everything else seems to be going. When numerical evidence indicates that all orbits of a system on an invariant interval converge to a periodic orbit $O(x_0)$ of period $p>1$, Sarkovskii's theorem ensures the presence of exceptions to this behavior.

Theorem 2.3.1 below gives sufficient conditions for the existence of a unique stationary state x_s and for the convergence to x_s of every orbit of the system. The result is fairly general and will be extended, with minor changes, to dynamical systems in $\mathbf{R}^q$. The proof of the theorem is based on a fundamental result from calculus, which we state without proof. Recall that a function h is **increasing** in an interval I if $h(x)\le h(y)$ for every pair of points $x,y\in I$, $x<y$, and **strictly increasing** if the inequality $h(x)\le h(y)$ can be replaced by the stronger inequality $h(x)<h(y)$. **Decreasing** and **strictly decreasing** are defined in a similar manner.

Lemma 2.3.1 *Let* I *be an interval and* $h: I\to\mathbf{R}$ *be continuous on* I *and differentiable at every interior point of* I. *Assume that* $h'(x)\ge 0$. *Then* h *is increasing on* I. *If, in addition,* $h'(x)$ *is zero only finitely many times on every bounded interval* J *contained in* I, *then* h *is strictly increasing on* I.

We can now establish the following result.

Theorem 2.3.1 *Let* I *be an interval and* $F: I\to I$ *be continuous on* I *and differentiable at every interior point of* I. *Assume that*

- F *has a fixed point* x_s.
- $|F'(x)|\le 1$.
- *There is an open interval* $J\subset I$, $x_s\in J$, *such that* $|F'(x)|<1$, $x\in J$, $x\ne x_s$.

Then x_s *is the only fixed point of* F *and every orbit converges to* x_s.

Proof. Define $h(x)=x-F(x)$ and notice that $h(x)=0$ if and only if x is a fixed point of F. Hence, $h(x_s)=0$. The inequality $|F'(x)|\leq 1$ implies that $h'(x)=1-F'(x)\geq 0$ on I and $h'(x)>0$ on J except possibly at x_s. Hence, the function h is increasing on I and strictly increasing on J. Since $h(x_s)=0$ we have $h(x)<0$ $[x<F(x)]$ for all $x<x_s$ and $h(x)>0$ $[F(x)<x]$ for all $x>x_s$. Consequently, x_s is the only fixed point of F.

We now establish that every orbit converges to x_s. To start, let $x_0>x_s$ and select a point $y\in(x_s,x_0)\cap J$. Then, by the MVT,

$$x_1-x_s=F(x_0)-F(x_s)=F(x_0)-F(y)+F(y)-F(x_s)=F'(d)(x_0-y)+F'(c)(y-x_s),$$

with $c\in(x_s,y)$ and $d\in(y,x_0)$. Consequently,

$$|x_1-x_s|<|x_0-y|+|y-x_s|=|x_0-x_s|.$$

In an analogous manner we establish that $|x_1-x_s|<|x_0-x_s|$ when $x_0<x_s$. Thus, the sequence $\{|x_n-x_s|,\ n=0,1,\dots\}$ is strictly decreasing, and from this point on, we can follow the same reasoning used in the proof of Theorem 2.2.8. QED

Example 2.3.1 Let $F(x)=\arctan x$. Then $x_s=0$ is a stationary state. Moreover, the derivative of $\arctan x$ is $1/(1+x^2)$ and it satisfies the inequality $0<1/(1+x^2)\leq 1$ with equality only at $x=0$. Hence, every orbit converges to 0 (see Fig. 2.3.1).

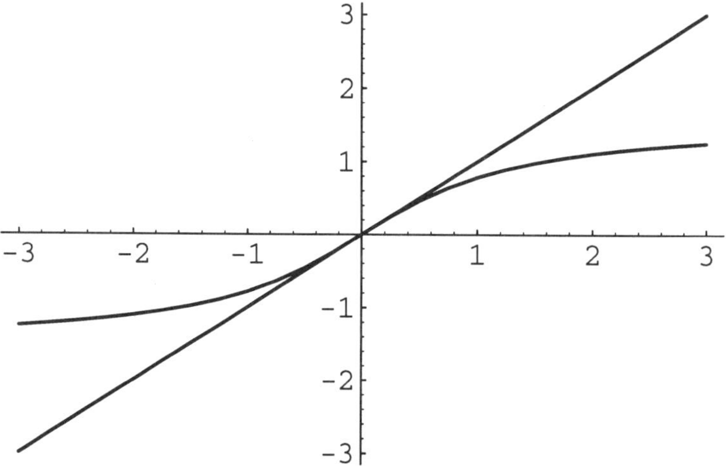

Fig. 2.3.1 Graph of $y=\arctan x$ and of $y=x$. The cobweb method could be used to graphically illustrate that every orbit converges to 0.

Example 2.3.2 Let $F(x)=\cos x$. There exists $x_s\in(0,1)$ such that $F(x_s)=x_s$ (see Fig. 2.3.2). The function satisfies all assumptions of Theorem 2.3.2. Hence, every orbit converges to x_s.

In some cases numerical evidence seems to indicate that all orbits of a dynamical system converge to a periodic orbit. Example 2.3.3 is typical.

Example 2.3.3 Let $F(x)=3.2x(1-x)$, $x\in[0,1]$. We have seen that

$$x_0 = \frac{2.1+\sqrt{.21}}{3.2}, \qquad x_1 = \frac{2.1-\sqrt{.21}}{3.2}$$

is a periodic orbit of period 2. Since

$$(F^2)'(x_0) = (-1-2\sqrt{.21})(-1+2\sqrt{.21}) = 1-.84 = .16$$

the periodic orbit $\{x_0,x_1\}$ is an attractor. Numerical evidence suggests that for every $y_0\in(0,1)$ the orbit $O(y_0)$ converges to the periodic orbit $\{x_0,x_1\}$ (see Remark 2.2.3). There are exceptions. For example, O(11/16) and O(5/16) do not converge to $\{x_0,x_1\}$.

There are no simple general criteria similar to the one provided by Theorem 2.3.1 to ensure that a periodic orbit behaves like the orbit $\{x_0, x_1\}$ of the preceding example.

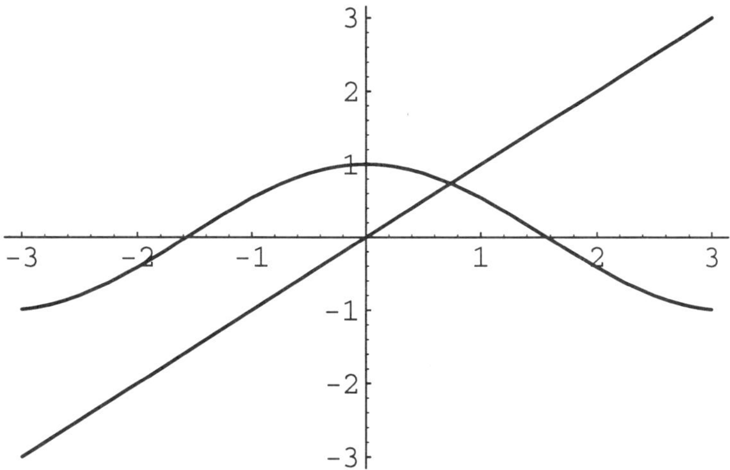

Fig. 2.3.2 Graph of $y=\cos x$ and $y=x$. Also in this case the cobweb method could be used to graphically illustrate the convergence of every orbit to x_s.

Recall that a set $A\subset\mathbf{R}$ is bounded if there exists $r>0$ such that $A\subset[-r,r]$. Consequently, an orbit $O(x_0)$ is **unbounded** if there exists a subsequence $\{x_{nk}\}$ of $O(x_0)$ such that $|x_{nk}|\to\infty$ as $k\to\infty$. A careful analysis of Theorem 2.3.1 shows that the assumption $|F'(x)|\le 1$ for all $x\in I$, without any additional condition, implies that either all orbits of F are bounded or all unbounded.

In fact, assume that at least one orbit $O(x_0)$ is bounded and let $y_0\neq x_0$. By the MVT we obtain

$$x_1 - y_1 = F'(c)(x_0-y_0) \quad \text{which implies that} \quad |x_1-y_1| \le |x_0-y_0|.$$

In an analogous manner we prove that $|x_n-y_n|\leq|x_0-y_0|$.

Consequently, $O(y_0)$ is bounded (see Problem 8). Since y_0 is arbitrary, every orbit of F is bounded. In the following example we have $|F'(x)|<1$ for every x, and every orbit is unbounded.

Example 2.3.4 Let $F(x)=x+2-\arctan x$. Then $F'(x)=1-1/(1+x^2)\in[0,1)<1$ for all $x\in\mathbf{R}$. Every orbit of the system defined by F goes to infinity (see Fig. 2.3.3).

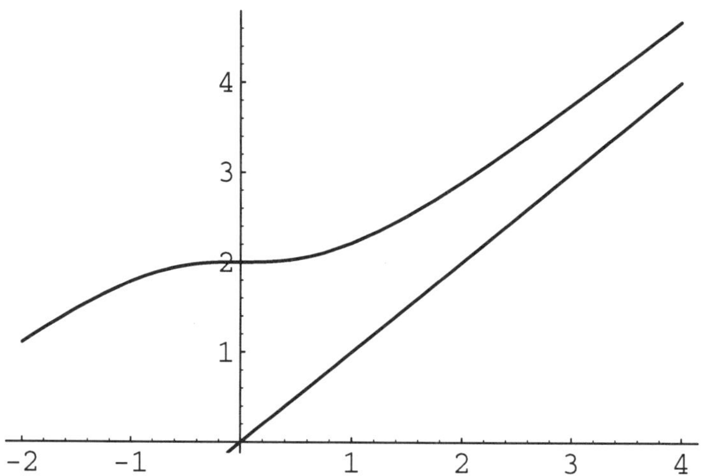

Fig. 2.3.3 Every orbit of the dynamical system generated by $F(x)=x+2-\arctan x$ goes to infinity. Use the cobweb method.

Problems

1. Prove that the equation $x=\cos(x+2)$ has one and only one solution.

2. * Prove that every orbit of the dynamical system governed by the function $F(x)=\cos(x+2)$ converges to the unique fixed point found in Problem 1. Illustrate this convergence numerically using the cobweb method and the Orbits program of Appendix 1.

3. * Prove that every trajectory of the dynamical system governed by the function $F(x) = \sin x$ converges to 0. Illustrate this convergence numerically using the Cobweb Method and the Orbits program of Appendix 1.

4. Let $F : [0,1]\rightarrow[0,1]$ be such that $F'(x)>0$ and $F''(x)>0$. Prove that $F'(0)<1$. [Hint : If $F'(0)\geq1$ then $F'(x)>1$ for all $x\in(0,1)$].

5. Let F be as in Problem 4. Assume that $F(0)=0$. Prove that every orbit with initial state $x_0\in(0,1)$ converges to 0.

6. * Let $F(x)=1+\arctan x^2-\ln(1+x^2)$. Find an interval of length not exceeding 1 which contains a fixed point of F. Show graphically (you can use suitable programs from Appendix 1) that $|F'(x)|<1$ for every $x\in\mathbf{R}$ and conclude that every orbit converges to the fixed point.

7. * Let $F(x)=\ln(1+x^2)-\arctan x^2+2$. Prove that F has a fixed point and every orbit converges to it (see Problem 6).

8. Assume that $O(x_0)$ is contained in the interval $[-r,r]$ and $|x_n-y_n|\le|x_0-y_0|$ for every $n\ge1$. Does $[-r,r]$ necessarily contain $O(y_0)$? If the answer is negative, find an interval that contains the orbit of $O(y_0)$.

Section 4. PARAMETER SPACE ANALYSIS

1. Fold, Transcritical, and Pitchfork Bifurcation

Let I be an interval, $a\in I$, and $F(a,x)$ be a **one-parameter family** of scalar maps. Assume that F is differentiable with continuous derivatives with respect to x and a. Let $J\subset I$ be an interval with more than one point and $x : J\to\mathbf{R}$ be a continuous function such that for every $a\in J$, $x(a)$ has the property

$$F(a,x(a)) = x(a),$$

i.e., $x(a)$ is a stationary state of $F(a,x)$ for every $a\in J$. Then the graph of the map $x(a)$, i.e., the curve $(a,x(a))$, $a\in J$, is called a **branch** of fixed points of the one-parameter family $F(a,x)$. For simplicity, $x(a)$ itself is sometimes called a branch. In a similar manner we define branches of periodic points of period p.

Example 2.4.1 Let $F(a,x)=ax(1-x)$. From $F(a,x)=x$ we derive $x=0$ and $x=(a-1)/a$. Thus, the graph of $x_1 : \mathbf{R}\to\mathbf{R}$, $x_1(a)=0$ is a branch of fixed points of F. Notice that the fixed point $x_1(a)$ is independent of a. The graph of $x_2(a)=(a-1)/a$ is another branch of fixed points of F. This time the fixed point changes with a.

Example 2.4.2 Recall that the function of Example 2.4.1 does not have any periodic point of period 2 for $a\in(-1,3)$ because the discriminant a^2-2a-3 is negative (see Example 1.2.6). For $a>3$ there are exactly two periodic points of period 2, namely

$$x_3(a) = \frac{a+1+\sqrt{a^2-2a-3}}{2a}, \qquad x_4(a) = \frac{a+1-\sqrt{a^2-2a-3}}{2a}.$$

Thus, we have two branches of periodic points of period 2 of F when $a>3$. At $a=3$ we obtain $x_2(a)=x_3(a)=x_4(a)$, where $x_2(a)$ is as defined in Example 2.4.1.

Definition 2.4.1
A value $a_0\in I$ is a **bifurcation point** of the one-parameter family F (or, briefly, of F) if there are intervals $J_i\subset I$, $a_0\in J_i$, $i=1,2$ or $i=1,2,3$, and continuous functions $x_i : J_i\to\mathbf{R}$, whose graphs are branches of fixed points or periodic points of F such that

$$x_i(a_0) = x_j(a_0) \quad \text{and} \quad x_i(a) \ne x_j(a), \quad a \ne a_0, \quad i\ne j.$$

We require J_i to have nonzero length, and we allow a_0 to be one of its endpoints. Hence, J_i may have the form (b,c), $[a_0,b)$, or $(b,a_0]$. When i=1,2, then $J_1=J_2$ and all points $x_i(a)$, i=1,2, have the same period. When i=1,2,3, then $J_1=(b,c)$, $J_2=J_3=[a_0,b)$ (or $(b,a_0]$) and the points of $x_i(a)$, i=2,3, may have period equal or double (period-doubling bifurcation) the period of $x_1(a)$.

Identify $\mathbf{R}^2$ with a plane and represent the parameter a along the horizontal axis. The pictorial display obtained by plotting all branches of stationary states and periodic points is called the **bifurcation diagram** of F on I. At a bifurcation point a_0 there is frequently an exchange of stability between the different branches. The following examples illustrate these ideas.

Example 2.4.3 Let $F(a,x)=ax(1-x)$. We have seen before that the graphs of $x_1(a)=0$ and $x_2(a)=(a-1)/a$ are branches of fixed points. They meet at (1,0). Hence, a=1 is a bifurcation point.

From $F_x(a,x)=a(1-2x)$ we see that the fixed point 0 is a sink as long as $a\in(-1,1)$ and a source for $a\notin[-1,1]$. Replacing x with $x_2(a)$ in $F_x(a,x)$ we obtain $F_x(a,x_2(a))=2-a$. Therefore, the fixed points $x_2(a)$ are sinks for $a\in(1,3)$ and sources for $a\notin[1,3]$. At $a_0=1$ there is an exchange of stability between $x_1(a)$ and $x_2(a)$.

Example 2.4.4 Let $F(a,x)=ax-x^3$. The stationary states of $F(a,x)$ are the solutions of the equation $x=ax-x^3$. Hence, $x_1(a)=0$ is always a stationary state of our system. No other fixed point exists for $a\leq 1$. When the parameter a crosses the value 1 (see Fig. 2.4.1), two more stationary states appear, namely

$$x_2(a)=\sqrt{a-1} \quad \text{and} \quad x_3(a)=-\sqrt{a-1}.$$

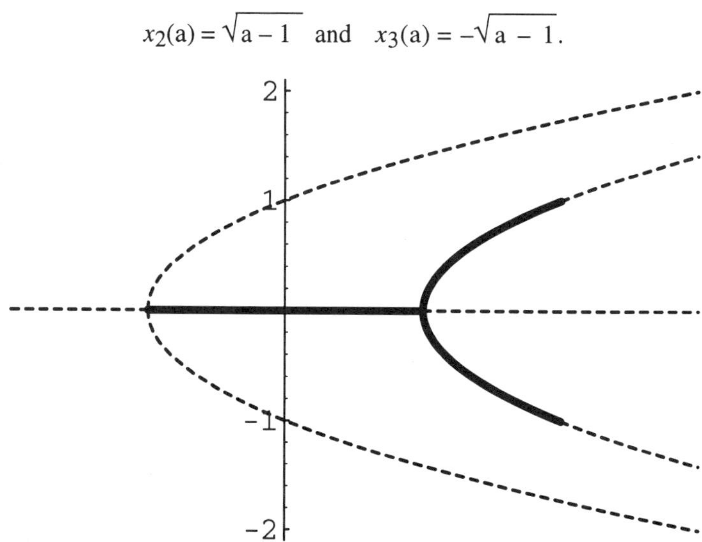

Fig. 2.4.1 Part of the bifurcation diagram for the map $F(a,x)=ax-x^3$.

From $F_x(a,x)=a-3x^2$ we derive that 0 is a repeller for $|a|>1$ and a sink for $|a|<1$. At a=1 we have $F_x(1,0)=1$. Replacing x with $x_2(a)$ [or $x_3(a)$] in $F_x(a,x)$ we obtain $F_x(a,x_2(a))=3-2a$. Since $-1<3-2a<1$ for $a\in(1,2)$, the points of the two branches (they exist only for $a\geq 1$) are sinks for $a\in(1,2)$ and sources for $a>2$. At $a_0=1$ we have an exchange of stability between $x_1(a)$ and $x_2(a)$, $x_3(a)$.

At $a_0=-1$ a periodic orbit of period 2 arises: $\{x_4(a)=(a+1)^{.5}, x_5(a)=-(a+1)^{.5}\}$. Since the derivative of the second iterate of F at $x_4(a)$ [or $x_5(a)$] is $(2a+3)^2$, which is larger than 1 for $a>-1$, the periodic orbit is always a source. There is no exchange of stability at $a=-1$.

Four types of bifurcation points are presented in the following pages. In three of the four types, all branches (two or three) which meet for $a=a_0$ are made of fixed points (or periodic points of the same period). In the fourth case, three branches meet for $a=a_0$: $x_1(a)$, $x_2(a)$, and $x_3(a)$, with $x_1(a)$ being periodic points of period p and $x_2(a)$, $x_3(a)$ being periodic points of period 2p. The first three types of bifurcation points are called **fold**, **transcritical**, and **pitchfork** bifurcation according to the different behavior of the one-parameter family $F(a,x)$ near $a=a_0$. The fourth is called, for obvious reasons, **period-doubling** bifurcation. The following examples illustrate the fold, transcritical, and pitchfork bifurcation.

Example 2.4.5 Let $F(a,x)=x^2-a$ (see Fig. 2.4.2). The stationary states of F are the solutions of the equation $x^2-a=x$ or

$$x_1(a) = \frac{1+\sqrt{1+4a}}{2} \qquad x_2(a) = \frac{1-\sqrt{1+4a}}{2}.$$

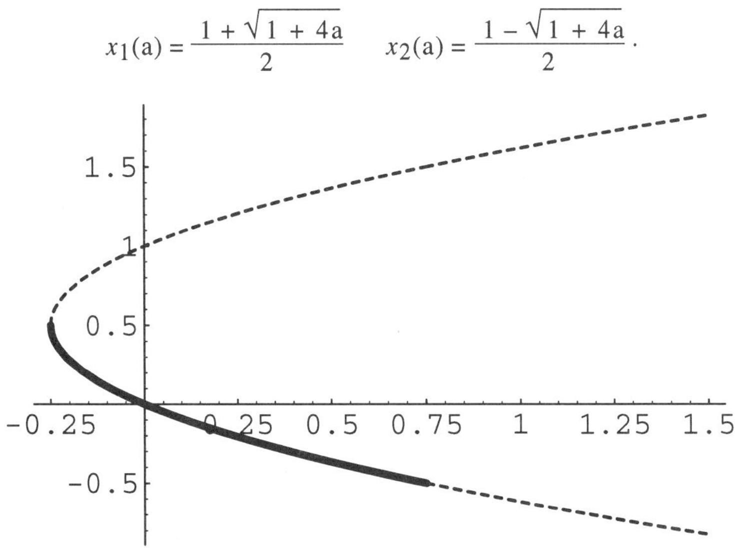

Fig. 2.4.2 Supercritical fold bifurcation for the one-parameter family $F(a,x)=x^2-a$ at $a_0=-1/4$. The upper branch is unstable and the lower is stable for $a<.75$.

The fixed points exist for $a\geq -1/4$. At $a_0=-1/4$ we have $x_1(-1/4)=x_2(-1/4)=.5$. Hence, $a_0=-1/4$ is a bifurcation point. The interval J is of the form $[a_0,b)$ since no branches of fixed points exist before a_0. The type of bifurcation taking place at

$-1/4$ is called **supercritical fold**. Notice that $F_x(a_0,x_1(a_0))=F_x(-.25,.5)=1$. Along the branch $(a,x_1(a))$ we have $F_x(a,x_1(a))=2x_1(a)=1+(1+4a)^{.5}>1$. Hence, the fixed points $x_1(a)$ are sources for all $a>-1/4$. Along $(a,x_2(a))$ we have $F_x(a,x_2(a))=1-(1+4a)^{.5}$, which implies that $x_2(a)$ are sinks for $a\in(-1/4,3/4)$ and sources for $a>3/4$.

Example 2.4.6 Let $F(a,x)=x^2+a$ (see Fig. 2.4.3). The stationary states of F are the solutions of the equation $x^2+a=x$ or

$$x_1(a) = [1 + (1 - 4a)^{.5}]/2, \quad x_2(a) = [1 - (1 - 4a)^{.5}]/2$$

The two branches of stationary states exist for $a\leq 1/4$. At $a_0=1/4$ we have $x_1(1/4)=x_2(1/4)$. The interval J is of the form $(b,a_0]$, and the type of bifurcation taking place at a_0 is called a **subcritical fold**. The fixed points of the lower branch are sinks for $a\in(-.75,.25)$ and sources for $a<-.75$. Those of the upper branch are always sources.

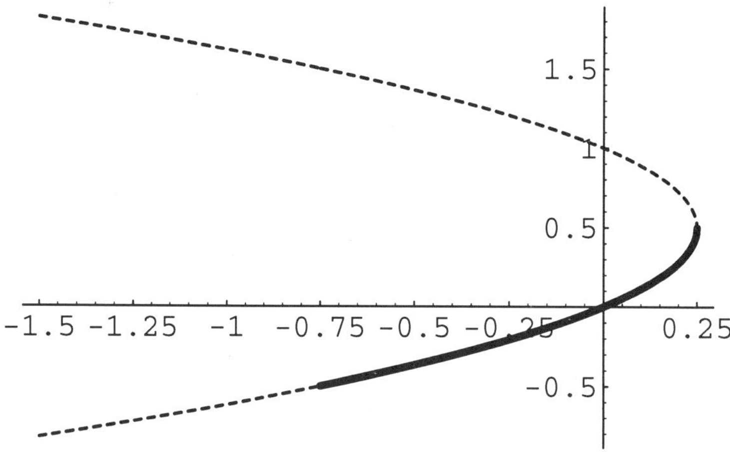

Fig. 2.4.3 Subcritical fold bifurcation for the family $F(a,x)=x^2+a$ at $a=1/4$.

Example 2.4.7 Let $F(a,x)=ax-x^3$ (see Fig. 2.4.4). We have seen that $x_1(a)=0$ is always a fixed point independently of a. No other fixed point exists for $a\leq 1$. When the parameter crosses the value 1, two more fixed points appear, $x_2(a)=(a-1)^{.5}$ and $x_3(a)=-(a-1)^{.5}$. Hence, $a_0=1$ is a bifurcation point. Since before $a_0=1$ there is only one branch of fixed points, and after there are three, the interval J is of the form $[a_0,b)$. The type of bifurcation taking place at a_0 is called **supercritical pitchfork**. The **subcritical** case is when $J=(b,a_0]$.

Example 2.4.8 The one-parameter family of maps $F(a,x)=ax(1-x)$ provides an example of another type of bifurcation (see Fig. 2.4.5). As mentioned before, the two branches of stationary states $(a,x_1(a))=(a,0)$ and $(a,x_2(a))=(a,(a-1)/a)$ meet at $(1,0)$. Hence, $a_0=1$ is a bifurcation point, called **transcritical**.

For a∈(0,1) the stationary state $x_2(a)$ is a repeller since $F_x(a,(a-1)/a)=2-a$ and 2–a>1 when a∈(0,1), while $x_1(a)=0$ is a sink. For a∈(1,3) the roles are reversed. The fixed point $x_1(a)=0$ is a source, while $x_2(a)=(a-1)/a$ is a sink.

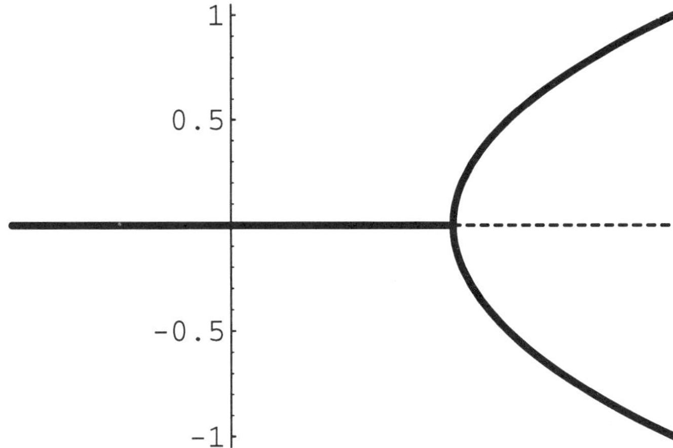

Fig. 2.4.4 Bifurcation diagram for $F(a,x)=ax-x^3$ (Example 2.4.7). As a crosses the value 1, the stationary state 0 becomes a source. Its stability is passed on to the pair of stationary states $(a-1)^{.5}$ and $-(a-1)^{.5}$

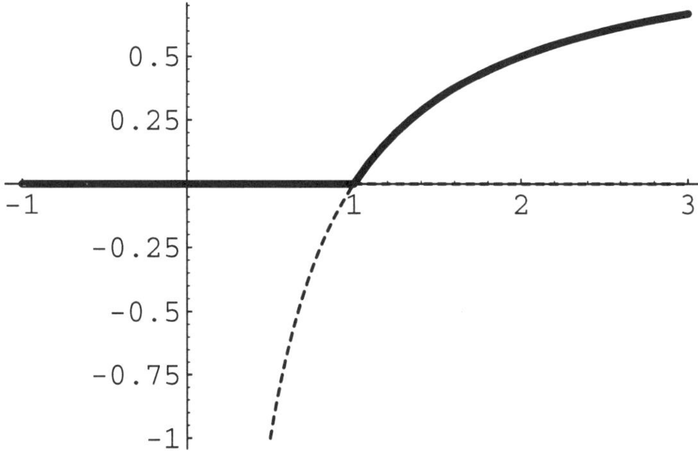

Fig. 2.4.5 Transcritical bifurcation at a=1 for the logistic map $ax(1-x)$ (Example 2.4.8). The stationary state $x=0$ loses its stability as a crosses the value 1.

From Example 2.4.8 we see that transcritical bifurcation takes place at $a=a_0$ when two branches of fixed points (or periodic points of the same period) exist in an open interval J, $a_0 \in J$, and they meet at $a=a_0$.

Fold bifurcation happens (see Examples 2.4.5 and 2.4.6) when two branches of fixed points (or periodic points of the same period) exist only in an interval $[a_0,b)$ (supercritical case) or $(b,a_0]$ (subcritical case) and they are equal at $a=a_0$.

Pitchfork bifurcation (see Example 2.4.7) requires three branches of fixed points (or periodic points of the same period). Two of them exist only in an interval $[a_0,b)$ (supercritical case) or $(b,a_0]$ (subcritical case). The third exists in an open interval J, $a_0 \in J$. The three branches meet at $a=a_0$.

In Subsection 3 we will show that $F_x(a_0,x(a_0))=1$ is necessary for a_0 to be a fold, transcritical, or pitchfork bifurcation. The condition is not sufficient, as Example 2.4.9 shows.

Example 2.4.9 Let $F(a,x)=(1+a^2)x+x^3$. The fixed point $x=0$ exists for all a. Along the branch (a,0) we have $F_x(a,0)=1+a^2$. Hence the derivative is 1 for $a_0=0$. However, a_0 is not a bifurcation point. In fact, from $(1+a^2)x+x^3=x$ we derive $a^2x+x^3=0$, of which $x=0$ is obviously a solution. Any solution with $x\neq 0$ would require that $a^2+x^2=0$, which is impossible if $x\neq 0$. Hence, $a_0=0$ is not a bifurcation point.

Sufficient conditions for the different types of bifurcation will be provided in Subsection 3. In the meantime we would like to encourage the reader to use a careful numerical and graphical investigation to detect the type of bifurcation taking place. The complexity of many processes frequently makes intelligent use of numerical and graphical techniques a good source of prompt and reliable information. We conclude this section with two additional examples.

Example 2.4.10 Consider the one-parameter family of maps $F(a,x)=x^3-ax$ (see Fig. 2.4.6). Notice that $x_1(a)\equiv 0$ is always a fixed point regardless of the value of a.

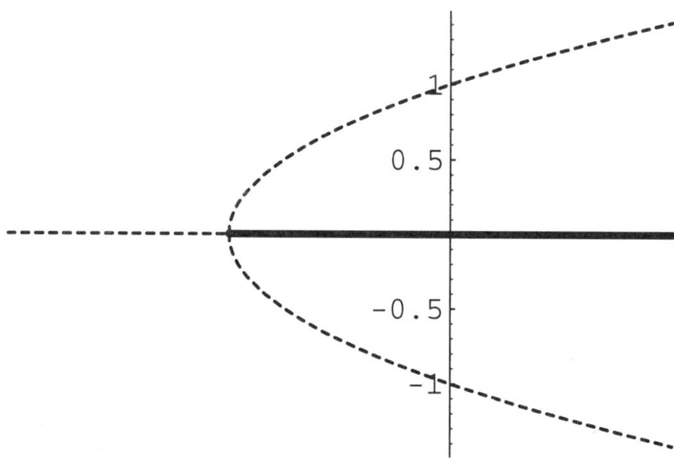

Fig. 2.4.6 Pitchfork bifurcation at a=–1 for the map $F(a,x)=x^3-ax$. The stationary state $x_s=0$ becomes a sink as a crosses the value –1.

For a>–1 there are two more stationary states, namely

$$x_2(a) = (a + 1)^{.5} \quad \text{and} \quad x_3(a) = -(a + 1)^{.5}$$

which are never stable. In fact, their stability requires that $a \in (-2,-1)$, but they exist only for $a \geq -1$. The point $a_0=-1$ is a supercritical pitchfork bifurcation. The stationary state $x_1(a)=0$ is a source for $|a|>1$ and a sink for $a \in (-1,1)$. This time, no exchange of stability is taking place at $a_0=-1$.

Example 2.4.11 Let $F(a,x)=ax+x^3$. For $a<1$ the one-parameter family of maps has the branches of stationary states $(a,x_2(a))$, $(a,x_3(a))$ with

$$x_2(a) = (1 - a)^{.5} \quad \text{and} \quad x_3(a) = -(1 - a)^{.5}.$$

The fixed points $x_2(a)$, $x_3(a)$ are always sources. The point $x_1(a)=0$ is a source for $|a|>1$ and a sink for $a \in (-1,1)$. The value $a_0=1$ is a subcritical pitchfork bifurcation. The bifurcation diagram is "symmetric" to the one in Example 2.4.10 with respect to the vertical axis.

Problems

1. * Study the bifurcation diagram of the map $F(a,x)=a-x^2$. In particular, determine the fold bifurcation and investigate the stability of the two branches of fixed points.

2. * Study the bifurcation diagram of $F(a,x)=ax-x^2$. In particular, determine the transcritical bifurcation and the stability of the branches of fixed points.

3. * Study the bifurcation diagram of $F(a,x)=-ax+x^2$. Determine the transcritical bifurcation and investigate the stability of the branches of fixed points.

4. * Study the bifurcation diagram of $F(a,x)=-ax+x^3$. In particular, determine the pitchfork bifurcation and the stability of the branches of fixed points.

5. * Study the bifurcation diagram of the map $F(a,x)=(x+a^2-1)(x^2-2x-a)+x$. Determine what type of bifurcation point is $a=-1$.

6. * Let $F(a,x)=ax+\arctan x$. Is $a=0$ a bifurcation point? Of what type?

7. * Let $F(a,x)=x-\arctan ax$. Is $a=0$ a bifurcation point?

8. * Let $F(a,x)=ax-1+\cos x$. Is $a=1$ a bifurcation point? Of what type?

2. Period-Doubling Bifurcation

The bifurcation taking place at a_0 is called **period-doubling** if the period of the points of one of the branches is double the period of the points of another.

Example 2.4.12 Let $F(a,x)=ax(1-x)$. At $a=3$ the branch of stationary states $(a,x_2(a))$, $x_2(a)=(a-1)/a$ (see Examples 2.4.3 and 2.4.8) meets the two branches of periodic points of period 2 $(a,x_3(a))$ and $(a,x_4(a))$:

$$x_3(a) = \frac{a + 1 + \sqrt{a^2 - 2a - 3}}{2a}, \quad x_4(a) = \frac{a + 1 - \sqrt{a^2 - 2a - 3}}{2a}.$$

Since $\{x_3(a),x_4(a)\}$ is a periodic orbit of period 2, period-doubling bifurcation is taking place at $a_0=3$. The interval J is of the form $[a_0,b)$ (**supercritical** case). There is an exchange of stability at 3. In fact, the fixed point $x_2(a)=(a-1)/a$ is a sink for $a\in(1,3)$ and a source for $a\notin[1,3]$. At a=3 we have

$$x_2(3) = 2/3 \text{ and } F_x(3,2/3) = -1.$$

The periodic orbit $\{x_3(a), x_4(a)\}$ is a sink for $a\in(3,1+\sqrt{6})$. This conclusion is arrived at as follows. Let

$$g(a) = \frac{d}{dx}F^2(a,x_3(a)) = F_x(a,x_4(a))\, F_x(a,x_3(a)) = 4 + 2a - a^2.$$

Since $g'(a)=2-2a$, we obtain that $g'(a)<0$ for $a>3$. Thus, g(a) is strictly decreasing for $a>3$ and it reaches the value -1 when $a=1+\sqrt{6}$. It follows that the periodic orbit $\{x_3(a), x_4(a)\}$ is a sink for $a\in(3,1+\sqrt{6})$. At $a_0=3$, we have an exchange of stability.

It can be shown that a periodic orbit of period 4 appears at $a=1+\sqrt{6}$ and this periodic orbit is a sink for $a\in(1+\sqrt{6}, a_3)$ with $a_3\approx3.54409...$. Thus, $a=1+\sqrt{6}$ is also a period-doubling bifurcation point. This behavior continues in the sense that there exists an increasing sequence $\{a_n: n=1,2,...\}$ such that $a=a_n$ is always a period-doubling bifurcation point with a periodic orbit of period 2^n appearing at that value of a. The sequence $\{a_1=3, a_2=1+\sqrt{6}, ...\}$ converges to $a_\infty\approx3.569946...$. For this value of a, F has aperiodic orbits.

The elements of the sequence $\{a_n\}$ are all **period-doubling bifurcation points** of the one-parameter family of maps $\{F(a,x)=ax(1-x) : a\in[0,4]\}$ (see Fig. 2.4.7), and the entire sequence $\{a_n\}$ is called a **cascade of bifurcations**.
M. J. Feigenbaum [Feigenbaum, 1983] noted that this period-doubling cascade of bifurcations appears in many other processes. Moreover, he observed that although the sequence $\{a_n: n=1,2,...\}$ may vary from one process to another, there exists a real number $\alpha\approx4.669201609...$ such that

$$\lim_{n\to\infty} \frac{a_n - a_{n-1}}{a_{n+1} - a_n} = \alpha \tag{2.4.1}$$

for many processes. For this reason α has been called the *Feigenbaum constant.* The sequence $\{a_n\}$ converges very quickly to a_∞. More precisely, we have the relation

$$a_n \approx a_\infty - c\alpha^{-n} \tag{2.4.2}$$

where c=2.6327... .

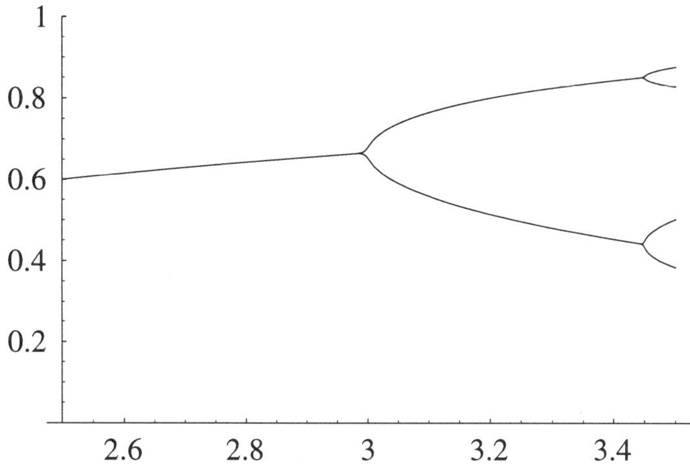

Fig. 2.4.7 Part of the bifurcation diagram of the one-parameter family of maps $F(a,x)=ax(1-x)$. The fixed point $(a-1)/a$ loses its stability for $a>3$, and an attractive periodic orbit of period 2 arises. For $a>1+(6)^{.5}$ the orbit becomes a repeller and its attractive character is passed on to a periodic orbit of period 4, and so on.

In summary, the sequence of period-doubling bifurcations for $ax(1-x)$ starts with two branches $(a,x_{21}(a))$, $(a,x_{22}(a))$ of periodic points of period 2 arising from a branch $(a,x_2(a))$ of stationary states when the parameter a crosses the value $a_1=3$. The derivative $F_x(a, x)$ computed at the point $(3,x_2(3)) =(3,2/3)$ is equal to -1, i.e.,

$$F_x(3,2/3) = -1.$$

The stationary states of the branch $x_2(a)$ are sinks when the parameter a ranges in an open interval $(3-\varepsilon, 3)$. For $a>3$ the stability is passed on to the periodic orbit of period 2:

$$x_{21}(a) = \frac{a + 1 + \sqrt{a^2 - 2a - 3}}{2a}, \quad x_{22}(a) = \frac{a + 1 - \sqrt{a^2 - 2a - 3}}{2a}.$$

This orbit is a sink until the parameter reaches $1+\sqrt{6}$. At $a_2=1+\sqrt{6}$ two branches of periodic points of period $4=2^2$ start from $(a,x_{21}(a))$ and two more from $(a,x_{22}(a))$. The four branches are

$$(a,x_{41}(a)), (a,x_{42}(a)) \text{ from } (a,x_{21}(a)); \quad (a,x_{43}(a)), (a,x_{44}(a)) \text{ from } (a,x_{22}(a)).$$

Moreover, the orbit $\{x_{41}(a),x_{44}(a),x_{42}(a),x_{43}(a)\}$, where $x_{41}(a)>x_{42}(a)>x_{43}(a) >x_{44}(a)$, is periodic of period 4. The derivative of the second iterate of F computed at

$$(a_2,x_{21}(a_2)) \quad \text{or} \quad (a_2,x_{22}(a_2))$$

is equal to -1. The periodic orbit of period 2 becomes a repeller for $a>1+\sqrt{6}$. The periodic orbit of period 4 is a sink for $a\in(1+\sqrt{6},a_3)$, $a_3\approx3.54409$.

At this point the bifurcation scheme is repeated with $F^4(a,x)$ replacing $F^2(a,x)$ and $F^8(a,x)$ replacing $F^4(a,x)$ until the parameter reaches a_4. The process just described continues as a goes through the sequence $\{a_n: n=1,2,...\}$ of period-doubling bifurcations.

Example 2.4.13 Let $F(a,x)=x(a-1)-x^3$. We see that for $a>0$ there is a periodic orbit of period 2 $\{x_2(a)=a^{.5},x_3(a)=-a^{.5}\}$. The two branches of periodic points of period 2, $(a,x_2(a))$ and $(a,x_3(a))$, meet the branch of fixed points $(a,0)$ at $a_0=0$. Hence, period-doubling bifurcation is taking place at 0. The interval J is of the form $[0,b)$ (**supercritical** case). At $(0,0)$ we have $F_x(0,0)=-1$. For $a\in(0,2)$ the stationary state $x_1(a)=0$ is a sink. The periodic orbit $\{a^{.5},-a^{.5}\}$ is never a sink. No exchange of stability is taking place at $a=0$.

In Subsection 3 we will show that $F_x(a_0,x(a_0))=-1$ is necessary for a_0 to be a period-doubling bifurcation. The condition is not sufficient as Example 2.4.14 shows. Recall that the analogous condition $F_x(a_0,x(a_0))=1$ is also not sufficient for the other three types of bifurcation (see Example 2.4.11) .

Example 2.4.14 Let $F(a,x)=a-x-x^3$. Notice that for every $a\in\mathbf{R}$ the equation $F(a,x)=x$ has one and only one solution. In fact, $F_x(a,0)=-3x^2-1\leq-1$ implies that F is strictly decreasing and crosses the line $y=x$ exactly once. The derivative is -1 only for $x=0$. The corresponding value of a is $a=0$, which is the only candidate for period-doubling bifurcation. Assume that $\{x_0=p,\ x_1=q\}$ is a periodic orbit of period 2. Hence,

$$a-p-p^3=q \quad \text{and} \quad a-q-q^3=p.$$

Subtracting the second equation from the first, we obtain

$$(q-p)+(q-p)(p^2+pq+q^2)=q-p.$$

Thus, $p^2+pq+q^2=0$. The equality requires that $pq<0$, which is impossible since adding pq to both sides of the equality we obtain $(p+q)^2=pq$. Hence $a=0$ is not a period-doubling bifurcation point.

Remark 2.4.1 Notice the difference between period-doubling and pitchfork. In both cases we have three branches coming together when $a=a_0$. In the case of pitchfork, all three are branches of fixed points (or periodic points of period p). In the case of period-doubling, one is a branch of fixed points (or periodic points of period p) and the other two are branches of periodic points of period 2 (or 2p). The derivative of the map (F or F^p) is

-1 when a_0 is a period-doubling bifurcation point,
1 when a_0 is a pitchfork bifurcation point.

Moreover, observe that in all four bifurcation cases we have $|F_x(a_0,x(a_0))| = 1$. More precisely, $F_x(a_0,x(a_0))=-1$ when a_0 is a period-doubling bifurcation, while in the three remaining cases, $F_x(a_0,x(a_0))=1$.

Problems

1. * Determine the period-doubling bifurcation point of $F(a,x)=a-x^2$. Investigate the stability of the periodic orbit of period 2.

2. * Determine the period-doubling bifurcation point of $F(a,x)=-a+x^2$. Investigate the stability of the periodic orbit of period 2.

3. * Determine the period-doubling bifurcation point of $F(a,x)=ax-x^2$. Investigate the stability of the periodic orbit of period 2.

4. * Determine the period-doubling bifurcation point of $F(a,x)=-ax+x^2$. Investigate the stability of the periodic orbit of period 2.

5. * Study the bifurcation diagram of the map $F(a,x)=-ax+x^3$. In particular, determine if there is any period-doubling bifurcation point.

6. * Study the bifurcation diagram of the map $F(a,x)=ax-x^3$. In particular, determine the period-doubling bifurcation points and the stability of the bifurcating branches.

7. * Let $F(a,x)=ax^3-(a-1)x$. Find the period-doubling bifurcation points and analyze the stability of the periodic orbits of period 2.

8. * Let $F(a,x)=ax-\arctan x$. Is a=0 a period-doubling bifurcation point?

9. * Let $F(a,x)=x-\sin ax$. Is a=2 a period-doubling bifurcation point?

3. **Bifurcation: A Theoretical Viewpoint**

We now present, perhaps in a slightly different form, some ideas which were introduced in calculus. Let $G : \mathbf{R}^2 \to \mathbf{R}$ be continuous together with its partial derivatives up to the order that will be needed. Consider the level curve of level 0 of G, i.e., the set $C=\{(x,y): G(x,y)=0\}$, and let $P=(x_0,y_0)$ be a point of C. Let L be the family of lines passing through P. The family depends on a parameter m, representing the slope of the line corresponding to it: $L=\{y=y_0+m(x-x_0): m\in \mathbf{R}\}$. To study the intersections between any line of L and the curve C, consider the zeros of the function $g(x)=G(x,y_0+m(x-x_0))$. As expected, x_0 is one of them, namely $g(x_0)=0$, since P belongs to C and to every line of L. The line tangent to C at P is given by the value of m for which x_0 is a double root of the equation $g(x)=0$. The root x_0 is double for $g(x)=0$ when, in addition to $g(x_0)=0$, we have $g'(x_0)=0$. Since

$$g'(x) = G_x(x,y)+ mG_y(x,y), \tag{2.4.3}$$

we obtain

$$m = -G_x(P) / G_y(P), \qquad P=(x_0,y_0). \tag{2.4.4}$$

The tangent line to C at P is the vertical line $x=x_0$ if and only if $G_y(P)=0$ and $G_x(P)\neq 0$.

Example 2.4.15 Let $G(x,y)=x^2-xy+y^2-1$. The level curve of level 0 of G is $C=\{(x,y) : x^2-xy+y^2-1=0\}$. $P=(1,1)\in C$. The generic line through P has equation $y=1+m(x-1)$. Substituting in $G(x,y)$, we obtain

$$g(x) = G(x,1 + m(x - 1)) = x^2 - x(1 + m(x - 1)) + (1 + m(x - 1))^2 - 1.$$

As expected, $g(1)=0$. Taking the derivative we obtain

$$g'(x) = 2x - 2mx + 3m - 1 + 2m^2(x - 1).$$

Hence $g'(1)=1+m$ and $g'(1)=0$ if and only if $m=-1$. Since $(\partial G/\partial x)(1,1)=1 =(\partial G/\partial y)(1,1)$, we see that (2.4.4) is satisfied. The tangent line is $y=1-(x-1)=-x+2$.

Another point on the curve C is $P=(2/\sqrt{3},1/\sqrt{3})$, for which we have

$$g(x) = x^2 - x(1/\sqrt{3} + m(x - 2/\sqrt{3})) + (1/\sqrt{3} + m(x - 2/\sqrt{3}))^2 - 1.$$

As expected, $g(2/\sqrt{3})=0$. Moreover,

$$g'(2/\sqrt{3}) = 4/\sqrt{3} -1/\sqrt{3} - 4m/\sqrt{3} + 4m/\sqrt{3} = 0$$

implies that $\sqrt{3}-0m=0$, which can be true only if $m=\infty$. The tangent line is $x=\sqrt{4/3}$. Notice that $(\partial G/\partial y)(2/\sqrt{3},1/\sqrt{3})=0$ and $(\partial G/\partial x)(2/\sqrt{3},1/\sqrt{3})\neq 0$.

It may happen that the curve C "passes through" $P=(x_0,y_0)$ twice (we assume that P is not an isolated point of C). In this case x_0 is a double root of $g(x)=0$ regardless of the value of m. For this to happen, (2.4.3) must have infinitely many solutions when the partial derivatives are evaluated at P. Thus,

$$G_x(P) = G_y(P) = 0. \tag{2.4.5}$$

There are now two tangent lines to the curve, one for each of the two branches of C passing through P. Their slopes are found by searching those values of m for which $g''(x_0)=0$ (triple root). This requires that

$$G_{xx}(P) + 2mG_{xy}(P) + m^2G_{yy}(P) = 0. \tag{2.4.6}$$

with $[G_{xy}(P)]^2-G_{xx}(P)G_{yy}(P)\geq 0$. The solutions of the quadratic equation (2.4.6) are the slopes of the two tangent lines. It may happen that (2.4.6) is a perfect square, in which case the two branches have the same tangent at P. When one of the tangent lines is vertical (i.e., one solution has to be $m=\infty$), the quadratic term $m^2G_{yy}(P)$ must disappear, i.e.,

$$G_{yy}(P) = 0. \tag{2.4.7}$$

To understand why this should happen, consider the quadratic equation $ax^2+2bx+c=0$. Keep b and c fixed, and select a sequence a_n such that $a_n \to 0$ as $n \to \infty$. Assuming that $b>0$, we obtain that the root $x_1=[-b-(b^2-a_nc)^{.5}]/a_n \to \infty$ as $a_n \to 0$.

Both tangent lines could be vertical. Then we must have

$$G_{yy}(P) = G_{xy}(P) = 0. \tag{2.4.8}$$

In any case, provided that P is only a double point for the curve C, (2.4.6) cannot be an identity since this would imply that P is at least a triple point. Hence, one (or more) of the three coefficients of (2.4.6) has to be different from 0.

Example 2.4.16 Let $G(x,y)=x^3+x^2-y^2$. The level curve of level 0 of G is the cubic curve C: $x^3+x^2-y^2=0$ (see Fig. 2.4.8), which intersects the x-axis at $x=0$ and $x=-1$. Moreover, x cannot be smaller than -1 since $x^3+x^2=y^2\geq 0$. In polar coordinates, the curve C is given by $\rho=-\cos 2\theta/\cos^3\theta$, $\theta \in (-\pi/2,\pi/2)$. It passes through the origin twice, once for $\theta=-\pi/4$ and a second time for $\theta=\pi/4$. Hence, the origin is a double point. This occurrence could have been recognized immediately by noticing that both partial derivatives of G at (0,0) are 0. Equation (2.4.6) becomes $2-2m^2=0$, which gives $m=\pm 1$. The two tangent lines are $y=x$ and $y=-x$. They can be obtained by setting equal to 0 the quadratic terms of G: $x^2-y^2=0$. This gives $(x-y)(x+y)=0$.

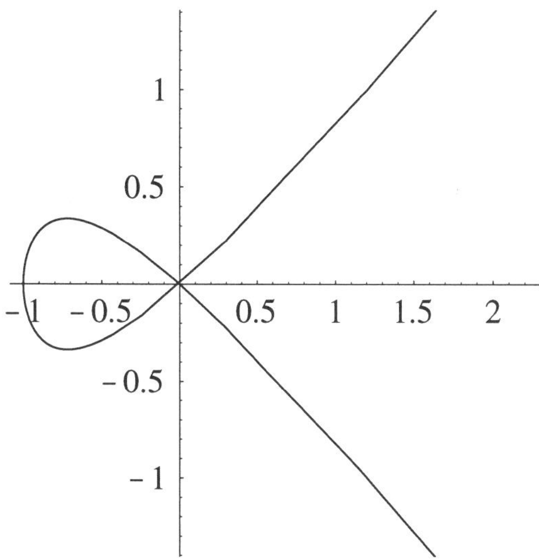

Fig. 2.4.8 The graph of $x^3+x^2-y^2=0$ shows that the curve passes through the origin twice. The tangent lines to the two different branches are $y=x$ and $y=-x$.

We now apply the ideas just outlined to bifurcation. The function F, its first and second partial derivatives, and F_{xxx} are assumed to be continuous. For each of the four types of bifurcation we first obtain the necessary condition(s). Using the simple geometric properties of plane curves mentioned above and the chain rule, we then derive additional conditions, that can be shown to be sufficient, although this result is not proved here. We conclude with an example of transcritical bifurcation in which one of the sufficient conditions is not verified. Its inclusion has the purpose of reminding our readers that for each type of bifurcation we have simply provided the necessary condition(s) and a set of possible sufficient conditions.

Let us first analyze the case of **fold** bifurcation. The function G is now $G(a,x)=x-F(a,x)$. The point P has coordinates $P=(a_0,x_0)$ with $x_0=x_1(a_0)=x_2(a_0)$, where $(a,x_1(a))$ and $(a,x_2(a))$ are the two branches of fixed points which start (supercritical case) or end at $a=a_0$. The tangent line at $P=(a_0,x_0)=(a_0,x_1(a_0))$ $=(a_0,x_2(a_0))$ is vertical and the point P is **not a double point** of the curve $G(a,x)=0$. Using x as the independent variable, the generic line through P has the form $a=a_0+m(x-x_0)$ (the horizontal axis is the a-axis). According to (2.4.3), the slope m of the tangent line must satisfy

$$0 = g'(x_0) = 1 - F_x(a_0,x_0) + mF_a(a_0,x_0). \tag{2.4.9}$$

The equation of the tangent line is $a=a_0$. Hence, m=0 is a solution of (2.4.9). Moreover, it is the only one since P is a simple point for the curve. Thus the necessary conditions for fold bifurcation are

$$1 - F_x(a_0,x_0) = 0,\ F_a(a_0,x_0) \neq 0. \tag{2.4.10}$$

We can go one step further. Assume that the fold bifurcation is **supercritical.** We can regard a as function of x, namely $a=a(x)$, along the curve $G(a,x)=0$. Since the tangent line is $a=a_0$ we have $a'(x_0)=0$. The supercritical fold is guaranteed by $a''(x_0)>0$. The function $a(x)$ satisfies the equality $x-F(a(x),x)=0$. Differentiating twice with respect to x gives

$$F_{xx} + 2a'(x)\, F_{xa} + (a'(x))^2\, F_{aa} + a''(x)F_a = 0. \tag{2.4.11}$$

Evaluating (2.4.11) at $P=(a_0,x_0)$, we obtain

$$F_{xx}(P) + a''(x_0)\, F_a(P) = 0. \tag{2.4.12}$$

Hence,

$$F_{xx}(P)F_a(P) < 0 \tag{2.4.13}$$

is the condition we look for. When $F_{xx}(P)F_a(P)>0$, we have **subcritical** fold.

Example 2.4.17 Let $F(a,x)=4a+x-x^2$. $a_0=0$ is a fold bifurcation since two branches of fixed points are $x_1(a)=2\sqrt{a}$ and $x_2(a)=-2\sqrt{a}$. The coordinates of P are (0,0). Since $F_a(0,0)=4$ and $F_x(0,0)=1$, we see that (2.4.10) is satisfied. Moreover, $F_{xx}(0,0)=-2$, which implies that (2.4.13) holds. At a=0 we have supercritical fold bifurcation.

Summarizing, a_0 is a fold bifurcation if

$$F_x(a_0,x_0) = 1, \quad F_a(a_0,x_0) \neq 0, \quad F_{xx}(P) \neq 0, \tag{2.4.14}$$

with $F_{xx}(P)F_a(P)<0$ giving a supercritical fold. In the analysis above F should be replaced by its pth iterate when $x_1(a)$ and $x_2(a)$ are branches of periodic points of period p.

We now examine the **transcritical** and **pitchfork** cases. Recall again that the horizontal axis is the a-axis, the vertical axis is the x-axis, and $G(a,x)=x-F(a,x)$. The point $P=(a_0,x_0)$ is a double point of the curve $G(a,x)=0$ in both cases. Hence, the two partial derivatives of G have to be 0 at P:

$$F_a(a_0,x_0) = 0, \qquad F_x(a_0,x_0) = 1. \tag{2.4.15}$$

Notice that in agreement with what we already knew, the derivative of F with respect to x at (a_0,x_0) is 1. Moreover, in the case of transcritical bifurcation, we can assume that neither one of the two tangent lines is vertical, while in the case of pitchfork one tangent line is vertical while the other is not. Hence, according to (2.4.7), we have

$$F_{xx}(a_0,x_0) \neq 0 \tag{2.4.16}$$

in the transcritical bifurcation, and

$$F_{xx}(a_0,x_0) = 0, \qquad F_{xa}(a_0,x_0) \neq 0 \tag{2.4.17}$$

in the pitchfork case. Notice that from (2.4.17) we derive that the two tangent lines do not coincide in the pitchfork case. We shall assume that the same is true in the transcritical case. Hence,

$$[F_{xa}(P)]^2 - F_{xx}(P)F_{aa}(P) > 0 \tag{2.4.18}$$

in both cases. The analysis of the pitchfork case can go one step further. Assume that a branch $x_1(a)$ of fixed points exists in an open interval J, $a_0 \in J$. Write

$$x - F(a,x) = G(a,x) = [x - x_1(a)][x - H(a,x)]. \tag{2.4.19}$$

It can be shown that H is continuous together with its first and second partial derivatives. Differentiating (2.4.19) with respect to x, and evaluating at P, we obtain

$$1 - F_x(P) = x_0 - H(P) + [x_0 - x_1(a_0)][1 - H_x(P)].$$

Hence, according to (2.4.14), $x_0-H(P)=0$. Differentiating with respect to x a second time, and evaluating at P, gives

$$-F_{xx}(P) = 2[1 - H_x(P)] - [x_0 - x_1(a_0)]H_{xx}(P).$$

From (2.4.17) we derive $1-H_x(P)=0$. Differentiating (2.4.19) first with respect ot x and then with respect to a, and evaluating at P, we obtain

$$-F_{xa}(P) = -H_a(P) - x_1'(a_0)[1 - H_x(P)] - [x_0 - x_1(a_0)]H_{xa}(P).$$

Hence, according to (2.4.17), $H_a(P)=F_{xa}(P)\neq 0$. We see that a_0 is a natural candidate for fold bifurcation for the function H. The only condition which remains to be required is $H_{xx}(P) \neq 0$. Differentiating (2.4.19) three times with respect to x and evaluating at P, gives

$$F_{xxx}(P) = 3H_{xx}(P) \neq 0. \quad (2.4.20)$$

The pitchfork bifurcation is supercritical [see (2.4.13)] when $F_{xa}(P)F_{xxx}(P) = H_{xx}(P)H_a(P) < 0$.

Example 2.4.18 Let $F(a,x)=ax(1-x)$. Recall that a=1 is a transcritical bifurcation point since the two branches of fixed points (a,0) and (a,(a–1)/a) intersect each other at (1,0). Since $F_a(a,x)=x(1-x)$, $F_x(a,x)=a-2ax$, $F_{xx}(a,x)=-2a$, we see that for a=1 and $x=0$, (2.4.15) and (2.4.16) are verified.

Example 2.4.19 Let $F(a,x)=ax-x^3$. The parameter value a=1 is a pitchfork bifurcation, since the branch of fixed points (a,0) is met at (1,0) by the two branches of fixed points $(a,x_2(a))$, $(a,x_3(a))$ with $x_2(a) = (a-1)^{.5}$ and $x_3(a) = -(a-1)^{.5}$. We can write $x-F(a,x)=x(1-a+x^2)$. We see that $H(a,x)=x-1+a-x^2$. As expected, $H(1,0)= H_x(1,0)-1=0$, while $H_a(1,0)=1$ and $H_{xx}(1,0)=-2$. Hence, we have a supercritical pitchfork. However, we can derive the same conclusion without using H. In fact, by means of F we obtain $F_a(1,0)=0$, $F_x(1,0)=1$, $F_{xx}(1,0)=0$, $F_{ax}(1,0)=1$, and $F_{xxx}(1,0)=-6$. The value a=–1 is a pitchfork bifurcation. The bifurcation is supercritical, since $F_{ax}(1,0)F_{xxx}(1,0)<0$.

Summarizing, a_0 is a transcritical bifurcation if at $P=(a_0,x_0)$ we have

$$F_a(P) = 0, \quad F_x(P) = 1, \quad F_{xx}(P) \neq 0, \quad [F_{xa}(P)]^2 - F_{xx}(P)F_{aa}(P) > 0. \quad (2.4.21)$$

The pitchfork bifurcation at a_0 takes place if

$$F_a(P) = 0, \quad F_x(P) = 1, \quad F_{xx}(P) = 0, \quad F_{xa}(P) \neq 0, \quad F_{xxx}(P) \neq 0. \quad (2.4.22)$$

The bifurcation is supercritical when $F_{xa}(P)F_{xxx}(P)<0$. In the analysis above, F should be replaced by its pht iterate when the bifurcation involves branches of periodic points of period p.

When a_0 is a **period-doubling** bifurcation, we can repeat the considerations made above for the pitchfork case by replacing $F(a,x)$ with $F(a,F(a,x))$. However, we should notice that $P=(a_0,x_0)$ is not a double point of the curve $G(a,x)=x-F(a,x)=0$. Hence, at least one of the partial derivatives of G has to be different from 0 at P. This implies that

$$[1 - F_x(P)]^2 + [F_a(P)]^2 > 0. \quad (2.4.23)$$

At the same time, P is at least a double point for the curve $K(a,x)=x-F(a,F(a,x))=0$. Hence, both partial derivatives of K are 0 at P, i.e.,

$$0 = 1 - (F_x(P))^2, \quad 0 = F_a(P)[1 + F_x(P)]. \quad (2.4.24)$$

To satisfy (2.4.23) and (2.4.24) we need $F_x(P)=-1$. In fact, if $F_x(P)=1$, then the second equality of (2.4.24) requires that $F_a(P)=0$. Both conditions imply that (2.4.23) is violated. Consequently, the tangent line to the branch of fixed points passing through P is not vertical since is slope m is given by

$$m = F_a(P)/2. \tag{2.4.25}$$

From (2.4.6), and knowing that P is a double point for the curve $K(a,x)=x-F(a,F(a,x))=0$, we obtain that the slopes of the tangent lines to $K(a,x)=0$ at the point P are given by the solutions of the quadratic equation

$$K_{aa}(P) + 2mK_{xa}(P) + m^2K_{xx}(P) = 0. \tag{2.4.26}$$

Since

$$K_{xx}(P) = -F_{xx}(P)[F_x(P)]^2 - F_x(P)F_{xx}(P) \qquad (2.4.26)$$

and $F_x(P)=-1$, we obtain $K_{xx}(P)=0$. Hence, one of the two tangent lines is vertical, i.e., its equation is $a=a_0$. From (2.4.25) we know that the other one is not. Thus, $K_{xa}(P)\neq 0$. This implies that

$$F_{xx}(P)F_a(P) + 2F_{xa}(P) \neq 0. \tag{2.4.27}$$

To complete the picture of what is taking place at P, let us write $x-F(a,F(a,x)= K(a,x)$ in the form

$$x - F(a,F(a,x) = K(a,x) = (x - x_1(a))(x - H(a,x)), \tag{2.4.28}$$

where $x_1(a)$ is the branch of fixed points of $F(a,x)$ passing through P. We can now repeat the procedure previously used for the pitchfork bifurcation. We obtain $x_0-H(P)=0$, $1-H_x(P)=0$, and $H_a(P)=-F_{xx}(P)F_a(P)-2F_{xa}(P)\neq 0$. Hence a_0 is a natural candidate for fold bifurcation for the function H. The only condition which remains to be required is $H_{xx}(P)\neq 0$. Differentiating (2.4.28) three times with respect to x and evaluating at P, gives

$$2F_{xxx}(P) + 3(F_{xx}(P))^2 = -3H_{xx}(P). \tag{2.4.29}$$

Summarizing, a_0 is a period-doubling bifurcation if

$$F_x(P) = -1, \quad F_{xx}(P)\, F_a(P) + 2F_{xa}(P) \neq 0;$$
$$2F_{xxx}(P) + 3(F_{xx}(P))^2 \neq 0, \tag{2.4.30}$$

with $[2F_{xxx}(P)+3(F_{xx}(P))^2][F_{xx}(P)F_a(P)+2F_{xa}(P)]<0$ giving the supercritical case. In the analysis above, F and F^2 should be replaced by F^p and F^{2p}, respectively, when $x_1(a)$ is a branch of periodic points of period p.

Example 2.4.20 Let $F(a,x)=ax(1-x)$. Since $F_x(a,x)=a-2ax$ and $x_1(a)=(a-1)/a$, we see that $F_x(a,x_1(a))=-1$ when $a_0=3$. This implies that $x_0=2/3$. Since $F_a(a,x)=x(1-x)$, $F_{xa}(a,x)=1-2x$, $F_{xx}(a,x)=-2a$, and $F_{xxx}(a,x)=0$, we obtain

$$F_{xx}(P)F_a(P) + 2F_{xa}(P) = -2, \qquad 2F_{xxx}(P) + 3[F_{xx}(P)]^2 = 108.$$

Hence, $a_0=3$ is a period-doubling bifurcation. From

$$[2F_{xxx}(P) + 3(F_{xx}(P))^2][F_{xx}(P)\ F_a(P) + 2F_{xa}(P)] = -2\times 108 < 0$$

we derive that the bifurcation is supercritical.

Example 2.4.21 Let $F(a,x)=ax-x-x^3$. Along the branch of fixed points (a,0) we have $F_x(a,0)=-1$ if and only if a=0. Since $F_{xx}(0,0)=0$, $F_{xa}(0,0)=1$, $F_{xxx}(0,0)=-6$, all conditions are satisfied and $a_0=0$ is a period-doubling bifurcation. In fact, the two branches of periodic points of period 2 are $(a,x_2(a))$ and $(a,x_3(a))$, $x_2(a)=a^{0.5}$ and $x_3(a)=-a^{0.5}$. They meet the branch (a,0) at (0,0). Since

$$[2F_{xxx}(P) + 3(F_{xx}(P))^2][F_{xx}(P)F_a(P) + 2F_{xa}(P)] = -12\times 2 < 0$$

the bifurcation is supercritical.

Example 2.4.22 Let $F(a,x)=a-x-x^3$. We have seen (Example 2.4.13) that a=0 is not a period-doubling bifurcation. Notice that $F_{xx}(0,0)=0$ and $F_{xa}(0,0)=0$. Hence the condition $F_{xx}(P)F_a(P)+2F_{xa}(P)\neq 0$ is violated.

We summarize the results in the following table, where $P=(a_0,x_0)$.

	$F_x(P)$	$F_a(P)$	$F_{xx}(P)$	Other
Fold	1	$\neq 0$	$\neq 0$	$F_{xx}(P)F_a(P)<0$ supercritical
Transcritical	1	0	$\neq 0$	$[F_{xa}(P)]^2-F_{xx}(P)F_{aa}(P)>0$
Pitchfork	1	0	0	$F_{xa}(P)\neq 0$, $F_{xxx}(P)\neq 0$ $F_{xa}(P)F_{xxx}(P)<0$ supercritical
Period-doubling	-1			$r=F_{xx}(P)F_a(P)+2F_{xa}(P)\neq 0$ $s=2F_{xxx}(P)+3(F_{xx}(P))^2\neq 0$ $rs<0$ supercritical

We now conclude with the example that was mentioned at the beginning of this theoretical overview on bifurcation.

Example 2.4.23 Let $F(a,x)=a^6+x-x^2$. Then $a_0=0$ is a transcritical bifurcation point. The two branches of fixed points are $x_1(a)=a^3$ and $x_2(a)=-a^3$. The necessary conditions are satisfied since $F_x(0,0)=1$, $F_a(0,0)=0$. Moreover, $F_{xx}(0,0)\neq 0$. However, the tangent lines to the two branches coincide. Hence, the sufficient condition $[F_{xa}(P)]^2-F_{xx}(P)F_{aa}(P)>0$ is not verified.

Problems

1. Let $F(a,x)=a^2+e^{-x}-1$. Verify that for each $a\in\mathbf{R}$ the equation $F(a,x)=x$ has one and only one solution $x(a)$. Verify that a=0 is the only potential candidate for period-doubling. Verify that a=0 is not a period-doubling bifurcation point.

2. Let $F(a,x)=a+e^{-x}-1$. Verify that a=0 is a period-doubling bifurcation point. Is the bifurcation supercritical or subcritical?

3. Let $F(a,x)=a^2x(1-x)$. Verify that $a=3^{.5}$ and $a=-3^{.5}$ are the potential candidates for period-doubling bifurcation. Are they period-doubling bifurcation points?

4. Let $F(a,x)=a^2x(1-x)$. Are a=1 and a=–1 transcritical bifurcation points?

5. Let $F(a,x)=ax-\arctan x$. Verify that $a_0=2$ is a subcritical pitchfork bifurcation.

6. Let $F(a,x)=x-\arctan ax$. Verify that a=0 is not a bifurcation point.

7. Let F be as in Problem 6. Is a=2 a period-doubling bifurcation?

8. Let $F(a,x)=ax-1+\cos x$. Is a=–1 a period-doubling bifurcation?

Section 5. CONJUGACY AND CHAOS

Orbits of conjugate systems

Let $F : J\rightarrow J$ and $G : I\rightarrow I$ be conjugate via a map $\phi : I\rightarrow J$. We have seen that x_s is a stationary state for G if and only if $y_s=\phi(x_s)$ is a stationary state for F. Similarly, x_0 is a periodic point of period p for G if and only if $y_0=\phi(x_0)$ is a periodic point of the same period for F.

We now analyze the stability of corresponding fixed points and periodic orbits. Let x_s, $y_s=\phi(x_s)$ be two corresponding stationary states. The orbit starting at $x_0\in I$ converges to x_s if and only if the orbit starting at $y_0=\phi(x_0)$ converges to y_s. The conclusion follows from the continuity of ϕ and ϕ^{-1}. In fact, from $G^n=\phi^{-1}F^n\phi$, we obtain that

$$x_n = G^n(x_0) = \phi^{-1}F^n\phi(x_0) = \phi^{-1}F^n(y_0) = \phi^{-1}(y_n).$$

Hence, the sequence $\{x_n\}$ converges to x_s if and only if $\{y_n\}$ converges to $y_s=\phi(x_s)$. A similar argument can be used for periodic orbits of period p, since F^p and G^p are also conjugate via the same map ϕ.

What about the set of limit points of corresponding orbits? A subsequence of $\{x_n\}$ converges to z if and only if the corresponding subsequence of $\{y_n\}$ converges to $\phi(z)$. Hence, the set $L(x_0)$ of limit points of $O(x_0)$ is mapped by ϕ in a one-to-one manner onto the set $L(y_0)$ of limit points of $O(y_0)$. Thus, $L(x_0)$ and $L(y_0)$ are either both finite or both infinite. All these comments are based on the continuity of ϕ.

Something more can be said when ϕ is differentiable. For example, assume that $\phi'(x_s)\neq 0$. From

$$\phi(G(x)) = F(\phi(x))$$

we derive

$$\phi'(G(x_s))G'(x_s) = \phi'(x_s)G'(x_s) = F'(\phi(x_s))\phi'(x_s).$$

Therefore, dividing by $\phi'(x_s)\neq 0$, we obtain

$$G'(x_s) = F'(\phi(x_s)). \tag{2.5.1}$$

Similarly, from $\phi(G^p(x))=F^p(\phi(x))$ we derive

$$\phi'(G^p(x_0))(G^p)'(x_0) = \phi'(x_0)(G^p)'(x_0) = (F^p)'(\phi(x_0))\phi'(x_0).$$

This implies that

$$(G^p)'(x_0) = (F^p)'(\phi(x_0)) \tag{2.5.2}$$

provided that $\phi'(x_0)\neq 0$. Equality (2.5.1) guarantees that if x_s is a sink (source) the same is true for $\phi(x_s)$. Equality (2.5.2) authorizes similar conclusions for the periodic orbit $O(x_0)$ and the corresponding orbit $O(\phi(x_0))$.

These comments, combined with the observations made at the beginning of this chapter, show the importance of conjugacy in the study of dynamical systems.

Example 2.5.1 Recall that the two maps

$$G(x) = 2x^2 - 1 \quad \text{and} \quad F(x) = 2|x| - 1$$

are conjugate in $[-1,1]$ via the map $\phi(x)=(2/\pi)\arcsin x$.

Notice that $|F'(x)|=2$ except at $x=0$, where F is not differentiable. The orbit starting from $x_0=0$ is eventually stationary, since $x_1=-1$, $x_2=x_3=\ldots=1$. The map F has the periodic orbit of period 3:

$$x_0 = -7/9 \qquad x_1 = 5/9 \qquad x_2 = 1/9.$$

Therefore, according to the theorem of Sarkovskii, the dynamical system defined by F has a periodic orbit of every period. All these orbits are repellers. Hence, the same is true for all stationary states and periodic orbits of G. A direct analysis of G is not as simple since $|G'(x)|<1$ for $|x|<.25$.

Example 2.5.2 One can verify that the maps $T(x)=4x^3-3x$ and the map

$$L(x)=\begin{cases}3x+2 & -1\le x\le -1/3\\ -3x & -1/3<x\le 1/3\\ 3x-2 & 1/3<x\le 1\end{cases}$$

with $x\in[-1,1]$ are conjugate via the same map ϕ of Example 2.5.1.

Since $|L_x(x)|=3$ at every point where L is differentiable, we obtain that all stationary states and periodic orbits of the dynamical system $x_{n+1}=L(x_n)$ in $[-1,1]$ are repellers. Consequently, all stationary states and periodic orbits of the dynamical system $x_{n+1}=T(x_n)$ in the same interval are repellers.

Chaos in the Li-Yorke sense

The result of Sarkovskii (see Theorem 2.2.2) and the companion result of Li-Yorke (see Theorem 1.3.1) can provide a preliminary definition of chaotic systems in the real line. The reader should keep in mind that this definition is not valid in higher dimensions and has also some additional drawbacks. Its main advantage is that it can easily be applied.

Consider a dynamical system $F: I\to I$, where F is continuous and I is a bounded interval. We shall say that F is **chaotic** in I in the **Li-Yorke sense** if F has a periodic point in I of period 3. Notice that as a consequence of Sarkovskii's theorem F has a periodic orbit of period p for every integer p. Moreover, Li and Yorke proved that there is an infinite (actually, uncountable) set $S\subset I$ such that for every $x\in S$ the orbit of x is aperiodic and unstable. Recall that these are the two conditions required in our preliminary definition of chaos mentioned in Section 3 of Chapter 1.

Example 2.5.3 Let $I=[-1,1]$ and $F(x)=2|x|-1$. Then F has a periodic point of period 3. In fact, $F(-7/9)=5/9$, $F(5/9)=1/9$, $F(1/9)=-7/9$. Hence F is chaotic in [0,1] in the Li-Yorke sense. In later chapters we prove that F is chaotic in [0,1] according to a definition more demanding than the one presented here.

Example 2.5.4 Let $B(x)=2x-[2x]$. B maps the interval [0,1] onto itself and has a periodic orbit of period 3 (Chapter 1, Section 2, Subsection 2, Problem 11). B is not continuous in [0,1] (see Fig. 2.5.1). Therefore, the Li-Yorke definition of chaos cannot be used in this case, since it requires the continuity of B. We shall see in later chapters that B is chaotic according to the definition we will propose.

After seeing Example 2.5.4 one could suggest removing the continuity restriction from the definition of Li-Yorke. This, however, creates other problems as the following example shows. We should keep in mind that for non continuous functions, the result of Sarkovskii also fails.

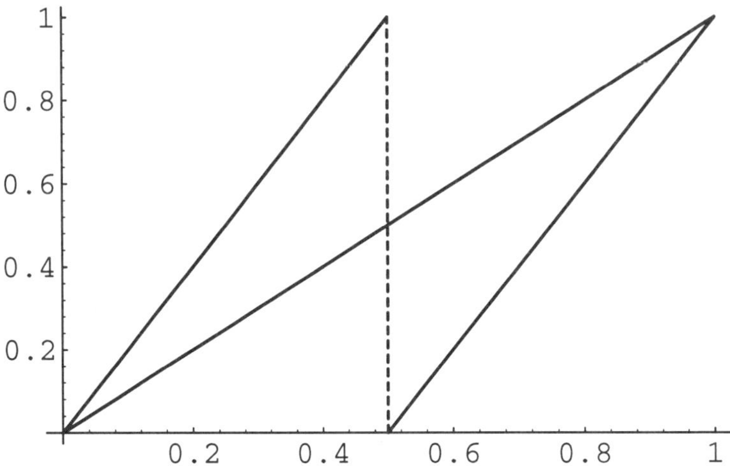

Fig. 2.5.1 Graph of the function $2x-[2x]$ for $x\in[0,1]$. The function is discontinuous at $x=0.5$ and $x=1$. By graphing the third iterate of $2x-[2x]$, one could verify that the function has a periodic orbit of period 3.

Example 2.5.5 Let

$$F(x) = \begin{cases} .5+x & 0\le x\le .5 \\ 0 & .5<x\le 1. \end{cases}$$

F maps the interval [0,1] into itself and F(0)=.5, F(.5)=1, F(1)=0. The reader can verify that for every $x\in[0,1]$, O(x) is eventually periodic of period 3. Therefore, the dynamical system generated by F in [0,1] cannot be considered chaotic. Notice that F has a nonremovable discontinuity at x=.5.

Using the intermediate value theorem it can be shown that F: I→I has a periodic orbit of period 3 if there are points a, b, c, d such that

$$F(c) = d < a < F(a) = b < F(b) = c \quad \text{or} \quad F(c) = d > a > F(a) = b > F(b) = c.$$

Example 2.5.6 Let $F(x)=.15x+4x^6e^{-2x}$ and let I=[0,8]. Then F(1.6)≅2.9755, F(2.9755)≅ 7.67291, F(7.67291)≅1.32763. Hence, the first condition stated above is verified and F has a periodic orbit of period 3.

Two systems F : J→J and G : I→I which are conjugate by ϕ : I→J are either both chaotic or not chaotic in the Li-Yorke sense in their respective intervals since x_0 is periodic of period 3 for G if and only if $y_0=\phi(x_0)$ is a periodic orbit of period 3 for F.

Example 2.5.7 Let I=J=[−1,1] and let $G(x)=2x^2-1$ and $F(x)=2|x|-1$. F and G are conjugate via the map $\phi(x)=(2/\pi)\arcsin x$ since $\phi(G(x))=F(\phi(x))$ for every $x\in[-1,1]$. The result was established in Example 2.1.5. F has the periodic orbit of period 3: $x_0=-7/9$, $x_1=5/9$, $x_2=1/9$. Thus, the dynamical system $x_{n+1}=F(x_n)$ is chaotic in the

Li-Yorke sense in [–1,1]. It follows that $x_{n+1}=G(x_n)$ is chaotic in the Li-Yorke sense in the same interval.

We could have obtained this result directly by plotting G(x), $G^3(x)=G(G(G(x)))$ and the line $y=x$. The points where the graph of G^3 intersects $y=x$ and which are not on the graph of G are fixed points of G^3 and not of G. Thus, they are periodic points of period 3 of G. We mention here that the two dynamical systems are chaotic in [–1,1] according to a more demanding definition of chaotic behavior we shall present in Chapter 6. In that case, as we shall see, the conjugacy will prove to be extremely useful.

Problems

1. * Let

$$K(x)=\begin{cases} 2x & 0\le x<.5 \\ 2-2x & .5\le x\le 1. \end{cases}$$

Verify that K maps the interval [0,1] onto itself. Find a periodic orbit of period 3 of K. Confirm your result graphically using the Graphing programs of Appendix 1, Section 1.

2. Find a periodic orbit of period 5 of the map K of Problem 1.

3. * Let $H(x)=4x(1-x)$. Verify that H maps the interval [0,1] onto itself. Find a conjugacy between H : $[0,1]\to[0,1]$ and G : $[-1,1]\to[-1,1]$, $G(x)=2x^2-1$. Verify the conjugacy using the programs of Appendix 1, Section 1.

4. Find a periodic orbit of period 3 of the map L of Example 2.5.2.

5. Using the conjugacy $\phi(x)=(2/\pi)\arcsin x$, find a periodic orbit of period 3 of the map T of Example 2.5.2.

6. * Let

$$F(a,x)=\begin{cases} ax & 0\le x<.5 \\ a-ax & .5\le x\le 1. \end{cases}$$

Show that for every $a\in[0,2]$ the function F(a,.) maps the interval [0,1] into itself. Find the values of a such that F(a,.) has a periodic orbit of period 3. Verify your conclusion using the programs of Appendix 1, Section 1.

7. Let $B(x)=3x-[3x]$, where $[x]$ denotes the greatest integer contained in x. Verify that B maps the interval [0,1] onto itself. Show that B has a periodic orbit of period 3. Can we conclude that B is chaotic in the Li-Yorke sense ?

8. * Provide graphical evidence that the maps T and L of Example 2.5.2 are conjugate in [–1,1].

9. * Using the conjugacy $\phi(x)=.5(1+x)$ construct a map in [0,1] that is conjugate to the map L of Example 2.5.2.

CHAPTER 3

$\mathbf{R}^q$, MATRICES, and FUNCTIONS

SUMMARY

The purpose of this chapter is to present a brief survey of those results, needed in Chapters 4 and 5, which are likely to be excluded in undergraduate courses on calculus of several variables and linear algebra. In Section 1 we introduce some algebraic and topological properties of $\mathbf{R}^q$ and discuss questions about continuity and equivalence of all norms. Section 2 deals with the operator norm of a matrix and it contains results on differentiable functions that will be needed in Chapter 5.

Section 1. STRUCTURE OF R^q AND CONTINUITY

1. Norms and Sets

Vectors and points

The elements of $\mathbf{R}^q$ are ordered q-tuples of real numbers: $\mathbf{R}^q=\{(x_1,...,x_q): x_i\in \mathbf{R}$ for all $i=1,2,...,q\}$. Each q-tuple is called a **point**. The name fits the task, since by means of a **coordinate system** every point of a given plane or of the three-dimensional space is identified with an element of $\mathbf{R}^2$ or $\mathbf{R}^3$, respectively. Therefore, we write $X=(x_1,...,x_q)$, since geometric points are usually denoted with upper-case letters. The equality sign stresses the identification. To reinforce the concept even further, we call $x_1,...,x_q$ the **coordinates** of X.

The idea of distance between geometric points in $\mathbf{R}^2$ or $\mathbf{R}^3$ and the way to measure it are extended to $\mathbf{R}^q$ in a straightforward manner. Given a pair of points $X=(x_1,...,x_q)$ and $Y=(y_1,...,y_q)$, the distance, d(X,Y), is usually defined by

$$d(X,Y)=\sqrt{(x_1-y_1)^2+...+(x_q-y_q)^2} \tag{3.1.1}$$

Formula (3.1.1) clearly reminds us of its geometric origin. In fact, in a plane and using the theorem of Pythagoras, the distance between $A=(a_1,a_2)$ and $B=(b_1,b_2)$ is the length of the hypotenuse of the right triangle with vertices A,B, and $C=(b_1,a_2)$ or $C=(a_1,b_2)$. Hence, $d(A,B)=[(a_1-b_1)^2+(a_2-b_2)^2]^{1/2}$.

This distance is called **Euclidean**. Although useful in many practical situations, it has limitations. For example, on the surface of our planet and for locations $X=(x_1,x_2,x_3)$ and $Y=(y_1,y_2,y_3)$ very far from each other, like Rome and San Francisco, we certainly cannot use $d(X,Y)=[(x_1-y_1)^2+(x_2-y_2)^2+(x_3-y_3)^2]^{1/2}$ to measure their distance! Hence, a more appropriate distance should be found: for example, the length of the shortest arc of the great circle passing through the two points. Similarly, the Euclidean distance is not very useful in a city, where a taxicab distance is certainly more appropriate, unless the taxi drivers are running up the tab! However, no matter what distance function we choose, it must satisfy the following three properties:

$$\begin{aligned}
&\text{i. } d(X,Y)\geq 0 \text{ and } d(X,Y)=0 \text{ if and only if } X=Y.\\
&\text{ii. } d(X,Y)=d(Y,X) \text{ (symmetric property).}\\
&\text{iii. } d(X,Y)\leq d(X,Z)+d(Z,Y) \text{ (triangular inequality).}
\end{aligned} \tag{3.1.2}$$

Almost daily in calculus we use the correspondence between points in a plane or in space with ordered pairs or triples of numbers. For example, we learned how to identify lines, planes, circles, spheres, curves, etc. with systems of equations in the variables x,y, and z. We learned how to use the same tool to decide if lines and/or planes were parallel or perpendicular, to study the motion of a point, to find areas, etc. We could safely say that among the challenging tasks of calculus, the following four ranked very high:

- Visualize a proposed problem geometrically.

- Convert the geometric interpretation into a system of equations.
- Solve the system.
- Interpret the meaning of the solution(s) for the proposed problem.

A quantum leap in the structure of $\mathbf{R}^q$ came with the course in linear algebra. We learned how to introduce in $\mathbf{R}^q$ two operations to make it into a **vector space**: the addition of q-tuples and the multiplication of q-tuples by real numbers (scalars):

$$(x_1,...,x_q) + (y_1,...,y_q) = (x_1 + y_1,...,x_q + y_q);$$
$$r(x_1,...,x_q) = (rx_1,...,rx_q). \tag{3.1.3}$$

Each q-tuple, being an element of a vector space, was called a **vector** and usually denoted with a lowercase boldfaced letter: $\mathbf{x}=(x_1,...,x_q)$. We learned how to choose a **basis** in $\mathbf{R}^q$, namely a set of q linearly independent vectors. Our preferred choice was

$$\mathbf{e}_1 = (1,0,...,0), \quad \mathbf{e}_2 = (0,1,0,...,0),..., \quad \mathbf{e}_q = (0,0,...,0,1). \tag{3.1.4}$$

It was established that they are a basis of $\mathbf{R}^q$, called the **standard basis**. We learned how to write each vector $\mathbf{x} \in \mathbf{R}^q$, $\mathbf{x}=(x_1,...,x_q)$ as a linear combination of the vectors $\mathbf{e}_1,...,\mathbf{e}_q$:

$$\mathbf{x} = x_1\mathbf{e}_1 + x_2\mathbf{e}_2 + ... + x_q\mathbf{e}_q. \tag{3.1.5}$$

Equality (3.1.5) suggested a new name for $x_1,...,x_q$: **components** of $\mathbf{x}$ in the standard basis $E=\{\mathbf{e}_1,...,\mathbf{e}_q\}$.

Problems to be solved with linear algebra techniques sometimes required the choice of a different basis $B=\{\mathbf{b}_1,\mathbf{b}_2,...,\mathbf{b}_q\}$. The components of $\mathbf{x}$ and of the vectors $\mathbf{b}_i$, i=1,2,...,q in the standard basis E allowed us to find the components of $\mathbf{x}$ in the new basis B. This topic, quite important in the study of discrete linear dynamical systems, will be discussed later (see Chapter 4).

The versatility and potential of $\mathbf{R}^q$ in the study of mathematics and its applications to other sciences were further enhanced with the introduction of the **dot** or **inner** product between pairs of vectors and the **norm** of a vector. $\mathbf{R}^q$ became an **inner product space** and a **normed vector space**.

This is our point of departure in the present chapter. In the sequel the elements of $\mathbf{R}^q$ will be denoted with boldfaced lowercase letters and called **vectors** or **points** to remind us of their geometric significance and their membership in the vector space $\mathbf{R}^q$. As the occasion demands, we will use the most convenient of the two representations

$$\mathbf{x} = (x_1,...,x_q), \quad \mathbf{x} = x_1\mathbf{e}_1 + x_2\mathbf{e}_2 + ... + x_q\mathbf{e}_q,$$

and call $x_1,...,x_q$ the **coordinates** of $\mathbf{x}$ or the **components** of $\mathbf{x}$ (in the standard basis).

Euclidean norm

The Euclidean norm of vectors in $\mathbf{R}^q$ is usually introduced by means of the **scalar** (or **dot**) product. Given $\mathbf{x}=(x_1,x_2,...,x_q)$ and $\mathbf{y}=(y_1,y_2,...,y_q)$, the scalar product $\mathbf{x}\bullet\mathbf{y}$ is defined as follows:

$$\mathbf{x}\bullet\mathbf{y} = x_1y_1 + ... + x_qy_q. \qquad (3.1.6)$$

The **Euclidean norm** (or **length** or **magnitude**) of $\mathbf{x}$, denoted by $\|\mathbf{x}\|$, is

$$\|\mathbf{x}\| = (\mathbf{x}\bullet\mathbf{x})^{1/2} = (x_1{}^2 +...+ x_q{}^2)^{1/2} \qquad (3.1.7)$$

According to what we said previously regarding distances, the number $\|\mathbf{x}\|$ in $\mathbf{R}^2$ and $\mathbf{R}^3$ is the length of the segment joining the origin with the point (x_1,x_2) or (x_1,x_2,x_3). The well-known Cauchy-Schwarz inequality can be easily established for every pair of vectors $\mathbf{x}$, $\mathbf{y}$:

$$|\mathbf{x}\bullet\mathbf{y}| \leq \|\mathbf{x}\| \, \|\mathbf{y}\|. \qquad (3.1.8)$$

With some help from the Cauchy-Schwarz inequality [for property (iii.)], it is not hard to show that the Euclidean norm has the following three properties:

$$\begin{aligned}
&\text{i. } \|\mathbf{x}\| \geq 0 \text{ and } \|\mathbf{x}\| = 0 \text{ if and only if } \mathbf{x} = \mathbf{0} = (0,...,0).\\
&\text{ii. } \|t\mathbf{x}\| = |t| \, \|\mathbf{x}\|. \qquad (3.1.9)\\
&\text{iii. } \|\mathbf{x} + \mathbf{y}\| \leq \|\mathbf{x}\| + \|\mathbf{y}\| \qquad \text{(triangular inequality).}
\end{aligned}$$

We recognize that $\|\mathbf{x}-\mathbf{y}\|$ is precisely the **Euclidean distance** between $(x_1,...,x_q)$ and $(y_1,...,y_q)$. The three properties (3.1.2) can easily be derived from (3.1.9).

From (3.1.8) we obtain that the ratio $\mathbf{x}\bullet\mathbf{y}/(\|\mathbf{x}\| \, \|\mathbf{y}\|) \in [-1,1]$. Hence, we can define the angle θ between $\mathbf{x}$ and $\mathbf{y}$ by setting

$$\theta = \arccos \frac{\mathbf{x}\bullet\mathbf{y}}{\|\mathbf{x}\| \, \|\mathbf{y}\|}, \quad \cos\theta = \frac{\mathbf{x}\bullet\mathbf{y}}{\|\mathbf{x}\| \, \|\mathbf{y}\|}. \qquad (3.1.10)$$

Using the cosine theorem from trigonometry, it can be verified that the number θ given by (3.1.10) coincides in $\mathbf{R}^2$ with the measurement, done with a protractor, of the angle between the segments joining (0,0) with (x_1,x_2) and (y_1,y_2). This fact motivates and justifies definition (3.1.10) for pairs of vectors in $\mathbf{R}^q$. The two vectors are orthogonal if and only if $\mathbf{x}\bullet\mathbf{y}=0$.

Example 3.1.1 Let $\mathbf{x}=(1,2)$ and $\mathbf{y}=(-3,1)$. Then $\|\mathbf{x}\|=(5)^{.5}$, $\|\mathbf{y}\|=(10)^{.5}$, $\mathbf{x}\bullet\mathbf{y}=-1$, $\cos\theta=-1/(5(2)^{.5})$, $\theta\approx 98.13^o$, $\|\mathbf{x}+\mathbf{y}\|=(13)^{.5}$, and $d(P,Q)=\|\mathbf{x}-\mathbf{y}\|=(17)^{.5}$. Notice that $\|\mathbf{x}\pm\mathbf{y}\|\leq\|\mathbf{x}\|+\|\mathbf{y}\|$.

Subsets of $\mathbf{R}^q$

The **ball** B, **sphere** S, and **closed ball** D, centered at $\mathbf{c}=(c_1,...,c_q)$ and with radius r, are defined as follows:

- $B(\mathbf{c},r) = \{\mathbf{x} \in \mathbf{R}^q : \|\mathbf{x} - \mathbf{c}\| < r\}$.

- $S(\mathbf{c},r) = \{\mathbf{x} \in \mathbf{R}^q : \|\mathbf{x} - \mathbf{c}\| = r\}$.
- $D(\mathbf{c},r) = \{\mathbf{x} \in \mathbf{R}^q : \|\mathbf{x} - \mathbf{c}\| \leq r\}$.

When $\mathbf{c}$ is the origin we frequently omit it. For example, $D(r) = D(\mathbf{0},r)$.

Example 3.1.2 Let $\mathbf{c}=(2,1)$, $r=3$. Then

$$B(\mathbf{c},3) = \{\mathbf{x}=(x,y) : \sqrt{(2-x)^2 + (1-y)^2} < 3\}.$$
$$S(\mathbf{c},3) = \{\mathbf{x}=(x,y) : \sqrt{(2-x)^2 + (1-y)^2} = 3\};$$
$$D(\mathbf{c},3) = \{\mathbf{x}=(x,y) : \sqrt{(2-x)^2 + (1-y)^2} \leq 3\}.$$

- A subset U of $\mathbf{R}^q$ is **open** if for every $\mathbf{x}\in U$ there is $r>0$ such that $B(\mathbf{x},r) \subset U$. A subset K of $\mathbf{R}^q$ is **closed** if its complement is open. Only two subsets of $\mathbf{R}^q$ are both open and closed: $\varnothing$ and $\mathbf{R}^q$. Most subsets of $\mathbf{R}^q$ are neither open nor closed.

- The **boundary**, ∂A, of $A\subset\mathbf{R}^q$ is the set of points $\mathbf{x}\in\mathbf{R}^q$ such that for every $r>0$ the ball $B(\mathbf{x},r)$ intersects both A and its complement A^c.

- The **closure** of A is the union of A and its boundary. It is denoted by $\bar{A}$.

- The **interior** of A is $\mathring{A}=\{\mathbf{x}\in A: \mathbf{x}\notin\partial A\}$.

Thus, $S(\mathbf{c},r)=\partial D(\mathbf{c},r)=\partial B(\mathbf{c},r)$; $\bar{B}(\mathbf{c},r)=D(\mathbf{c},r)$; and $\mathring{D}(\mathbf{c},r)=B(\mathbf{c},r)$.

- A set $A\subset\mathbf{R}^q$ is **bounded** if there exists $r>0$ such that $\|\mathbf{x}\|\leq r$ for every $\mathbf{x}\in A$. A set K which is closed and bounded is said to be **compact**.

Remark 3.1.1 It can be shown that $\bar{A}$ is a closed set and $\mathring{A}$ is an open set. Moreover, it is not difficult to prove that a set A is closed if and only if $\|\mathbf{x}_n-\mathbf{x}\|\to 0$ as $n\to\infty$ and $\mathbf{x}_n\in A$ for all $n=1,2,\ldots$ imply that $\mathbf{x}\in A$.

The following examples illustrate how one can verify that a set is open, closed, or neither and compute its boundary and closure.

Example 3.1.3 Let $U=\{(x,y)\in\mathbf{R}^2: x<y\}$, i.e., U is the region of the xy plane which is above the diagonal $y=x$. U is open. In fact, let $\mathbf{x}_0=(x_0,y_0)\in U$. Then $y_0-x_0=d>0$. Consider the ball $B(\mathbf{x}_0,r)$, where $r=d/3$. We want to show that for every $\mathbf{x}\in B(\mathbf{x}_0,r)$, we have $x<y$. Among all elements of $D(\mathbf{x}_0,r)$, the one with largest x-coordinate is (x_0+r,y_0) and the one with smallest y-coordinate is (x_0,y_0-r). Therefore, it is enough to show that $x_0+r<y_0-r$, or $y_0-x_0>2r$. This is obviously true since $y_0-x_0=d=3r$. The closure of U is the set $\{(x,y) : x\leq y\}$. It is a closed set. Its boundary is the same as the boundary of U and we have $\partial U=\{(x,y): x=y\}$. The set U is unbounded.

Now consider the closed set $V=\{(x,y)\in\mathbf{R}^2: x+y\leq 0\}$ and let $W=U\cap V$. Then $W=\{(x,y)\in\mathbf{R}^2: x+y\leq 0 \text{ and } x-y<0\}$. The set W is neither open nor closed. In fact, the point $(-1,1)\in W$, but for every $r>0$ the point $(-1+r,1)\notin W$. Hence W is not open. The complement of W is the set $W^c=\{(x,y)\in\mathbf{R}^2: x+y>0 \text{ or } x-y\geq 0\}$. This set is not open.

In fact, the point $(-1,-1)\in W^c$, but for every $r>0$ the point $(-1-r,-1)\notin W^c$. Hence, W is not closed.

In Example 3.1.3 the set U is defined by a strict inequality. U is an open set. The set V is defined by a nonstrict inequality. V is a closed set. The set W is defined by a system of a strict inequality and a nonstrict inequality. W is neither open nor closed. Any subset of $\mathbf{R}^q$ defined exclusively by strict inequalities is open. Likewise, any subset of $\mathbf{R}^q$ defined exclusively by nonstrict inequalities is closed. When strict inequalities are mixed to nonstrict inequalities in the definition of a set, we usually obtain a set which is neither open nor closed. The boundary of a set defined by inequalities is usually equal to the set of points satisfying the corresponding equalities.

Example 3.1.4 Let $U=\{\mathbf{x}=(x,y)\in \mathbf{R}^2: |x|+|y|<2 \text{ and } |x|+|y|\leq 2\}$. Notice that the vectors which satisfy the first inequality (strict) also satisfy the second (nonstrict). Hence, the set is defined only by the first, and it is an open and bounded set with r=2. Its boundary is the set of points $\partial U=\{\mathbf{x}=(x,y): |x|+|y| =2\}$. The complement of U is the set $U^c=\{\mathbf{x}=(x,y): |x|+|y|\geq 2\}$. It is a closed and unbounded set.

Example 3.1.5 Let $U=\{\mathbf{x}=(x,y,z): xyz>0\}$. U is an open set. Its complement is the set $U^c=\{\mathbf{x}=(x,y,z) : xyz\leq 0\}$. U^c is closed. The boundary of U, which is also the boundary of U^c, is the set $\partial U=\{\mathbf{x}=(x,y,z):xyz=0\}$. Both U and its complement are unbounded.

Other norms

The Euclidean norm is just an example of a norm, since every function f $:\mathbf{R}^q\to\mathbf{R}$ which satisfies (i.), (ii.) and (iii.) of (3.1.9) is called a **norm** in $\mathbf{R}^q$. Two examples of such functions are

- The taxicab norm $\|\mathbf{x}\|_1 = |x_1| + |x_2| + ... + |x_q|$.
- The maximum norm $\|\mathbf{x}\|_\infty = \max\{|x_i| : i =1,2,...,q\}$.

(3.1.11)

Each norm defines a corresponding distance between pairs of points. For example, the taxicab norm defines the taxicab distance, and the maximum norm defines the maximum distance. However, our readers should be aware that not all distance functions can be obtained in this way. For example, the distance between pairs of points on the surface of our planet, defined as the length of the arc of the great circle passing through them, is a distance that cannot be obtained from a norm. In the following example we compute the distance between two points in the plane using the two norms mentioned above and the Euclidean norm.

Example 3.1.6 The Euclidean distance between the two points (3,4) and (7,2) is $(20)^{.5}=2(5)^{.5}$. In the maximum norm the distance is $\max\{|7-3|,|2-4|\}=4$. In the taxicab norm the distance is $|3-7|+|4-2|=6$. This result explains the name taxicab norm. In fact, a taxi driver going from (3,4) to (7,2) most likely cannot travel along the line segment of length $2(5)^{.5}$ joining (3,4) with (7,2), but may have to go from (3,4) to (7,4) and from (7,4) to (7,2). Thus, the distance traveled by the taxi driver will be 4+2=6 rather than the Euclidean distance $2(5)^{.5}$.

Two norms $||.||_a$ and $||.||_b$ are said to be **equivalent** if there are two positive constants m and M such that for every $\mathbf{x} \in \mathbf{R}^q$ we have

$$m||\mathbf{x}||_b \le ||\mathbf{x}||_a \le M||\mathbf{x}||_b. \tag{3.1.12}$$

From (3.1.12) we obtain

$$(1/M)||\mathbf{x}||_a \le ||\mathbf{x}||_b \le (1/m)||\mathbf{x}||_a. \tag{3.1.13}$$

Example 3.1.7 Let us verify that the Euclidean norm and the maximum norm are equivalent. Since

$$\max\{|x_i|^2 : i=1,2,...,q\} \le x_1{}^2 +...+ x_q{}^2 \le q \max\{|x_i|^2 : i = 1,2,...,q\},$$

we obtain $||\mathbf{x}||_\infty \le ||\mathbf{x}|| \le (q)^{.5}\, ||\mathbf{x}||_\infty$. Hence, $M=(q)^{.5}$ and $m=1$.

Later in this section we will show that all norms in $\mathbf{R}^q$ are equivalent. Thus, a set $U \subset \mathbf{R}^q$ which is open in the Euclidean norm, is open in any other norm. The interior, the boundary, and the closure of a set are the same in every norm. Spheres, balls, and closed balls change with the norm.

For example, the unit sphere of $\mathbf{R}^2$ in the maximum norm is the square with vertices (1,1), (–1,1), (–1,–1), (1,–1), since we require that $\max\{|x|, |y|\}=1$. Unless otherwise stated, we shall always use the Euclidean norm and the corresponding Euclidean distance. The equivalence of all norms implies that all results we mention are valid in every norm, although statements and proofs may need some adjustments.

The following useful inequality is valid in every norm. Given any pair of vectors **x** and **y** and denoting by $||.||_0$ an arbitrary norm in $\mathbf{R}^q$, we have

$$|\ ||\mathbf{x}||_0 - ||\mathbf{y}||_0\ | \le ||\mathbf{x} - \mathbf{y}||_0. \tag{3.1.14}$$

In fact, by (3.1.9) iii. we have $||\mathbf{x}||_0=||\mathbf{x}-\mathbf{y}+\mathbf{y}||_0 \le ||\mathbf{x}-\mathbf{y}||_0+||\mathbf{y}||_0$. Hence, $||\mathbf{x}||_0-||\mathbf{y}||_0 \le ||\mathbf{x}-\mathbf{y}||_0$. Reversing the roles of **x** and **y**, we obtain $||\mathbf{y}||_0-||\mathbf{x}||_0 \le ||\mathbf{y}-\mathbf{x}||_0$. Since $||\mathbf{x}-\mathbf{y}||_0=||\mathbf{y}-\mathbf{x}||_0$, we have

$$||\mathbf{x}||_0 - ||\mathbf{y}||_0 \le ||\mathbf{x} - \mathbf{y}||_0 \quad \text{and} \quad ||\mathbf{y}||_0 - ||\mathbf{x}||_0 \le ||\mathbf{x} - \mathbf{y}||_0$$

from which (3.1.14) follows.

Problems

1. Find the boundary of the set $U=\{(x,y): |x|+|y|<1\}$. Provide a graphical description of the boundary.

2. Does the point $\mathbf{x}=(1,2,1)$ belong to $B(\mathbf{c},2)$ where $\mathbf{c}=(1,-1,0)$?

3. Let $\mathbf{c}=(1,2,-1)$. Produce four points that belong to $S(\mathbf{c},3)$.

4. Show that a set $A \subset \mathbf{R}^q$ is not closed if there is a sequence $\mathbf{x}_n \in A$ such that $\|\mathbf{x}_n - \mathbf{x}\| \to 0$ as $n \to \infty$ and $\mathbf{x} \notin A$.

5. Let $A \subset \mathbf{R}^q$ be not closed. Describe how to construct a sequence $\mathbf{x}_n \in A$ such that $\|\mathbf{x}_n - \mathbf{x}\| \to 0$ and $\mathbf{x} \notin A$.

6. Decide whether or not this statement is correct: A set B that does not contain its boundary is open. Explain your conclusions.

7. Show that the interior of a set is an open set.

8. Let $A = \{(x,y): |x|+|y| = 4\}$. Show that A is bounded and closed.

9. Show that in $\mathbf{R}^q$ the complement of a bounded set is unbounded.

10. Verify that the taxicab norm is a norm.

11. Verify that the maximum norm is a norm.

12. Show that the closed ball of radius 1 of $\mathbf{R}^2$ in the taxicab norm is contained in the closed ball of radius 1 in the Euclidean norm.

13. Show that the closed ball of radius 1 of $\mathbf{R}^2$ in the Euclidean norm is contained in the closed ball of radius 1 in the maximum norm.

14. Show that in $\mathbf{R}^2$ the taxicab distance between two points is never smaller than the Euclidean distance.

15. Draw a picture of the closed ball of radius 1 in the taxicab norm and in the maximum norm.

2. Continuity

A function $F : \mathbf{R}^q \to \mathbf{R}^p$ is an action on the elements of $\mathbf{R}^q$, which maps each element of $\mathbf{R}^q$ into an element of $\mathbf{R}^p$. F acts on q-tuples of real numbers. Given the q-tuple $\mathbf{x} = (x_1, x_2, \ldots, x_q)$ we shall use either one of the two symbols $F(x_1, \ldots, x_q)$ and $F(\mathbf{x})$ to denote the action of F on $(x_1, \ldots, x_q)$. Every function $F : \mathbf{R}^q \to \mathbf{R}^p$ is called a **vector function of a vector variable.**

Example 3.1.8 Let $F : \mathbf{R}^3 \to \mathbf{R}^4$ be defined by $F(x,y,z) = (x-y, y^2, y+z^2, x+z)$. We have $F(1,1,1) = (0,1,2,2)$, $F(2,-1,0) = (3,1,-1,2)$, etc.

From Example 3.1.8 we see that a function $F : \mathbf{R}^q \to \mathbf{R}^p$ is completely determined by its **component functions** $F_i : \mathbf{R}^q \to \mathbf{R}$, $i = 1, 2, \ldots, p$. In Example 3.1.8 the four component functions are

$F_1(x,y,z) = x - y, \quad F_2(x,y,z) = y^2, \quad F_3(x,y,z) = y + z^2, \quad F_4(x,y,z) = x + z.$

Example 3.1.9 Let $F(x,y)=(xy,x-y^2)$. Then $F : \mathbf{R}^2 \to \mathbf{R}^2$, and its components are $F_1(x,y)=xy$, $F_2(x,y)=x-y^2$.

A function F may be defined only on a proper subset U of $\mathbf{R}^q$. Unless otherwise stated, we assume that U is open and we write $F : U \to \mathbf{R}^p$.

Example 3.1.10 $F(x,y)=(x/(x-y),x+y)$ is defined only on the open subset $U \subset \mathbf{R}^2$, $U=\{(x,y) : x-y \neq 0\}$. Its component functions are $F_1(x,y)=x/(x-y)$ and $F_2(x,y)=x+y$.

We now give the definition of continuity using the Euclidean norm. Similar definitions can be given using any other norm. The equivalence of all norms implies the equivalence of all these definitions.

Definition 3.1.1
Let U be an open set of $\mathbf{R}^q$. A function $F : U \to \mathbf{R}^p$ is said to be **continuous** at $\mathbf{x} \in U$ if for every $r>0$ there exists $d>0$ such that $\|\mathbf{y}-\mathbf{x}\| \leq d$ implies that $\|F(\mathbf{y})-F(\mathbf{x})\| \leq r$.

Continuity can be checked using sequences. To explain how this can be done, we need the definition of convergent sequence.

Definition 3.1.2
A sequence of vectors $\{\mathbf{x}_n \in \mathbf{R}^q : n=1,2,...\}$ converges to $\mathbf{x} \in \mathbf{R}^q$, written $\mathbf{x}_n \to \mathbf{x}$, if for every $i=1,2,...,q$ we have $|x_{in}-x_i| \to 0$ as $n \to \infty$.

In other words, if $\mathbf{x}_n=(x_{1n},x_{2n},...,x_{qn})$ and $\mathbf{x}=(x_1,x_2,...,x_q)$, then $\mathbf{x}_n \to \mathbf{x}$ if for each $i=1,2,...,q$ the numerical sequence of the ith components of $\mathbf{x}_n$ converges to the ith component of $\mathbf{x}$.

Example 3.1.11 Let $\mathbf{x}_n=(n/(n+1),1/n)$. The sequence of the first components converges to 1 and the sequence of the second components converges to 0. Hence, $\mathbf{x}_n \to \mathbf{x}=(1,0)$.

Clearly, $\mathbf{x}_n \to \mathbf{x}$ if and only if $\|\mathbf{x}_n-\mathbf{x}\| \to 0$ as $n \to \infty$. The following lemma shows how sequences can be used to prove continuity.

Lemma 3.1.1 $F : U \to \mathbf{R}^p$ *is continuous at* $\mathbf{x} \in U$ *if and only if for every sequence* $\mathbf{x}_n \to \mathbf{x}$ *we have* $F(\mathbf{x}_n) \to F(\mathbf{x})$.

Proof. Assume that F is continuous at $\mathbf{x}$ and let $\mathbf{x}_n \to \mathbf{x}$. Choose $r>0$. By the definition of continuity there exists $d>0$ such that $\|F(\mathbf{y})-F(\mathbf{x})\| \leq r$ for every $\|\mathbf{y}-\mathbf{x}\| \leq d$. Since $\|\mathbf{x}_n-\mathbf{x}\| \to 0$ there exists an index n_0 such that $\|\mathbf{x}_n-\mathbf{x}\| \leq d$ for all $n \geq n_0$. Hence, $\|F(\mathbf{x}_n)-F(\mathbf{x})\| \leq r$ for every $n \geq n_0$. Since r is arbitrary, we conclude that the sequence $\|F(\mathbf{x}_n)-F(\mathbf{x})\| \to 0$ as $n \to \infty$.

Conversely, assume that F is not continuous at $\mathbf{x}$. Then there exists $r>0$ such that for every $d_n=1/n$ we can find $\mathbf{x}_n$ with the two properties $\|\mathbf{x}_n-\mathbf{x}\|\le 1/n$ and $\|F(\mathbf{x}_n)-F(\mathbf{x})\|>r$. Clearly, $\mathbf{x}_n\to\mathbf{x}$ but $F(\mathbf{x}_n)\not\to F(\mathbf{x})$. QED

Let $F_1,\ldots,F_p$ be the component functions of F. Since $F(\mathbf{x}_n)\to F(\mathbf{x})$ if and only if $F_j(\mathbf{x}_n)\to F_j(\mathbf{x})$ for every $j=1,2,\ldots,p$, the continuity of F at $\mathbf{x}$ is equivalent to the continuity of all its component functions at $\mathbf{x}$. In other words, $F(\mathbf{x}_n)\to F(\mathbf{x})$ if and only if the numerical sequences $F_j(\mathbf{x}_n)-F_j(\mathbf{x})\to 0$ for every $j=1,2,\ldots,p$. This observation makes it easier to check the continuity of F at $\mathbf{x}$.

Definition 3.1.3
$F: U\to\mathbf{R}^p$ is said to be **continuous in** U if F is continuous at every $\mathbf{x}\in U$.

Checking the continuity of a function is sometimes a demanding task. We mention here a few helpful principles:

- *A sum or a difference of continuous functions is continuous.*
- *A composition of continuous functions is continuous.*
- *A polynomial function is continuous.*

In the following examples we show how to apply the principles above.

Example 3.1.12 Let $F(x,y,z)=(x^2-y,\cos xz)$. F is continuous, since its first component is a polynomial of degree 2 in the two variables x and y, and its second component is the composition of a polynomial of degree 2 with the continuous trigonometric function cosine.

Example 3.1.13 Let $F(x,y,z)=(xyz,x-\arctan z,\sqrt{x+y})$. Then $F: U\to\mathbf{R}^3$, where $U=\{(x,y,z): x+y\ge 0\}$ and F is continuous since the arctan and the square root functions are continuous.

We now prove that every norm function is continuous.

Lemma 3.1.2 *Assume that* $\|.\|_a$ *is a norm in* $\mathbf{R}^q$. *Then* $F:\mathbf{R}^q\to\mathbf{R}$, $F(\mathbf{x})=\|\mathbf{x}\|_a$ *is continuous.*

Proof. We show that if $\mathbf{x}_n\to\mathbf{x}$, then $F(\mathbf{x}_n)\to F(\mathbf{x})$. Let $\mathbf{x}_n=(x_{1n},\ldots,x_{qn})$, $\mathbf{x}=(x_1,\ldots,x_q)$. From

$$\mathbf{x}=(x_1,\ldots,x_q)=x_1\mathbf{e}_1+x_2\mathbf{e}_2+\ldots+x_q\mathbf{e}_q,\quad \mathbf{x}_n=(x_{1n},\ldots,x_{qn})=x_{1n}\mathbf{e}_1+\ldots+x_{qn}\mathbf{e}_q,$$

and using inequality (3.1.14) in conjunction with the second and third properties of a norm we obtain

$$|F(\mathbf{x}_n)-F(\mathbf{x})| = |\,\|\mathbf{x}_n\|_a-\|\mathbf{x}\|_a\,| \le \|\mathbf{x}_n-\mathbf{x}\|_a$$

$$\le |x_{1n}-x_1|\,\|\mathbf{e}_1\|_a+\ldots+|x_{qn}-x_q|\,\|\mathbf{e}_q\|_a.$$

Since $|x_{in}-x_i|\to 0$ as $n\to\infty$, for all $i=1,\ldots,q$, we obtain $|\, \|\mathbf{x}_n\|_a - \|\mathbf{z}\|_a \,| \to 0$ as $n\to\infty$. Hence, F is continuous. QED

Before establishing the equivalence of all norms, which is one of the key results needed in the following chapters, we need to extend to $\mathbf{R}^q$ the extreme value theorem (EVT). The EVT in $\mathbf{R}$ states that given a continuous function $g: K\to\mathbf{R}$, where $K\subset\mathbf{R}$ is a bounded and closed set, there are $x_m, x_M\in K$ such that $g(x_m)\leq g(x)\leq g(x_M)$ for every $x\in K$. In other words, g has in K absolute maximum and absolute minimum value. A similar result holds when K is a compact subset of $\mathbf{R}^q$ and $g: K\to\mathbf{R}$ is continuous. Here is the result, stated more formally (for a proof, see [Marsden-Hoffman, 1993, pg. 214]).

Theorem 3.1.1 (Extreme Value Theorem) *Let* $K\subset\mathbf{R}^q$ *be compact and* $g: K\to\mathbf{R}$ *be continuous. Then* g *has maximum and minimum value in* K, *i.e., there exist* $\mathbf{x}_m, \mathbf{x}_M\in K$ *such that* $g(\mathbf{x}_m)\leq g(\mathbf{x})\leq g(\mathbf{x}_M)$ *for every* $\mathbf{x}\in K$.

We can now prove the following result.

Theorem 3.1.2 *All norms in* $\mathbf{R}^q$ *are equivalent.*

Proof. Let $\|.\|_a$, $\|.\|_b$ be two norms. We first establish the equivalence when $\|\mathbf{x}\|_b=\|\mathbf{x}\|$, the Euclidean norm of $\mathbf{x}$. Consider the closed and bounded set $S(1)=\{\mathbf{x}\in\mathbf{R}^q: \|\mathbf{x}\|=1\}$. By Lemma 3.1.2 the function $g(\mathbf{x})=\|\mathbf{x}\|_a$ is continuous. Hence, it has in S(1) a minimum m_1 and a maximum M_1. The minimum m_1 has to be strictly positive since $\|\mathbf{x}\|_a=0$ only if $\mathbf{x}=\mathbf{0}$. Thus, $0<m_1\leq g(\mathbf{x})\leq M_1$ for all $\mathbf{x}\in S(1)$. For every other $\mathbf{y}\in\mathbf{R}^q$, $\mathbf{y}\neq\mathbf{0}$, let $\mathbf{x}=\mathbf{y}/\|\mathbf{y}\|$. Then $\|\mathbf{x}\|=1$, $\|\mathbf{x}\|_a=\|\mathbf{y}\|_a/\|\mathbf{y}\|$, and

$$m_1 \leq g(\mathbf{x}) = \|\mathbf{x}\|_a = \|\mathbf{y}\|_a/\|\mathbf{y}\| \leq M_1.$$

It follows that $m_1\|\mathbf{y}\|\leq\|\mathbf{y}\|_a\leq M_1\|\mathbf{y}\|$. This inequality, which is now true for every vector $\mathbf{y}\in\mathbf{R}^q$, shows that the a-norm and the Euclidean norm are equivalent. In a similar manner it can be shown that the b-norm and the Euclidean norm are equivalent, i.e.,

$$m_2\|\mathbf{y}\| \leq \|\mathbf{y}\|_b \leq M_2\|\mathbf{y}\|$$

for some constants $0<m_2<M_2$. From the two inequalities we obtain

$$(m_2/M_1)\|\mathbf{y}\|_a \leq \|\mathbf{y}\|_b \leq (M_2/m_1)\|\mathbf{y}\|_a,$$

which implies that the two norms are equivalent. QED

Remark 3.1.2 As a consequence of Theorem 3.1.2, a function $F: U\to\mathbf{R}^p$, $U\subset\mathbf{R}^q$, which is continuous in one norm is continuous in every other norm. In fact, assume that F is continuous when $\|.\|_a$ is used in both $\mathbf{R}^q$ and $\mathbf{R}^p$. Then $\|\mathbf{x}_n-\mathbf{x}\|_a\to 0$ as $n\to\infty$ implies that $\|F(\mathbf{x}_n)-F(\mathbf{x})\|_a\to 0$ as $n\to\infty$. From

$$\|\mathbf{x}_n-\mathbf{x}\|_b \leq (M_2/m_1)\|\mathbf{x}_n-\mathbf{x}\|_a \qquad \text{and} \qquad (m_2/M_1)\|\mathbf{x}_n-\mathbf{x}\|_a \leq \|\mathbf{x}_n-\mathbf{x}\|_b$$

we see that $\|\mathbf{x}_n-\mathbf{x}\|_b\to 0$ if and only if $\|\mathbf{x}_n-\mathbf{x}\|_a\to 0$. Likewise, $\|F(\mathbf{x}_n)-F(\mathbf{x})\|_b\to 0$ if and only if $\|F(\mathbf{x}_n)-F(\mathbf{x})\|_a\to 0$. Hence, F is continuous in the b-norm if and only if F is continuous in the a-norm.

F remains continuous when either the a-norm or the b-norm is used in $\mathbf{R}^q$ and the other is used in $\mathbf{R}^p$. Thus, the continuity of F can be verified or disproved by using an arbitrary norm in $\mathbf{R}^q$, the space where the domain U of F is contained, and a different norm in the space $\mathbf{R}^p$ where the image (or range) of F is contained.

Problems

1. Let $F(x,y)=(x-y^2, x^2-y)$. What are the component functions of F? Is the function F continuous in $\mathbf{R}^2$?

2. Let $F(x,y)=(x/y,y/x)$. Find the component functions of F and its domain. Is the function continuous?

3. Let $F(x,y,z)=(x-y, xy-z^2, \sin(x+z))$. Find the component functions of F and show that F is continuous.

4. Let $F(x,y,z)=(x-y,\tan(x+z),x-z)$. Find the component functions of F and its domain. Is F continuous?

5. Let $\mathbf{x}_0=(2,3)$. Show that $F : \mathbf{R}^2\to\mathbf{R}$, $F(\mathbf{x})=\mathbf{x}\bullet\mathbf{x}_0$ is continuous.

6. Let $\mathbf{x}_0\in\mathbf{R}^q$. Define $F : \mathbf{R}^q\to\mathbf{R}$ by $F(\mathbf{x})=\mathbf{x}\bullet\mathbf{x}_0$. Show that F is continuous.

7. Let $\mathbf{x}_0\in\mathbf{R}^q$. Define $F : \mathbf{R}^q\to\mathbf{R}$ by $F(\mathbf{x})=\mathbf{x}\bullet\mathbf{x}_0$. Find the maximum and minimum value of F on D(1).

8. Let $F(x)=1/x$ for $x\neq 0$ and $F(0)=2$. Notice that F does not have a maximum in [0,1]. Which assumption of Theorem 3.1.1 is not satisfied?

9. Let $F(x)=\sin x/x$ for $x\neq 0$ and $F(0)=1$. Show that F is continuous in [0,1] and find its maximum and minimum value in [0,1].

10. Let $F(x,y)=x^2-xy+y^2$. Consider the closed and bounded set $D=\{(x,y): x^2+y^2\leq 1\}$. Find the maximum and the minimum of F on D.

11. Let F be as in Problem 10. Find its maximum and minimum on the closed and bounded set $S(1)=\{(x,y): x^2+y^2=1\}$.

12. Find two constants a and A such that for every $\mathbf{x}\in\mathbf{R}^q$ we have $a\|\mathbf{x}\|_1\leq\|\mathbf{x}\|\leq A\|\mathbf{x}\|_1$, where the 1-norm is the taxicab norm.

13. Find two constants a and A such that for every $\mathbf{x}\in\mathbf{R}^q$ we have $a\|\mathbf{x}\|_\infty\leq\|\mathbf{x}\|_1\leq A\|\mathbf{x}\|_\infty$, where the ∞- and the 1-norm are the maximum and the taxicab norm, respectively.

Section 2. **OPERATOR NORM AND DERIVATIVE**

1. **Operator Norm**

Let M be a square matrix of size q. We define an action of M : $\mathbf{R}^q \to \mathbf{R}^q$ by setting

$$M\mathbf{x} = (\mathbf{m}_1 \cdot \mathbf{x})\mathbf{e}_1 + \ldots + (\mathbf{m}_q \cdot \mathbf{x})\mathbf{e}_q = (\mathbf{m}_1 \cdot \mathbf{x}, \ldots, \mathbf{m}_q \cdot \mathbf{x}), \tag{3.2.1}$$

where $\mathbf{m}_i$, i=1,2,...,q, represents the ith row of M. Since each row of M can be regarded as a vector in $\mathbf{R}^q$, the scalar product $\mathbf{m}_i \cdot \mathbf{x}$ is defined for all i=1,2,...,q. Uppercase letters of the alphabet will usually denote matrices.

Example 3.2.1 Let

$$M = \begin{pmatrix} 2 & 3 \\ 1 & 5 \end{pmatrix}.$$

Then M : $\mathbf{R}^2 \to \mathbf{R}^2$; $M\mathbf{x} = (\mathbf{m}_1 \cdot \mathbf{x})\mathbf{e}_1 + (\mathbf{m}_2 \cdot \mathbf{x})\mathbf{e}_2 = (2x+3y, x+5y)$.

Example 3.2.2 Let

$$M = \begin{pmatrix} 2 & 3 & 1 \\ 5 & 4 & 2 \\ 4 & 3 & 7 \end{pmatrix}.$$

Then M : $\mathbf{R}^3 \to \mathbf{R}^3$; $M\mathbf{x} = (\mathbf{m}_1 \cdot \mathbf{x})\mathbf{e}_1 + (\mathbf{m}_2 \cdot \mathbf{x})\mathbf{e}_2 + (\mathbf{m}_3 \cdot \mathbf{x})\mathbf{e}_3 = (2x+3y+z,\ 5x+4y+2z,\ 4x+3y+7z)$.

The following properties of the action of M are important.

1. For every pair of real numbers a, b and for every pair of elements $\mathbf{x}$, $\mathbf{y} \in \mathbf{R}^q$, we have $M(a\mathbf{x}+b\mathbf{y}) = aM\mathbf{x} + bM\mathbf{y}$.
2. For every element $\mathbf{e}_i$ of the standard basis E we have $M\mathbf{e}_i = (m_{1i}, \ldots, m_{qi})$; i.e., the components of the element $M\mathbf{e}_i$ in the standard basis are the entries of the ith column of M.
3. For every pair of matrices M, P of size q, and for every element $\mathbf{x}$, we have $P(M\mathbf{x}) = (PM)\mathbf{x}$, where PM is the row by column product of P and M.

Property 1 is referred to as the **linearity** of the action of M. Property 3 tells us that the action of M followed by the action of P has the same effect as the action of the matrix PM. The justification of properties 1 and 2 is straightforward and is based on the properties of the inner product. Property 3 is proven by writing the coordinates of $P(M\mathbf{x})$, $(PM)\mathbf{x}$ and by verifying that they are equal.

In the following example we verify properties 1, 2, and 3 using specific numbers a and b, and specific matrices M and P.

Example 3.2.3 Let

$$M = \begin{pmatrix} 2 & 3 \\ 1 & 5 \end{pmatrix}$$

and let $\mathbf{x}=(2,3)$, $\mathbf{y}=(4,-2)$, a=5, and b=4. Then a$\mathbf{x}$+b$\mathbf{y}$=(26,7). Hence, M(a$\mathbf{x}$+b$\mathbf{y}$)=(73,61). Since 5M$\mathbf{x}$=5(13,17)=(65,85) and 4M$\mathbf{y}$=4(2,–6)=(8,–24), we see that 5M$\mathbf{x}$+4M$\mathbf{y}$=(73,61)=M(5$\mathbf{x}$+4$\mathbf{y}$). Let

$$P = \begin{pmatrix} 1 & -2 \\ 0 & 2 \end{pmatrix}.$$

Then (PM)$\mathbf{x}$=(–21,34). Since M$\mathbf{x}$=(13,17), we obtain P(M$\mathbf{x}$)=(13–34,34))=(–21,34)=(PM)$\mathbf{x}$.

The action of M can also be written in the form

$$M\mathbf{x} = x_1 M\mathbf{e}_1 + ... + x_q M\mathbf{e}_q. \tag{3.2.2}$$

The equality above can be obtained as follows. From $\mathbf{x}=(x_1,x_2,...,x_q)=x_1\mathbf{e}_1+...+x_q\mathbf{e}_q$ and using property 1 we have $M\mathbf{x}=M(x_1\mathbf{e}_1+...+x_q\mathbf{e}_q)=x_1M\mathbf{e}_1+...+x_qM\mathbf{e}_q$. Hence, the image M$\mathbf{x}$ of $\mathbf{x}$ is a linear combination of the vectors $M\mathbf{e}_1$, ..., $M\mathbf{e}_q$ using the components of $\mathbf{x}$ in the standard basis of $\mathbf{R}^q$ as coefficients.

By property 2 the components of the vector $M\mathbf{e}_i$, i=1,...,q are the entries of the ith column of M. Hence, the image of M is the subspace of $\mathbf{R}^q$ spanned by the columns of M. Moreover, (3.2.2) tells us that M is entirely determined by its action on the elements of the standard basis. The following examples illustrate these remarks.

Example 3.2.4 Let

$$M = \begin{pmatrix} 2 & 3 \\ 1 & 5 \end{pmatrix}$$

and let $\mathbf{x}=(x_1,x_2)=x_1\mathbf{e}_1+x_2\mathbf{e}_2$. From (3.2.1) we obtain $M\mathbf{x}=(2x_1+3x_2,x_1+5x_2)$. From (3.2.2) we obtain

$$M\mathbf{x} = x_1M\mathbf{e}_1 + x_2M\mathbf{e}_2 = x_1(2,1) + x_2(3,5) = (2x_1 + 3x_2, x_1 + 5x_2).$$

The two results coincide and M is completely determined by its action on $\mathbf{e}_1$ and $\mathbf{e}_2$.

Example 3.2.5 Let M(1,0)=(3,–2), M(0,1)=(–1,5). For every $\mathbf{x} \in \mathbf{R}^2$ we have M$\mathbf{x}$=$(3x-y,-2x+5y)$. In fact,

$$\mathbf{x} = x\mathbf{e}_1 + y\mathbf{e}_2 \quad \text{and} \quad M\mathbf{x} = xM\mathbf{e}_1 + yM\mathbf{e}_2.$$

Therefore,

$$\mathbf{M}\mathbf{x} = x(3,-2) + y(-1,5) = (3x - y, -2x + 5y).$$

Since $(3x-y,-2x+5y)=((3,-1)\bullet\mathbf{x}, (-2,5)\bullet\mathbf{x})$, we obtain

$$\mathrm{M} = \begin{pmatrix} 3 & -1 \\ -2 & 5 \end{pmatrix}.$$

The entries of the first column of M are the components of $\mathrm{M}\mathbf{e}_1$, and the entries of the second column are the components of $\mathrm{M}\mathbf{e}_2$.

Example 3.2.6 Let M be a 3×3 matrix such that $\mathrm{M}\mathbf{e}_1=(1,2,-2)$, $\mathrm{M}\mathbf{e}_2=(2,3,-4)$, and $\mathrm{M}\mathbf{e}_3=(3,-2,5)$. Then

$$\mathrm{M} = \begin{pmatrix} 1 & 2 & 3 \\ 2 & 3 & -2 \\ -2 & -4 & 5 \end{pmatrix}.$$

Operator norm of a matrix

From (3.2.1) and the Cauchy-Schwarz inequality we derive

$$\|\mathrm{M}\mathbf{x}_0\|^2 = (\mathbf{m}_1\bullet\mathbf{x}_0)^2 + \ldots + (\mathbf{m}_q\bullet\mathbf{x}_0)^2 \le (\|\mathbf{m}_1\|^2 + \ldots + \|\mathbf{m}_q\|^2)\,\|\mathbf{x}_0\|^2 = p^2\|\mathbf{x}_0\|^2$$

with $p^2=\|\mathbf{m}_1\|^2+\ldots+\|\mathbf{m}_q\|^2$. Let $\mathbf{x}$ be given and set $\mathbf{y}=\mathbf{x}+\mathbf{x}_0$. Then, using the linearity of the action of M and the inequality above, we obtain

$$\|\mathrm{M}\mathbf{y} - \mathrm{M}\mathbf{x}\| = \|\mathrm{M}(\mathbf{y} - \mathbf{x})\| \le p\|\mathbf{y} - \mathbf{x}\|. \qquad (3.2.3)$$

The following theorem is an important consequence of the inequality above.

Theorem 3.2.1 M *is continuous at every* $\mathbf{x}\in\mathbf{R}^q$ *regardless of the norm used in* $\mathbf{R}^q$.

Proof. Assume that $\mathbf{x}_n\to\mathbf{x}$. Then $\|\mathbf{x}_n-\mathbf{x}\|\to 0$ as $n\to\infty$. From inequality (3.2.3) we obtain $\|\mathrm{M}\mathbf{x}_n-\mathrm{M}\mathbf{x}\|\to 0$ as $n\to\infty$. Hence, M is continuous at $\mathbf{x}$. Since $\mathbf{x}$ is an arbitrary element of $\mathbf{R}^q$ we conclude that M is continuous in the entire space $\mathbf{R}^q$. From Remark (3.1.2) we obtain that M is continuous in every norm. QED

From the continuity of the Euclidean norm and Theorem 3.2.1, we derive that the function $g(\mathbf{x})=\|\mathrm{M}\mathbf{x}\|$ is continuous. By Theorem 3.1.1, F has a maximum on $S(\mathbf{0},1)$. This maximum, called the **Euclidean operator norm** of M, is denoted by $\|\mathrm{M}\|$. Hence,

$$\|\mathrm{M}\| = \max\{\|\mathrm{M}\mathbf{x}\| : \|\mathbf{x}\|=1\}. \qquad (3.2.4)$$

For every vector $\mathbf{y}\in\mathbf{R}^q$ we have

$$\|\mathrm{M}\mathbf{y}\| \le \|\mathrm{M}\|\,\|\mathbf{y}\|. \qquad (3.2.5)$$

The inequality is obvious if $\mathbf{y}=\mathbf{0}$ since $M\mathbf{0}=\mathbf{0}$. For every other vector $\mathbf{y}$ consider the vector $\mathbf{x}=\mathbf{y}/\|\mathbf{y}\|$. Then $\|\mathbf{x}\|=1$, and according to (3.2.4), $\|M\mathbf{x}\|\le\|M\|$. Inequality (3.2.5) follows from property (ii) of the norm, since

$$\|M\mathbf{y}\| = \|\,\|\mathbf{y}\|\, M\mathbf{x}\| = \|M\mathbf{x}\|\,\|\mathbf{y}\| \le \|M\|\,\|\mathbf{y}\|.$$

Let M and P be two matrices of size q. The following basic inequality is easily established from the definition of the Euclidean operator norm:

$$\|PM\| \le \|P\|\,\|M\|. \tag{3.2.6}$$

In particular, for P = M we obtain $\|M^2\| \le \|M\|^2$. In general,

$$\|M^n\| \le \|M\|^n. \tag{3.2.7}$$

The operator norm defined in (3.2.4) depends on the Euclidean norm of $\mathbf{R}^q$. The use of a different norm in $\mathbf{R}^q$, for example the a-norm, produces the a-operator norm of M:

$$\|M\|_a = \max\{\|M\mathbf{x}\|_a\!: \|\mathbf{x}\|_a=1\}. \tag{3.2.4'}$$

The inequalities (3.2.5), (3.2.6), and (3.2.7) remain valid in the a-operator norm.

Recall that the **eigenvalues** of a square matrix M of size q are the (real or complex) roots of the **characteristic polynomial** of M

$$P(r) = \det(rI-M). \tag{3.2.8}$$

Since P(r) has degree q, the fundamental theorem of algebra implies that there are r_1, $r_2,\dots,r_k$ real or complex numbers and $m_1,m_2,\dots,m_k$ positive integers such that $m_1+\dots+m_k=q$ and

$$P(r) = \det(rI-M) = (r-r_1)^{m_1}\dots(r-r_k)^{m_k}. \tag{3.2.9}$$

The number m_i is the **algebraic multiplicity** of the eigenvalue r_i. Since the coefficients of the polynomial P(r) are real, complex roots of (3.2.8) come in pairs: a_j+ib_j and a_j-ib_j with the same algebraic multiplicity m_j.

The set of all eigenvalues of M, called the **spectrum** of M, is denoted by $\sigma(M)$. The number

$$\rho(M) = \max\{|r| : r\in\sigma(M)\} \tag{3.2.10}$$

is the **spectral radius** of M. In (3.2.10) the positive number $|r|$ is simply the absolute value of r if r is real, and $|r|=(a^2+b^2)^{1/2}$ when r=a+ib.

Example 3.2.7 Let

$$M = \begin{pmatrix} 2 & 0 & 0 \\ 0 & 3 & -2 \\ 0 & 2 & 3 \end{pmatrix}.$$

The reader can verify that $\sigma(M)=\{2,3+2i,\ 3-2i\}$. Hence, $\rho(M)=\sqrt{13}$.

The following theorem, presented without proof, will be needed in Chapters 4 and 5.

Theorem 3.2.2 $\lim_{n\to\infty} \sqrt[n]{\|M^n\|_a} = \rho(M)$.

Theorem 3.2.2 states that regardless of the a-norm used in $\mathbf{R}^q$, the sequence of the nth roots of the a-operator norm of the nth power M^n of M converges to the spectral radius of M. From Theorem 3.2.2 we derive the following result.

Theorem 3.2.3 *Assume that* $\rho(M)<1$. *Then there exists a norm* $\|.\|_a$ *in* $\mathbf{R}^q$ *such that* $\|M\|_a<1$.

Proof. Using the Euclidean operator norm, we have

$$\lim_{n\to\infty} \sqrt[n]{\|M^n\|} = \rho(M) < 1.$$

Hence, there exists a positive integer p such that $\|M^p\|<1$. Define the a-norm as follows:

$$\|\mathbf{x}\|_a = \|\mathbf{x}\| + \|M\mathbf{x}\| + \ldots + \|M^{p-1}\mathbf{x}\|. \tag{3.2.11}$$

We leave to the reader the verification that the a-norm just defined satisfies all properties of a norm. To find $\|M\|_a$, select any $\mathbf{x}$ such that $\|\mathbf{x}\|_a = \|\mathbf{x}\| + \|M\mathbf{x}\| + \ldots + \|M^{p-1}\mathbf{x}\| = 1$. Then

$$\|M\mathbf{x}\|_a = \|M\mathbf{x}\| + \ldots + \|M^p\mathbf{x}\| = \|\mathbf{x}\| + \|M\mathbf{x}\| + \ldots + \|M^p\mathbf{x}\| - \|\mathbf{x}\|$$

$$= 1 - \|\mathbf{x}\| + \|M^p\mathbf{x}\| \le 1 - \|\mathbf{x}\|(1-\|M^p\|) .$$

Since $\|\mathbf{x}\|_a=1$ we obtain from (3.2.11) and (3.2.5)

$$1 \le (1 + \|M\| + \ldots + \|M^{p-1}\|)\, \|\mathbf{x}\| = K\|\mathbf{x}\| ,$$

with $K=1+\|M\|+\ldots+ \|M^{p-1}\|$. Thus

$$\|M\mathbf{x}\|_a \le 1 - \frac{1}{K}(1 - \|M^p\|)$$

for every $\mathbf{x}$ such that $\|\mathbf{x}\|_a=1$. Therefore, the a-operator norm $\|M\|_a$, associated with the a-norm of $\mathbf{R}^q$ satisfies the inequality

$$\|M\|_a \le 1 - \frac{1}{K}(1 - \|M^p\|) < 1. \text{ QED} \tag{3.2.12}$$

Example 3.2.8 Let

$$M = \begin{pmatrix} 1/4 & 1 \\ 0 & 1/4 \end{pmatrix}.$$

Then $\|M\|>1$. Since

$$M^2 = \begin{pmatrix} 1/16 & 1/2 \\ 0 & 1/16 \end{pmatrix}$$

we see that $\|M^2\|<1$. Hence, we define

$$\|\mathbf{x}\|_a = \|\mathbf{x}\| + \|M\mathbf{x}\| = (x^2 + y^2)^{1/2} + [(.25x+y)^2 + (.25y)^2]^{1/2}.$$

From $\|M\|<1.2$ and $\|M^2\|<.52$ we obtain $K<2.2$ and $\|M\|_a \leq .8$.

Problems

1. Let $M\mathbf{e}_1=(2,3,2)$, $M\mathbf{e}_2=(2,-2,3)$, and $M\mathbf{e}_3=(-1,0,-2)$. Find $M\mathbf{x}$ for every $\mathbf{x} \in \mathbf{R}^3$.

2. Let $M\mathbf{e}_1=(1,2,-1)$, $M\mathbf{e}_2=(1,1,0)$, and $M\mathbf{e}_3=(2,3,0)$. Find $M\mathbf{x}$ and verify that $M(M\mathbf{x})=M^2\mathbf{x}$.

3. * Let

$$M = \begin{pmatrix} 2 & 1 \\ 0 & 3 \end{pmatrix}.$$

Estimate the operator norm of M and M^2.

4. Find the eigenvalues of the matrix M in Problem 3. Compute the spectral radius of M.

5. * Estimate the operator norm of the matrix of Problem 3 using the taxicab norm and the maximum norm.

6. Let

$$M = \begin{pmatrix} 1/4 & -\sqrt{3}/4 \\ \sqrt{3}/4 & 1/4 \end{pmatrix}.$$

Find the spectral radius of M.

7. The matrix M of the Problem 6 can be rewritten as

$$M = .5\begin{pmatrix} 1/2 & -\sqrt{3}/2 \\ \sqrt{3}/2 & 1/2 \end{pmatrix}.$$

Recall that $1/2=\cos\pi/3$ and $\sqrt{3}/2=\sin\pi/3$. Use this information to find the operator norm of M.

8. * Estimate the operator norm of the matrix in Example 3.2.7 using the maximum norm, the Euclidean norm, and the taxicab norm.

9. * Let

$$M = \begin{pmatrix} -18 & 10 & 20 \\ -10 & 7 & 10 \\ -10 & 5 & 12 \end{pmatrix}.$$

Find the spectral radius of M.

10. * Estimate the operator norm of the matrix of Problem 9 using the maximum norm.

11. * Let

$$M = \begin{pmatrix} 3 & -3 & 8 \\ 2 & 1 & -3 \\ 4 & 2 & -3 \end{pmatrix}.$$

Find the spectral radius of M and estimate its operator norm in the maximum norm.

2. **Derivative and Mean Value Inequality**

Let U be an open subset of $\mathbf{R}^q$, $F : U \to \mathbf{R}^q$ be continuous, and let $\mathbf{x} \in U$. Assume that all partial derivatives of the component functions of F exist at $\mathbf{x} \in U$. The matrix

$$J_{\mathbf{x}} = \begin{pmatrix} \frac{\partial F_1}{\partial x_1} & \frac{\partial F_1}{\partial x_2} & \cdots & \frac{\partial F_1}{\partial x_q} \\ \cdots & \cdots & \cdots & \cdots \\ \frac{\partial F_q}{\partial x_1} & \frac{\partial F_q}{\partial x_2} & \cdots & \frac{\partial F_q}{\partial x_q} \end{pmatrix},$$

where all partial derivatives are evaluated at **x**, is called the **Jacobian** matrix of F at **x**. Since U is open there exists r>0 such that $B(\mathbf{x},r) \subset U$. Hence, F is defined for every $\mathbf{y} \in B(\mathbf{x},r)$ and we can consider the ratio

$$\frac{F(\mathbf{y}) - F(\mathbf{x}) - J_{\mathbf{x}}(\mathbf{y} - \mathbf{x})}{\|\mathbf{y} - \mathbf{x}\|}.$$

Definition 3.2.1
We say that F is **differentiable** at **x** if

$$\lim_{\mathbf{y} \to \mathbf{x}} \frac{F(\mathbf{y}) - F(\mathbf{x}) - J_{\mathbf{x}}(\mathbf{y} - \mathbf{x})}{\|\mathbf{y} - \mathbf{x}\|} = \mathbf{0}. \qquad (3.2.13)$$

In this case, the Jacobian matrix of F at **x** is called the **derivative** of F at **x** and is denoted by $F'(\mathbf{x})$. Equality (3.2.13) says that the norm of $\mathbf{w}=F(\mathbf{y})-F(\mathbf{x})-$

$J_\mathbf{x}(\mathbf{y}-\mathbf{x})$ goes to zero faster than the norm of $\mathbf{y}-\mathbf{x}$. A sufficient condition that ensures the validity of (3.2.13) is the **continuity** at $\mathbf{x}$ of all **partial derivatives** $\partial F_i/\partial x_j$, i,j=1,2,...,q.

Example 3.2.9 Let $F(\mathbf{x})=F(x,y)=(x^2+y, y^2-2x+1)$. Then, $F_1(x,y)=x^2+y$, $F_2(x,y)=y^2-2x+1$, and

$$J_\mathbf{x} = \begin{pmatrix} 2x & 1 \\ -2 & 2y \end{pmatrix}.$$

Since the partial derivatives are continuous, F is differentiable at every point. Let $\mathbf{x}_0=(1,1)$. The derivative of F at $\mathbf{x}_0$ is

$$F'(\mathbf{x}_0) = \begin{pmatrix} 2 & 1 \\ -2 & 2 \end{pmatrix}.$$

Consequently,

$$F(\mathbf{y}) - F(1,1) - F'(\mathbf{x}_0)(\mathbf{y} - \mathbf{x}_0) = ((x - 1)^2, (y - 1)^2).$$

We see that as $\mathbf{y}\to\mathbf{x}_0$ the norm of the vector $\mathbf{w}=((x-1)^2, (y-1)^2)$ goes to 0 faster than $\|\mathbf{y}-\mathbf{x}_0\|=[(x-1)^2+(y-1)^2]^{1/2}$.

The following example shows that equality (3.2.13) may fail at a point $\mathbf{x}$ where $J_\mathbf{x}$ exists.

Example 3.2.10 Let $F(x,y)=(xy,xy)$ if $xy=0$ and (1,1) if $xy\neq0$. At $\mathbf{x}=(0,0)$ the Jacobian matrix of F is

$$J_\mathbf{0} = \begin{pmatrix} 0 & 0 \\ 0 & 0 \end{pmatrix}.$$

Equality (3.2.13) fails since $((1,1)-J_\mathbf{0}(x,y))/\|\mathbf{x}\|$ does not go to $\mathbf{0}$ as $\mathbf{x}\to\mathbf{0}$.

First-order approximation

Assume that F is differentiable at $\mathbf{x}$ and let $M=F'(\mathbf{x})$. The difference

$$F(\mathbf{y}) - F(\mathbf{x}) - M(\mathbf{y} - \mathbf{x})$$

is usually denoted by $o(\mathbf{y} - \mathbf{x})$ and is called a **higher-order term**. Hence,

$$F(\mathbf{y}) = F(\mathbf{x}) + M(\mathbf{y} - \mathbf{x}) + o(\mathbf{y} - \mathbf{x}). \tag{3.2.14}$$

This formula is useful for approximation purposes, since the size of the higher-order term $o(\mathbf{y}-\mathbf{x})$ becomes negligible and $o(\mathbf{y}-\mathbf{x})$ can frequently be omitted when $\mathbf{y}$ is very close to $\mathbf{x}$. Its omission in (3.2.14) destroys the equality and we have

$$F(\mathbf{y}) \neq F(\mathbf{x}) + M(\mathbf{y} - \mathbf{x}). \tag{3.2.15}$$

However, the right-hand side is a good approximation of $F(\mathbf{y})$ when $\|\mathbf{y}-\mathbf{x}\|$ is very close to 0. The right-hand side of (3.2.15) is called the **first-order approximation** of F at the point **x**. We write

$$F(\mathbf{y}) \approx F(\mathbf{x}) + M(\mathbf{y} - \mathbf{x}), \quad \text{with} \quad M = F'(\mathbf{x}). \tag{3.2.16}$$

Example 3.2.11 With F and **x** as in Example 3.2.9, we obtain

$$F(\mathbf{y}) \approx (2,0) + M(x-1,y-1) = (2x+y-1,-2x+2y).$$

The right-hand side of the equality is the first-order approximation of F at $\mathbf{x}_0=(1,1)$.

The following list summarizes those properties of the derivative that we shall need in the future.

(a) $(F \pm G)'(\mathbf{x}) = F'(\mathbf{x}) \pm G'(\mathbf{x})$;
(b) $(rF)'(\mathbf{x}) = rF'(\mathbf{x})$, where r is a real number;
(c) $(G\circ F)'(\mathbf{x}) = G'(F(\mathbf{x}))\, F'(\mathbf{x})$, i.e., the derivative of the composition is the product of the two matrices representing the derivative of the two functions, with the derivative of G evaluated at $F(\mathbf{x})$. (3.2.17)
(d) Let $\mathbf{r} : \mathbf{R} \to \mathbf{R}^q$ and let **a** be a vector of $\mathbf{R}^q$. Then $(\mathbf{a}\bullet(F\circ\mathbf{r}))'(t) = \mathbf{a}\bullet F'(\mathbf{r}(t))\mathbf{r}'(t)$, i.e., the derivative of $\mathbf{a}\bullet F(\mathbf{r}(t))$ is the scalar product of **a** with the vector obtained by applying the matrix $F'(\mathbf{r}(t))$ to the vector $\mathbf{r}'(t)$.

Equality (3.2.13) may hold even in cases when some or all partial derivatives of the component functions are not continuous at **x**. Regardless of the reason behind its validity, we say that F is **differentiable** at **x** whenever (3.2.13) holds, and we call $F'(\mathbf{x})$ the derivative of F at **x**. As we move **x** in U we normally find points where (3.2.13) holds and points where it does not. We say that F is **differentiable** in U if (3.2.13) holds at every point of U.

As **x** is moved in U the matrix $F'(\mathbf{x})$ changes. We say that $F'(\mathbf{x})$ is **continuous** at **x** if for every real number r>0 we can find s>0 such that the operator norm of the matrix $F'(\mathbf{y})-F'(\mathbf{x})$ does not exceed r whenever $\|\mathbf{y}-\mathbf{x}\| \le s$, i.e.,

$$\|\mathbf{y} - \mathbf{x}\| \le s \quad \text{implies that} \quad \|F'(\mathbf{y}) - F'(\mathbf{x})\| \le r.$$

Theorem 3.2.4 *Assume that the partial derivatives of the component functions of* F *are continuous at* **x**. *Then* F′ *is continuous at* **x**.

Proof. Since a square matrix of size q can be regarded as an element of $\mathbf{R}^{q^2}$ and all norms in $\mathbf{R}^{q^2}$ are equivalent, it is enough to prove continuity in the maximum norm. In other words, it is enough to show that given r>0 there exists s>0 such that

$$\|\mathbf{y}-\mathbf{x}\| \le s \quad \text{implies that} \quad \max\left\{ \left| \frac{\partial F_i}{\partial x_j}(\mathbf{y}) - \frac{\partial F_i}{\partial x_j}(\mathbf{x}) \right| : i,j=1,\ldots,q \right\} \le r.$$

The continuity of $\partial F_i/\partial x_j$ at $\mathbf{x}$ implies that given r>0 there exists s_{ij} such that

$$\|\mathbf{y}-\mathbf{x}\| \le s_{ij} \quad \text{implies that} \quad \left| \frac{\partial F_i}{\partial x_j}(\mathbf{y}) - \frac{\partial F_i}{\partial x_j}(\mathbf{x}) \right| \le r.$$

Selecting s=minimum$\{s_{ij}: i,j,=1,\ldots,q\}$ achieves the desired goal. QED

It may happen that $F'(\mathbf{x})$ is continuous at every $\mathbf{x}\in U$. In this case we say that F is (of class) C^1 in U. One condition which ensures this property is the continuity in U of all partial derivatives of the component functions.

Example 3.2.12 Let $F(x,y)=(x\cos y,\ \sin x+\sin y)$. The matrix $F'(\mathbf{x})$ is

$$F'(\mathbf{x}) = \begin{pmatrix} \cos y & -x\sin y \\ \cos x & \cos y \end{pmatrix}.$$

The function F is C^1 since all partial derivatives of the component functions are continuous everywhere. The first-order approximation of F at $\mathbf{0}=(0,0)$ is

$$(x\cos y,\ \sin x + \sin y) \approx (0,0) + F'(\mathbf{0})(x,y) = (x,\ x + y).$$

The first-order approximation of a function at some point $\mathbf{x}$ of its domain is an important tool for analyzing the behavior of the function close to $\mathbf{x}$. As we shall see later certain features of F are completely determined by its first-order approximation. The technique of replacing F with its first-order approximation in order to detect these features is called **linearization**.

Mean value inequality

The following result plays an important role in Chapter 5.

Theorem 3.2.5 (Mean Value Inequality in $\mathbf{R}^q$) *Let* r>0 *and* $D=D(\mathbf{c},r)\subset\mathbf{R}^q$. *Assume that* $F: D\to\mathbf{R}^q$ *is continuous on* D *and differentiable on* $B(\mathbf{c},r)$. *Then, for every* $\mathbf{x},\mathbf{y}\in D$ *there is a point* $\mathbf{c}$ *in the open line segment joining* $\mathbf{x}$ *with* $\mathbf{y}$ *such that*

$$\|F(\mathbf{y}) - F(\mathbf{x})\| \le \|F'(\mathbf{c})\|\ \|\mathbf{y}-\mathbf{x}\|. \tag{3.2.18}$$

Proof. Let $\mathbf{r}: [0,1]\to\mathbf{R}^q$ be defined by $\mathbf{r}(t)=t\mathbf{y}+(1-t)\mathbf{x}$ and let $\mathbf{a}$ be a vector of $\mathbf{R}^q$ which will be specified later. Notice that $\mathbf{r}(t)$ for $t\in[0,1]$ describes the segment joining $\mathbf{x}$ with $\mathbf{y}$. Consider the function $g(t)=\mathbf{a}\bullet F(\mathbf{r}(t))$. Then $g(0)=\mathbf{a}\bullet F(\mathbf{x})$, $g(1)=\mathbf{a}\bullet F(\mathbf{y})$, and by part (d) of (3.2.17), $g'(t)=\mathbf{a}\bullet F'(\mathbf{r}(t))\mathbf{r}'(t)=\mathbf{a}\bullet F'(\mathbf{r}(t))(\mathbf{y}-\mathbf{x})$. The mean value theorem implies that $g(1)-g(0)=g'(t_0)$ for some $t_0\in(0,1)$. Hence,

$$|\mathbf{a}\bullet F(\mathbf{y}) - \mathbf{a}\bullet F(\mathbf{x})| = |\mathbf{a}\bullet F'(\mathbf{r}(t_0))(\mathbf{y}-\mathbf{x})|$$

$$\text{or} \quad |\mathbf{a}\bullet(F(\mathbf{y}) - F(\mathbf{x}))| = |\mathbf{a}\bullet F'(\mathbf{r}(t_0))(\mathbf{y}-\mathbf{x})|. \tag{3.2.9}$$

From the Cauchy-Schwarz inequality we derive

$$|\mathbf{a}\bullet F'(\mathbf{r}(t_0))(\mathbf{y}-\mathbf{x})| \le \|\mathbf{a}\| \, \|F'(\mathbf{r}(t_0))(\mathbf{y}-\mathbf{x})\|.$$

Moreover, from the properties of the operator norm of a matrix, we have

$$\|F'(\mathbf{r}(t_0))(\mathbf{y}-\mathbf{x})\| \le \|F'(\mathbf{r}(t_0))\| \, \|\mathbf{y}-\mathbf{x}\|.$$

Let $\mathbf{c}=\mathbf{r}(t_0)$ and select $\mathbf{a}=F(\mathbf{y})-F(\mathbf{x})$. Since $t_0\in(0,1)$, $\mathbf{c}=t_0\mathbf{y}+(1-t_0)\mathbf{x}$ belongs to the open line segment joining $\mathbf{x}$ with $\mathbf{y}$. Moreover, (3.2.19) implies that

$$\|F(\mathbf{y})-F(\mathbf{x})\|^2 \le \|F(\mathbf{y})-F(\mathbf{x})\| \, \|F'(\mathbf{c})\| \, \|\mathbf{y}-\mathbf{x}\|. \tag{3.2.20}$$

Dividing by $\|F(\mathbf{y})-F(\mathbf{x})\|$ gives the mean value inequality. QED

Remark 3.2.1 The proof above can be adjusted so that the mean value inequality holds in every norm. The closed ball D should be considered in the new norm, and the operator norm of the derivative should be defined using the new norm. Moreover, we should make sure that the open line segment joining $\mathbf{x}$ with $\mathbf{y}$ is entirely contained in the interior of D. Inequality (3.2.18) becomes

$$\|F(\mathbf{x})-F(\mathbf{y})\|_a \le \|F'(\mathbf{c}_0)\|_a \, \|\mathbf{x}-\mathbf{y}\|_a. \tag{3.2.21}$$

The points $\mathbf{c}$ [see (3.2.18)] and $\mathbf{c}_0$ [see (3.2.21)] are not necessarily the same, although both belong to the open line segment joining $\mathbf{x}$ with $\mathbf{y}$. For a version of the mean value inequality which does not require differentiability at every point of B($\mathbf{c}$,r) see [Furi-Martelli, 1991].

Problems

1. Let $F(x,y)=(x-y^2,x^2-y)$. Find the derivative of F and its first-order approximation at (0,0). Verify that the derivative of F is continuous.

2. Let $F(x,y,z)=(x-y+z,\ x+z,y+z)$. Verify that F can be represented by a matrix, in the sense that there exists a matrix M such that $M\mathbf{x}=F(\mathbf{x})$ for all $\mathbf{x}\in\mathbf{R}^3$. Verify that the derivative of F is exactly the matrix M.

3. Let $F(x,y)=(x/y,y/x)$. Find the derivative of F in general and its first-order approximation at $\mathbf{x}=(2,3)$.

4. * Given $\mathbf{x}=(1,1)$ and $\mathbf{y}=(3,3)$, find $\mathbf{c}$ in the open line segment joining $\mathbf{x}$ with $\mathbf{y}$ such that the mean value inequality is satisfied for $F(x,y)=(x-y^2, x^2-y)$.

5. * Let F be the map of Problem 1 and let $\mathbf{x}=(0,1)$, $\mathbf{y}=(1,2)$. Find $\mathbf{c}$ in the open line segment joining $\mathbf{x}$ with $\mathbf{y}$ which satisfies the mean value inequality using the maximum norm.

6. * Do Problem 5 using the Euclidean norm.

7. * Do Problem 5 using the taxicab norm.

CHAPTER 4

DISCRETE LINEAR DYNAMICAL SYSTEMS

SUMMARY

The purpose of this chapter is to study linear dynamical systems in a dimension higher than 1. The first section contains the definition and properties of linear dynamical systems. In the second section we study the stability of the origin, which is always an equilibrium point of a linear system. Using the eigenvalues of a matrix, we determine under which conditions all orbits converge to the origin (all eigenvalues have modulus smaller than 1), or all nonstationary orbits diverge to infinity (all eigenvalues have modulus larger than 1). The third section contains the spectral decomposition theorem which is necessary for the analysis of all linear systems not studied in Section 2. In Section 4 we examine the case in which some eigenvalues have modulus smaller than 1 and the others have modulus larger than 1. In this section we introduce the idea of stable and unstable subspaces. In Section 5 we analyze the case when there are eigenvalues with modulus 1. Finally, in Section 6 we study affine systems using conjugacy with suitable linear systems.

Section 1. **ORBITS OF LINEAR PROCESSES**

An operator L : $\mathbf{R}^q \to \mathbf{R}^q$ is said to be **linear** if

$$L(a\mathbf{x}+b\mathbf{y}) = aL\mathbf{x}+bL\mathbf{y} \tag{4.1.1}$$

for every pair of vectors $\mathbf{x},\mathbf{y} \in \mathbf{R}^q$ and every pair a and b of real numbers. Let M be the square matrix of size q such that the entries of the ith column of M are the components of the vector $L\mathbf{e}_i$, i=1,2,...,q. In other words, if $L\mathbf{e}_i=(e_{1i}, e_{2i},...,e_{qi})$, then $m_{1i}=e_{1i},..., m_{qi}=e_{qi}$. The matrix M has the important property

$$L\mathbf{x} = M\mathbf{x}, \tag{4.1.2}$$

for every $\mathbf{x} \in \mathbf{R}^q$, when the standard basis is used. In fact, from $\mathbf{x}=x_1\mathbf{e}_1+...+x_q\mathbf{e}_q$ and (4.1.1) we have

$$L\mathbf{x} = x_1L\mathbf{e}_1 + ... + x_qL\mathbf{e}_q.$$

Likewise, the action of M can be written [see (3.2.2)]

$$M\mathbf{x} = x_1M\mathbf{e}_1 + ... + x_qM\mathbf{e}_q.$$

Since $M\mathbf{e}_i=(e_{1i},e_{2i},...,e_{qi})=L\mathbf{e}_i$ (see Chapter 3, Section 2) we see that equality (4.1.2) holds.

Example 4.1.1 Let $L\mathbf{x}=L(x,y)=(3x+2y,-x+y)$. Then $L\mathbf{e}_1=(3,-1)$ and $L\mathbf{e}_2=(2,1)$. Hence, M is the matrix

$$M = \begin{pmatrix} 3 & 2 \\ -1 & 1 \end{pmatrix}.$$

It is easily seen that $L\mathbf{x}=M\mathbf{x}$ for every $\mathbf{x} \in \mathbf{R}^2$, $\mathbf{x}=(x,y)=x\mathbf{e}_1+y\mathbf{e}_2$.

A **linear** dynamical system in $\mathbf{R}^q$ has the form

$$\mathbf{x}_{n+1} = L\mathbf{x}_n \tag{4.1.3}$$

where L : $\mathbf{R}^q \to \mathbf{R}^q$ is a linear operator. Using (4.1.2) and the standard basis of $\mathbf{R}^q$ we can write (4.1.3) in the form

$$\mathbf{x}_{n+1} = M\mathbf{x}_n \tag{4.1.4}$$

where M is the matrix defined above. Thus, the study of the discrete linear dynamical system $\mathbf{x}_{n+1}=L\mathbf{x}_n$ in $\mathbf{R}^q$ can be performed with the analysis of $\mathbf{x}_{n+1}=M\mathbf{x}_n$, where M is the matrix representing L in the standard basis.

Example 4.1.2 Let L be as in Example 4.1.1. The dynamical system $\mathbf{x}_{n+1}=L\mathbf{x}_n$ can be written in the form

$$\mathbf{x}_{n+1} = \begin{pmatrix} 3 & 2 \\ -1 & 1 \end{pmatrix}\mathbf{x}_n,$$

where $\mathbf{x}_n=(x_n,y_n)$, $\mathbf{x}_{n+1}=(x_{n+1},y_{n+1})$.

One of the important consequences of (4.1.1) and (4.1.2) is that the action of L on any vector $\mathbf{x}\in \mathbf{R}^q$ is determined once the images $L\mathbf{e}_i$, i=1,2,...,q are known. More precisely, if

$$\mathbf{x}_0 = \sum_{i=1}^{q} a_i\mathbf{e}_i, \tag{4.1.5}$$

then

$$\mathbf{x}_1 = L\mathbf{x}_0 = M\mathbf{x}_0 = \sum_{i=1}^{q} a_i M\,\mathbf{e}_i. \tag{4.1.6}$$

Using the property $M(M\mathbf{x})=M^2\mathbf{x}$ (see Chapter 3, Section 2) we obtain the equality

$$\mathbf{x}_n = L\mathbf{x}_{n-1} = M\mathbf{x}_{n-1} = M^n\mathbf{x}_0 = \sum_{i=1}^{q} a_i M^n\mathbf{e}_i, \tag{4.1.7}$$

where M^n is the product of M with itself n times. Hence, each state $\mathbf{x}_n$ of the orbit $O(\mathbf{x}_0)$ can be obtained from the corresponding states $M^n\mathbf{e}_i$, i=1,2,...,q. Once the q orbits $\{M^n\mathbf{e}_i$: n=1,2,...$\}$, i=1,2,...,q are known, the orbit $O(\mathbf{x}_0)$ of every initial state $\mathbf{x}_0$ is completely determined.

Example 4.1.3 Let $L(x,y,z)=(x+y,x-z,-y+z)$, and $\mathbf{x}_0=(1,2,-2)$, i.e., $\mathbf{x}_0=\mathbf{e}_1+2\mathbf{e}_2-2\mathbf{e}_3$. The action of L coincides with the action of the matrix

$$M = \begin{pmatrix} 1 & 1 & 0 \\ 1 & 0 & -1 \\ 0 & -1 & 1 \end{pmatrix}.$$

Moreover, $\mathbf{x}_n=M^n\mathbf{x}_0=M^n\mathbf{e}_1+2M^n\mathbf{e}_2-2M^n\mathbf{e}_3$. Unfortunately, the three orbits $\{M^n\mathbf{e}_i$: n= 1,2,...$\}$, i=1,2,3 do not have a simple form.

Example 4.1.3 shows that (4.1.7) may not be very advantageous when the orbits $O(\mathbf{e}_i)$, i=1,2,...,q are complicated. Writing $\mathbf{x}_0$ as a linear combination of the elements of the standard basis is certainly helpful, but it may not be the best approach. We now illustrate with Example 4.1.4 a more efficient strategy which will be fully discussed in Section 3 and used in Sections 4 and 5.

Example 4.1.4 With reference to Example 4.1.3, the eigenvalues of M (M is the matrix representing L) are $r_1=1$, $r_2=2$, and $r_3=-1$. Corresponding eigenvectors are $\mathbf{u}_1=(1,0,1)$, $\mathbf{u}_2=(1,1,-1)$, and $\mathbf{u}_3=(-1,2,1)$. They form a basis of $\mathbf{R}^3$. Select a vector $\mathbf{x}_0\in\mathbf{R}^3$. Assume that $\mathbf{x}_0=(1,2,-2)$. There exist three constants b_1, b_2, and b_3 such that

$$\mathbf{x}_0 = (1,2,-2) = b_1\mathbf{u}_1 + b_2\mathbf{u}_2 + b_3\mathbf{u}_3. \tag{4.1.8}$$

They can be determined by solving (4.1.8), i.e., by finding the solution of the linear system of three equations in three unknowns,

$$\begin{cases} 1 = b_1 + b_2 - b_3 \\ 2 = b_2 + 2b_3 \\ -2 = b_1 - b_2 + b_3. \end{cases} \tag{4.1.9}$$

We obtain $b_1=-1/2$, $b_2=5/3$, and $b_3=1/6$. Hence $\mathbf{x}_0=-\mathbf{u}_1/2 + 5\mathbf{u}_2/3 + \mathbf{u}_3/6$. Since $L^n\mathbf{u}_1=\mathbf{u}_1$, $L^n\mathbf{u}_2=2^n\mathbf{u}_2$, and $L^n\mathbf{u}_3=(-1)^n\mathbf{u}_3$, we arrive at

$$\mathbf{x}_n = L^n\mathbf{x}_0 = -(1/2)\mathbf{u}_1 + (5/3)2^n\mathbf{u}_2 + (1/6)(-1)^n\mathbf{u}_3. \tag{4.1.10}$$

In this case the orbits of $\mathbf{u}_1$, $\mathbf{u}_2$, and $\mathbf{u}_3$ can be determined easily since they are eigenvectors of M. Hence, (4.1.10) provides the easiest way of computing the orbit of $\mathbf{x}_0$.

In Section 3 we will show that the analysis presented in Example 4.1.4 can be extended to every linear operator L acting on $\mathbf{R}^q$ whenever the eigenvalues of L are real and to each eigenvalue r_i there corresponds a number of linearly independent eigenvectors equal to its algebraic multiplicity m_i. We shall also analyze more complex situations. However, before we embark on this adventure, we shall study in Section 2 some cases in which the behavior of all orbits can be established easily without using the eigenvalues and eigenvectors of M.

Problems

1. Let $L(x,y)=(3x-2y,5x+7y)$. Find the matrix M representing L in the standard basis.

2. Let $L(x,y,z)=(2x-y,x-y+z,2x-z)$. Find the matrix M representing L in the standard basis.

3. Assume that $L(1,0)=(3,2)$ and $L(0,1)=(-2,-1)$. Find $L(x,y)$ and find the matrix M representing L in the standard basis.

4. Assume that $L(1,0,0)=(2,1,0)$, $L(0,1,0)=(1,-2,3)$, and $L(0,0,1)=(0,-1,2)$. Find $L(x,y,z)$ and find the matrix M representing L in the standard basis.

5. Let $F(x,y,z)=(x+y,y+z+1,x+z)$. Is F linear?

6. Let $F(x,y,z) = (|x|,y+z,x-z)$. Is F linear?

7. * Let

$$M = \begin{pmatrix} 1 & 2 \\ 1 & 3 \end{pmatrix}.$$

Find the eigenvalues and eigenvectors of M. Find the form of O(1,1).

8. * Let

$$M = \begin{pmatrix} 1 & 2 \\ 4 & 3 \end{pmatrix}.$$

Find the eigenvalues and eigenvectors of M. Find the form of O(2,3).

9. * Let

$$M = \begin{pmatrix} .5 & 0 & 3 \\ -.5 & 1 & 1 \\ 0 & 0 & 2 \end{pmatrix}.$$

Find the eigenvalues and eigenvectors of M. Find the form of O(1,1,–1).

10. * Let

$$M = \begin{pmatrix} 1 & 0 & 0 \\ 1 & -1 & 0 \\ 1 & 1 & 2 \end{pmatrix}.$$

Find the eigenvalues and eigenvectors of M. Determine O(0,1,1).

Section 2. STABILITY AND INSTABILITY OF THE ORIGIN

In this section we present conditions which ensure that every nonstationary orbit of the dynamical system (4.1.4) (or (4.1.3)) either converges to **0** or diverges to infinity. Our goal is to prove that if $\rho(M)<1$, then every orbit converges to **0**, and if all eigenvalues of M have modulus larger than 1, then every nonstationary orbit goes to infinity.

The origin as an attractor

Recall that $M\mathbf{0}=\mathbf{0}$ for every matrix M. Hence **0** is always a fixed point for the dynamical system $\mathbf{x}_{n+1}=M\mathbf{x}_n$. The condition $\rho(M)<1$ is necessary and sufficient for obtaining that every orbit of

$$\mathbf{x}_{n+1} = M\mathbf{x}_n \tag{4.2.1}$$

goes to **0** as $n\to\infty$. However, in Theorem 4.2.1 we simply establish that the condition is sufficient.

Theorem 4.2.1 *Assume that* $\rho(M)<1$. *Then for every* $\mathbf{x}_0 \in \mathbf{R}^q$ *we have* $\|\mathbf{x}_n\| \to 0$ *as* $n \to \infty$.

Proof. Define a new norm $\|.\|_a$ in $\mathbf{R}^q$ such that the operator norm of M in this new norm is smaller than 1: $\|M\|_a<1$. Theorem 3.2.3 provides a strategy for finding a new norm with the required property. Let $\mathbf{x}_0 \in \mathbf{R}^q$. From $\mathbf{x}_n=M^n\mathbf{x}_0$ we derive

$$\|\mathbf{x}_n\|_a = \|M^n\mathbf{x}_0\|_a \leq \|M^n\|_a\|\mathbf{x}_0\|_a \leq \|M\|_a^n\|\mathbf{x}_0\|_a.$$

The last step in the inequality above is written by taking into account that $\|M^n\| \leq \|M\|^n$, a result which is true in every operator norm [see (3.2.7)].

Since $\|M\|_a<1$ we obtain $\|\mathbf{x}_n\|_a \to 0$ as $n \to +\infty$. The equivalence of all norms implies that $\|\mathbf{x}_n\| \to 0$. QED

Example 4.2.1 Let

$$M = \begin{pmatrix} 17/20 & -1/20 \\ 1/10 & 7/10 \end{pmatrix}.$$

The eigenvalues of M are 3/4 and 4/5. Hence the spectral radius is 4/5 and, according to Theorem 4.2.1, every orbit of the dynamical system $\mathbf{x}_{n+1}=M\mathbf{x}_n$ converges to $\mathbf{0}$.

Example 4.2.2 Let

$$M = \begin{pmatrix} -1.7 & 1.4 & .8 \\ .9 & .5 & -.9 \\ -2.5 & 1.4 & 1.6 \end{pmatrix}.$$

The eigenvalues of M are –.9, .8, and .5. Hence, the spectral radius of M is .9 and, according to Theorem 4.2.1, every orbit of the dynamical system $\mathbf{x}_{n+1}=M\mathbf{x}_n$ converges to $\mathbf{0}$.

In the two preceding examples we found the spectral radius of M by determining all eigenvalues of M. This task may be quite difficult when the dimension q of the space is large. However, only the knowledge that $\rho(M)<1$ is needed. There are efficient numerical methods for determining $\rho(M)$, and they do not require the knowledge of all eigenvalues. One of these methods, called the **power method**, is presented in Chapter 6.

Remark 4.2.1 From the proof of Theorem 4.2.1 we derive that $\|\mathbf{x}_n\| \leq Kh^n\|\mathbf{x}_0\|$ where $h \in (0,1)$ and K is a constant determined by the equivalence between the Euclidean norm and the a-norm. Hence, every orbit converges to $\mathbf{0}$ exponentially.

The origin as a repeller

We now assume that all eigenvalues of M have modulus larger than 1. To establish the theorem regarding the behavior of all orbits of a system $\mathbf{x}_{n+1}=M\mathbf{x}_n$, with M satisfying the condition above, we need two preliminary results.

Lemma 4.2.1 M *is invertible and* $\rho(M^{-1})<1$.

Proof. Since 0 is not an eigenvalue we have det(0I–M)=detM≠0. Hence, M is invertible. Let r be an eigenvalue of M. Define s=1/r. The equality

$$(sI-M^{-1})M = s\,M - I = s(M - rI)$$

implies that $\det((sI-M^{-1})M)=0$, since $\det(M-rI)=0$. Recall that the determinant of a product of two square matrices is the product of their determinants. Hence,

$$\det((sI - M^{-1})M) = \det(sI - M^{-1})\,\det M.$$

This equality implies that $\det(sI-M^{-1})=0$, since $\det M\neq 0$. Consequently, s=1/r is an eigenvalue of M^{-1} and $\rho(M^{-1})<1$. QED

We need an additional result.

Lemma 4.2.2 *Let* M *be invertible. Then* M^n *is invertible and* $(M^n)^{-1}=(M^{-1})^n$.

Proof. M^n is invertible since $\det(M^n)=(\det M)^n\neq 0$. From

$$M^2(M^{-1})^2 = MMM^{-1}M^{-1} = MM^{-1} = I$$

we derive that $(M^{-1})^2$ is the inverse of M^2, i.e., $(M^2)^{-1}=(M^{-1})^2$. An induction argument gives $(M^n)^{-1}=(M^{-1})^n$. QED

From Lemmas 4.2.1 and 4.2.2 we derive the behavior of all orbits of a system $\mathbf{x}_{n+1}=M\mathbf{x}_n$ when all eigenvalues of M have modulus larger than 1.

Theorem 4.2.2 *Assume that* $\rho(M^{-1})<1$. *Then for every* $\mathbf{x}_0\neq\mathbf{0}$ *we have*

$$\|\mathbf{x}_n\| = \|M^n\mathbf{x}_0\| \to \infty \quad as \quad n \to \infty. \tag{4.2.2}$$

Proof. Choose a norm in $\mathbf{R}^q$ such that $\|M^{-1}\|_a<1$ and let $\mathbf{x}_0 \neq \mathbf{0}$. From Lemma 4.2.2 and the equality $\mathbf{x}_n=M^n\mathbf{x}_0$ we obtain

$$\mathbf{x}_0 = M^{-1}M\mathbf{x}_0 = (M^{-1})^nM^n\mathbf{x}_0 = (M^{-1})^n\mathbf{x}_n.$$

Consequently,

$$\|\mathbf{x}_0\|_a = \|(M^{-1})^nM^n\mathbf{x}_0\|_a \leq \|(M^{-1})^n\|_a\,\|M^n\mathbf{x}_0\|_a = \|(M^{-1})^n\|_a\,\|\mathbf{x}_n\|_a. \tag{4.2.3}$$

Notice that

$$\|(M^{-1})^n\| \leq \|M^{-1}\|_a^n.$$

Therefore, $\|(M^{-1})^n\|_a \|\mathbf{x}_n\|_a \le \|M^{-1}\|_a^n \|\mathbf{x}_n\|_a$. Since $\|M^{-1}\|_a^n \to 0$ as $n\to\infty$, we must have $\|\mathbf{x}_n\|_a \to \infty$, for otherwise the product cannot remain larger that $\|\mathbf{x}_0\|_a > 0$, as required by (4.2.3). The equivalence of all norms implies the theorem. QED

Remark 4.2.2 Observe that the divergence of $\|\mathbf{x}_n\|$ is exponential, since $\|M^{-1}\|_a^n$ goes to 0 in an exponential manner.

Example 4.2.3 Let

$$M = \begin{pmatrix} 2 & 2 \\ 0 & 3 \end{pmatrix}.$$

The eigenvalues of M are 2 and 3. Thus every nonstationary orbit of the dynamical system defined by M goes to infinity.

Example 4.2.4 Let

$$M = \begin{pmatrix} -1 & 6 & 9 \\ -1 & 4.5 & 3.5 \\ -1 & -.5 & 2.5 \end{pmatrix}.$$

The eigenvalues of M are 2, 2–3**i** and 2+3**i**. Hence, the spectral radius of M^{-1} is 1/2. According to Theorem 4.2.2, every nonstationary orbit of the dynamical systems defined by M goes to infinity.

Problems

1. Let

$$M = \begin{pmatrix} 1 & -1 \\ .5 & -1 \end{pmatrix}.$$

Find the eigenvalues of M. Derive that every orbit of the dynamical system governed by M goes to **0**.

2. Let

$$M = \begin{pmatrix} 1 & 2 \\ -1 & 3 \end{pmatrix}.$$

Prove that every nonstationary orbit of the system governed by M goes to infinity.

3. Let

$$M = \begin{pmatrix} 2^{-1} & 2 \\ 0 & 3^{-1} \end{pmatrix}.$$

Study the orbits of the system governed by M.

4. * Let

$$M = \begin{pmatrix} 1/2 & 0 & 1 \\ 0 & 1/2 & 3/4 \\ 0 & -3/4 & 1/2 \end{pmatrix}.$$

Show that all orbits of the dynamical system defined by M converge to **0**.

5. * Let M be the matrix of Problem 4. Find M^{-1} and $\rho(M^{-1})$. Study the dynamical system $\mathbf{x}_{n+1}=M^{-1}\mathbf{x}_n$.

6. * Let

$$M = \begin{pmatrix} 5/2 & 0 & -1/2 \\ 0 & 2 & 0 \\ -1/2 & 0 & 5/2 \end{pmatrix}.$$

Verify that the characteristic polynomial of M is $P(r)=(r-3)(r-2)^2$. Find the eigenvectors of M and study the dynamical system governed by M. In particular, find the form of the orbit of (1,–1,2).

7. * Let M be the matrix of Problem 6. Find M^{-1} and determine the orbit of (1,–1,2) under the action of M^{-1}.

Section 3. SPECTRAL DECOMPOSITION THEOREM

The purpose of this section is to present the spectral decomposition theorem (SDT), which plays a pivotal role here and in Sections 4 and 5. The goal of the theorem is to identify subspaces of $\mathbf{R}^q$ where the action of a square matrix M of size q is particularly simple and derive conclusions on the action of M in the entire space by understanding it in these subspaces. We present three versions of the theorem. First, we discuss the easiest case (see Theorem 4.3.2), namely when all eigenvalues are real and semisimple (see the definition below). Then we present the case when the eigenvalues are real but not necessarily semisimple (see Theorem 4.3.4). Finally, we discuss the case when some eigenvalues are complex (see Theorem 4.3.5). The brief preliminary discussion below serves as an introduction to all three cases.

Definition 4.3.1
$\mathbf{R}^q$ is the **direct sum** of its subspaces $\mathbf{X}_1,\ldots,\mathbf{X}_k$, written $\mathbf{R}^q=\mathbf{X}_1\oplus\mathbf{X}_2\oplus\ldots\oplus\mathbf{X}_k$, if the following two conditions are verified:

(i) For every vector $\mathbf{x}\in\mathbf{R}^q$ there exist vectors $\mathbf{x}_i\in\mathbf{X}_i$, i=1,...,k such that $\mathbf{x}=\mathbf{x}_1+\mathbf{x}_2+\ldots+\mathbf{x}_k$ (we can call this a "decomposition" of **x**).
(ii) The decomposition of **x** written in (i) is unique.

Notice that the **0** vector can always be obtained by selecting the zero vector in each subspace. Condition 2 states that in a direct sum this should be the only way to get the zero vector. Once we have verified condition 2 for the zero vector, it will hold true for any other vector. In fact, the presence of a vector $\mathbf{x}_0$ with two different decompositions

$$\mathbf{x}_0 = \mathbf{x}_1 + \mathbf{x}_2 + \ldots + \mathbf{x}_k \quad \text{and} \quad \mathbf{x}_0 = \mathbf{y}_1 + \mathbf{y}_2 + \ldots + \mathbf{y}_k$$

would imply that $\mathbf{0}=\mathbf{x}_1-\mathbf{y}_1+\ldots+\mathbf{x}_k-\mathbf{y}_k$.

We see that both conditions are verified if we select a basis $B=\{\mathbf{x}_1,\ldots,\mathbf{x}_q\}$ of $\mathbf{R}^q$, we partition B into $k\leq q$ subsets $S_1,\ldots,S_k$, and we define $\mathbf{X}_1,\ldots,\mathbf{X}_k$ to be the subspaces spanned by the vectors of $S_1,\ldots,S_k$, respectively. This is exactly the strategy that will be used in the spectral decomposition theorem.

Example 4.3.1 Let $E=\{\mathbf{e}_1=(1,0,0),\ \mathbf{e}_2=(0,1,0),\ \mathbf{e}_3=(0,0,1)\}$. Then $\mathbf{R}^3= \mathbf{X}_1\oplus\mathbf{X}_2 \oplus\mathbf{X}_3$, where $\mathbf{X}_i=\text{Span}\{\mathbf{e}_i\}$, i=1,2,3. Also, $\mathbf{R}^3=\mathbf{Y}_1\oplus\mathbf{X}_3$, where $\mathbf{Y}_1= \text{Span}\{\mathbf{e}_1,\mathbf{e}_2\}$.

Example 4.3.2 Let $B=\{\mathbf{u}=(1,1),\mathbf{v}=(1,-1)\}$. Then $\mathbf{R}^2=\mathbf{X}_1\oplus\mathbf{X}_2$, where $\mathbf{X}_1=\text{Span}\{\mathbf{u}\}$ and $\mathbf{X}_2=\text{Span}\{\mathbf{v}\}$.

Example 4.3.3 Let $B=\{\mathbf{x}_1=(1,1,0),\ \mathbf{x}_2=(0,1,1),\ \mathbf{x}_3=(1,0,1)\}$. The three vectors are linearly independent. Hence, B is a basis of $\mathbf{R}^3$ and $\mathbf{R}^3=\mathbf{X}_1\oplus\mathbf{X}_2\oplus\mathbf{X}_3$, where $\mathbf{X}_i=\text{Span}\{\mathbf{x}_i\}$, i=1,2,3. Also, $\mathbf{R}^3=\mathbf{Y}_1\oplus\mathbf{X}_3$, where $\mathbf{Y}_1=\text{Span}\{\mathbf{x}_1,\mathbf{x}_2\}$. Notice that the vectors of the plane $\mathbf{Y}_1$ have the form a(1,1,0)+b(0,1,1)=(a,a+b,b). Hence, for these vectors we have $x+z=y$, which is the Cartesian equation of $\mathbf{Y}_1$. The vectors of $\mathbf{X}_3$ have the form t(1,0,1), and parametric equations of the line containing all these vectors are x=t, y=0, z=t.

An essential step toward the SDT is the identification of all eigenvectors of M. Below we discuss the case of matrices M with all real eigenvalues. The analysis of matrices M with some complex eigenvalues will be presented before Theorem 4.3.5.

Recall that given a real eigenvalue r_i we have $\det(r_iI-M)=0$. Consequently, the homogeneous system

$$(r_iI-M)\mathbf{x} = \mathbf{0} \tag{4.3.1}$$

has nonzero solutions, $\mathbf{x}\neq\mathbf{0}$, which are called **eigenvectors** of M corresponding to the eigenvalue r_i. How many linearly independent eigenvectors correspond to the eigenvalue r_i? Equality (4.3.1) requires **x** to be orthogonal to all row vectors of the matrix r_iI-M. Let **X** be the subspace spanned by the row vectors of r_iI-M. Reduce r_iI-M to a row echelon form Q. The nature of the operations performed in this reduction imply that **X** is spanned by the rows of Q. Moreover, the dimension of **X** is equal to the number p_i of leading ones of Q. Since there is at least one vector orthogonal to all rows of r_iI-M, we know that $p_i\leq q-1$. The number of linearly independent eigenvectors corresponding to the eigenvalue r_i is $q-p_i$. Such a number is called the **geometric multiplicity** of r_i, $gm(r_i)=q-p_i$.

Recall that the **algebraic multiplicity** of r_i, denoted by $am(r_i)$ is the exponent with which $(r-r_i)$ appears in the factorization of the characteristic polynomial of M.

Example 4.3.4 Let

$$M = \begin{pmatrix} 1 & 0 & 1 \\ 0 & 1 & 0 \\ 0 & 1 & 2 \end{pmatrix}.$$

From $\det(rI-M)=(r-1)^2(r-2)$ we obtain that $r_1=1$ and $r_2=2$ are the eigenvalues of M with algebraic multiplicity 2 and 1, respectively. Since

$$I - M = \begin{pmatrix} 0 & 0 & -1 \\ 0 & 0 & 0 \\ 0 & -1 & -1 \end{pmatrix}$$

we see that a row echelon form of M can be obtained by bringing the third row to position 1 and changing all signs. The number of leading ones is 2. Thus, the geometric multiplicity of 1 is 3–2=1.

Example 4.3.5 Let

$$M = \begin{pmatrix} 2 & -3 & -3 \\ 2 & -3 & -2 \\ -2 & 2 & 1 \end{pmatrix}.$$

Then $\det(rI-M)=(r-2)(1+r)^2$, $am(2)=1$ and $am(-1)=2$. A row echelon form of $-I-M$ is

$$\begin{pmatrix} 1 & -1 & -1 \\ 0 & 0 & 0 \\ 0 & 0 & 0 \end{pmatrix}.$$

We see that -1 has geometric multiplicity 2. Two linearly independent eigenvectors corresponding to -1 are $\mathbf{u}=(1,1,0)$ and $\mathbf{v}=(1,0,1)$.

We now state without proof a result that establishes a useful relation (see Remark 4.3.1 below) between the algebraic and geometric multiplicity of an eigenvalue r_i.

Theorem 4.3.1 *Assume that* r_i *is a real eigenvalue of* M *with algebraic multiplicity* m_i. *Then there are exactly* m_i *linearly independent vectors which satisfy the system*

$$(r_i I - M)^{m_i}\mathbf{x} = \mathbf{0}. \tag{4.3.2}$$

Example 4.3.6 According to Theorem 4.3.1 and using the matrix of Example 4.3.4 it must be the case that the system $(I-M)^2\mathbf{x}=\mathbf{0}$ has two linearly independent solutions. Since

$$(I-M)^2 = \begin{pmatrix} 0 & 1 & 1 \\ 0 & 0 & 0 \\ 0 & 1 & 1 \end{pmatrix},$$

two linearly independent solutions of $(I-M)^2\mathbf{x}=\mathbf{0}$ are $\mathbf{x}_1=(1,0,0)$, which is also a solution of (4.3.1) (hence, $\mathbf{x}_1$ is an eigenvector), and $\mathbf{x}_2=(0,1,-1)$, which solves (4.3.2) and not (4.3.1) (not an eigenvector).

Remark 4.3.1 Given a square matrix A and a nonzero vector $\mathbf{x}$ such that $A\mathbf{x}=\mathbf{0}$, we have $A^2\mathbf{x}=A(A\mathbf{x})=A\mathbf{0}=\mathbf{0}$. It follows that $\mathrm{Ker}(r_iI-M)\subset\mathrm{Ker}(r_iI-M)^{m_i}$. Hence, the geometric multiplicity of an eigenvalue never exceeds its algebraic multiplicity. When the two multiplicities are equal the eigenvalue is said to be **semisimple**.

SDT*: *real, semisimple eigenvalues

Theorem 4.3.2 below is the first version of the spectral decomposition theorem. Although not the most general case, it is useful for many applications.

Theorem 4.3.2 *Assume that all eigenvalues* $r_1,\ldots,r_k$ *of* M *are real and semisimple. Then to each eigenvalue* r_i *there corresponds a subspace* $\mathbf{X}_i$ *such that*:

(i) *For every* $\mathbf{x}\in\mathbf{X}_i$, *we have* $M\mathbf{x}\in\mathbf{X}_i$.
(ii) $\mathbf{X}_1\oplus\mathbf{X}_2\oplus\ldots\oplus\mathbf{X}_k=\mathbf{R}^q$.

Proof. Step 1. Let m_i be the algebraic multiplicity of r_i, $\mathbf{x}_1,\ldots,\mathbf{x}_{m_i}$ be linearly independent eigenvectors corresponding to r_i, and $\mathbf{X}_i=\mathrm{Span}\{\mathbf{x}_1,\ldots,\mathbf{x}_{m_i}\}$. Given $\mathbf{x}=a_1\mathbf{x}_1+\ldots+a_{m_i}\mathbf{x}_{m_i}$ (notice that $\mathbf{x}$ is an arbitrary element of $\mathbf{X}_i$ since it is a linear combination of a basis of $\mathbf{X}_i$) we have

$$M\mathbf{x} = a_1M\mathbf{x}_1 + \ldots + a_{m_i}M\mathbf{x}_{m_i} = a_1r_i\mathbf{x}_1 + \ldots + a_{m_i}r_i\mathbf{x}_{m_i}.$$

Hence, $M\mathbf{x}\in\mathbf{X}_i$.

Step 2. Let $m_1,\ldots,m_k$ be the algebraic multiplicities of $r_1,\ldots,r_k$, respectively. Since $m_1+\ldots+m_k=q$ and each eigenvalue is semisimple, we can produce k families, $S_1,\ldots,S_k$ of vectors such that each S_i is made of m_i linearly independent eigenvectors corresponding to the eigenvalue r_i. Hence, the vectors of S_1 are linearly independent. Assume that the same result is true for the vectors of $S_1 \cup S_2 ... \cup S_{j-1}$. If the vectors of $S_1 \cup S_2 ... \cup S_j$ are linearly independent, then an induction argument implies that the q vectors of $S_1 \cup S_2 ... \cup S_k$ are a basis of $\mathbf{R}^q$. We select as $\mathbf{X}_i$ the span of the vectors of S_i and taking into account step 1, we obtain Theorem 4.3.2.

To make the induction argument less cumbersome from a notational point of view, assume that the result is true for $S_1 \cup S_2$. We want to prove it for $S_1 \cup S_2 \cup S_3$. To be even more specific, assume that S_1 has one vector $\mathbf{x}_1$, S_2 has three vectors $\mathbf{x}_2$, $\mathbf{x}_3$, $\mathbf{x}_4$, and S_3 has two vectors $\mathbf{x}_5$ and $\mathbf{x}_6$. Let

$$\mathbf{0} = a_1\mathbf{x}_1 + a_2\mathbf{x}_2 + a_3\mathbf{x}_3 + a_4\mathbf{x}_4 + a_5\mathbf{x}_5 + a_6\mathbf{x}_6. \tag{4.3.3}$$

Apply M to both sides of (4.3.3) to obtain

$$\mathbf{0} = a_1r_1\mathbf{x}_1 + a_2r_2\mathbf{x}_2 + a_3r_2\mathbf{x}_3 + a_4r_2\mathbf{x}_4 + a_5r_3\mathbf{x}_5 + a_6r_3\mathbf{x}_6. \tag{4.3.4}$$

Multiply (4.3.3) by r_3 to get

$$\mathbf{0} = a_1r_3\mathbf{x}_1 + a_2r_3\mathbf{x}_2 + a_3r_3\mathbf{x}_3 + a_4r_3\mathbf{x}_4 + a_5r_3\mathbf{x}_5 + a_6r_3\mathbf{x}_6. \tag{4.3.5}$$

Subtracting (4.3.5) from (4.3.4), we arrive at

$$\mathbf{0} = a_1(r_1 - r_3)\mathbf{x}_1 + a_2(r_2 - r_3)\mathbf{x}_2 + a_3(r_2 - r_3)\mathbf{x}_3 + a_4(r_2 - r_3)\mathbf{x}_4. \tag{4.3.6}$$

Since the four vectors are linearly independent and neither r_1 nor r_2 is equal to r_3, we obtain $a_1=...=a_4=0$. Substituting these values into (4.3.3) gives $\mathbf{0}=a_5\mathbf{x}_5+a_6\mathbf{x}_6$, which implies $a_5=a_6=0$, since $\mathbf{x}_5$ and $\mathbf{x}_6$ are linearly independent. QED

We now apply Theorem 4.3.2 to linear dynamical systems in the case when all eigenvalues are real and semisimple. We start with an example.

Example 4.3.7 Let M be the matrix of Example 4.3.5. We have seen that two linearly independent eigenvectors corresponding to the eigenvalue –1 are $\mathbf{u}=(1,1,0)$ and $\mathbf{v}=(1,0,1)$. An eigenvector corresponding to the eigenvalue 2 is $\mathbf{w}=(-3,-2,2)$. Hence, $\mathbf{R}^3=\mathbf{X}_1 \oplus \mathbf{X}_2$ where $\mathbf{X}_1 = \text{Span}\{\mathbf{u},\mathbf{v}\}$ and $\mathbf{X}_2=\text{Span}\{\mathbf{w}\}$.

The study of the dynamical system $\mathbf{x}_{n+1}=M\mathbf{x}_n$ is now greatly simplified. Given a vector $\mathbf{x}_0$ there are constants a,b, and c such that $\mathbf{x}_0=a\mathbf{u}+b\mathbf{v}+c\mathbf{w}$. Consequently, $\mathbf{x}_1=M(a\mathbf{u}+b\mathbf{v}+c\mathbf{w})=-a\mathbf{u}-b\mathbf{v}+2c\mathbf{w}$ and $\mathbf{x}_n=(-1)^n(a\mathbf{u}+b\mathbf{v})+2^nc\mathbf{w}$.

The analysis presented in Example 4.3.7 can be used for every matrix M whenever the eigenvalues of M are real and semisimple. To avoid messy notation let us see how this can be done in a particular case. Assume that M is of size 5, with three real eigenvalues r_1, r_2, and r_3 , r_1 and r_2 being semisimple with multiplicity 2. Let $\mathbf{u}_1$ and $\mathbf{u}_2$ be linearly independent eigenvectors corresponding to r_1, $\mathbf{u}_3$ and $\mathbf{u}_4$ be linearly independent eigenvectors corresponding to r_2, and $\mathbf{u}_5$ be an eigenvector corresponding to r_3. Given $\mathbf{x}_0 \in \mathbf{R}^5$, find $b_1, b_2, ..., b_5$ such that $\mathbf{x}_0 = b_1\mathbf{u}_1 + b_2\mathbf{u}_2 + ... + b_5\mathbf{u}_5$. Then

$$\mathbf{x}_n = L^n\mathbf{x}_0 = M^n\mathbf{x}_0 = r_1^n(b_1\mathbf{u}_1 + b_2\mathbf{u}_2) + r_2^n(b_3\mathbf{u}_3 + b_4\mathbf{u}_4) + r_3^n b_5\mathbf{u}_5. \quad (4.3.7)$$

A problem that needs to be addressed is how to find the constants $b_1, ..., b_q$ which are the components of $\mathbf{x}_0$ in the eigenvectors basis. This is a typical change of basis problem. Assume that the components of the vector $\mathbf{x}_0$ in the standard basis are known, $\mathbf{x}_0 = x_1\mathbf{e}_1 + ... + x_q\mathbf{e}_q$. We need the ones in the "eigenvectors basis," namely $b_1, ..., b_q$ such that $\mathbf{x}_0 = b_1\mathbf{u}_1 + ... + b_q\mathbf{u}_q$.

Consider the matrix T whose columns are the entries of the eigenvectors $\mathbf{u}_1, ..., \mathbf{u}_q$. The components of the vector $\mathbf{u}_i$ in the basis $B = \{\mathbf{u}_1, ..., \mathbf{u}_q\}$ are all 0 except the one in position i, which is 1. In other words,

$$\mathbf{u}_i = (0,0,...,0,1,0,...,0)_B, \text{ where 1 is in position i.} \quad (4.3.8)$$

Since T(0,0,...,0,1,0,...,0) is the ith column of T (see Chapter 3, Section 2) and its entries are the components of $\mathbf{u}_i$ in the standard basis, we see that T is the transition matrix between the basis B (eigenvectors) and the standard basis E. Hence, by applying T^{-1} to the components of $\mathbf{x}_0$ in the standard basis we get its components in the basis B. Let us illustrate the strategy with an example.

Example 4.3.8 Let

$$M = \begin{pmatrix} 1.25 & .75 & -.75 \\ .75 & 1.25 & -.75 \\ 0 & 0 & .5 \end{pmatrix}.$$

The eigenvalues of M are $r_1 = r_2 = .5$ and $r_3 = 2$. The first is semisimple with multiplicity 2. Two eigenvectors corresponding to .5 are $\mathbf{u}_1 = (-1,1,0)$ and $\mathbf{u}_2 = (1,1,2)$. An eigenvector corresponding to 2 is $\mathbf{u}_3 = (1,1,0)$. The matrix T and its inverse are

$$T = \begin{pmatrix} -1 & 1 & 1 \\ 1 & 1 & 1 \\ 0 & 2 & 0 \end{pmatrix}, \qquad T^{-1} = \frac{1}{2}\begin{pmatrix} -1 & 1 & 0 \\ 0 & 0 & 1 \\ 1 & 1 & -1 \end{pmatrix}.$$

Given the vector $\mathbf{x}_0 = (4,6,4)$ we have $T^{-1}(4,6,4) = (1,2,3)_B$. Hence, $\mathbf{x}_0 = \mathbf{u}_1 + 2\mathbf{u}_2 + 3\mathbf{u}_3$ and $\mathbf{x}_n = M^n\mathbf{x}_0 = (0.5)^n(\mathbf{u}_1 + 2\mathbf{u}_2) + 2^n 3\mathbf{u}_3$.

Once the orbit is known in the B basis we can "go back" to the standard basis by applying the matrix T to the components of $\mathbf{x}_n$ in the basis B.

The strategy outlined in equality (4.3.7) provides a really powerful and simple way to find the orbit of every point $\mathbf{x}_0 \in \mathbf{R}^q$ whenever the eigenvalues of M are real and semisimple. However, determining the eigenvalues of M may not be an easy task in high-dimensional spaces since we need to solve the algebraic equation of degree q,

$$\det(rI - M) = 0. \tag{4.3.9}$$

Whenever (4.3.9) can be solved, and all eigenvalues are real and semisimple, we can use the strategy of (4.3.7) to determine the orbit of every vector $\mathbf{x}_0 \in \mathbf{R}^q$. Here is an additional example.

Example 4.3.9 Let $L\mathbf{x}=(x+z,y-z,2z)$. The matrix

$$M = \begin{pmatrix} 1 & 0 & 1 \\ 0 & 1 & -1 \\ 0 & 0 & 2 \end{pmatrix}$$

has the same action as L, and the eigenvalues are $r_1=r_2=1$ and $r_3=2$. The first is semisimple and $\mathbf{e}_1$, $\mathbf{e}_2$ are two linearly independent eigenvectors corresponding to it. The remaining eigenvalue is simple and $\mathbf{u}_3=(1,-1,1)$ is an eigenvector corresponding to it. Equality (4.3.7) becomes

$$\mathbf{x}_n = b_1\mathbf{e}_1 + b_2\mathbf{e}_2 + b_3 2^n \mathbf{u}_3,$$

where b_1, b_2, and b_3 are the components of $\mathbf{x}_0$ in the basis $\mathbf{e}_1$, $\mathbf{e}_2$, and $\mathbf{u}_3$. In this case the matrices T and T^{-1} are

$$T = \begin{pmatrix} 1 & 0 & 1 \\ 0 & 1 & -1 \\ 0 & 0 & 1 \end{pmatrix}, \qquad T^{-1} = \begin{pmatrix} 1 & 0 & -1 \\ 0 & 1 & 1 \\ 0 & 0 & 1 \end{pmatrix}.$$

Hence, given $\mathbf{x}_0=(x_0,y_0,z_0)$ we have $T^{-1}(x_0,y_0,z_0)=(x_0-z_0,y_0+z_0,z_0)_B$. Thus, $\boldsymbol{x}_0=(x_0-z_0)\mathbf{e}_1+(y_0+z_0)\mathbf{e}_2+z_0\mathbf{u}_3$ and $\mathbf{x}_n=(x_0-z_0)\mathbf{e}_1+(y_0+z_0)\mathbf{e}_2+2^n z_0\mathbf{u}_3$.

SDT*: *real, not semisimple eigenvalues (*)

Assume that r_i is not semisimple. There are $q-p_i=gm(r_i)$ linearly independent eigenvectors corresponding to r_i that are solutions of (4.3.1) (are eigenvectors). In addition, there are $am(r_i)-gm(r_i)$ linearly independent vectors that solve (4.3.2) and not (4.3.1) (not eigenvectors). According to Theorem 4.3.1, the number of linearly independent solutions of (4.3.2) is $m_i=am(r_i)$. Hence, they span (each one of them is a vector of $\mathbf{R}^q$) a m_i-dimensional subspace $\mathbf{X}_i$ of $\mathbf{R}^q$.

Theorem 4.3.3 *The subspace* $\mathbf{X}_i$ *is invariant under the action of* M.

Proof. Let $\mathbf{x} \in \mathbf{X}_i$. Since $\mathbf{x}$ is a linear combination of solutions of (4.3.2), $\mathbf{x}$ itself is a solution, i.e., $(r_iI-M)^{m_i}\mathbf{x}=\mathbf{0}$. Hence, $(r_iI-M)^{m_i+1}\mathbf{x}=\mathbf{0}$. Since

$$(r_iI - M)^{m_i+1}\mathbf{x} = (r_iI - M)^{m_i}(r_i\mathbf{x} - M\mathbf{x})$$

and $(r_iI-M)^{m_i}r_i\mathbf{x}=\mathbf{0}$, we obtain $\mathbf{0}=(r_iI-M)^{m_i}M\mathbf{x}$. Thus, $M\mathbf{x}$ is a solution of (4.3.2) and $M\mathbf{x}\in\mathbf{X}_i$. QED

At this point, we can obtain the second version of the spectral decomposition theorem, valid in the case when all eigenvalues are real but not necessarily semisimple. We choose as $\mathbf{X}_i$ the subspaces of Theorem 4.3.3. They are invariant, and the only item of business left out is to prove that $S_1\cup S_2...\cup S_k$ is again a family of linearly independent vectors. We shall not prove this result, and we summarize the situation for semisimple and not semisimple real eigenvalues in Theorem 4.3.4.

The subspaces $\mathbf{X}_i$ of Theorem 4.3.2, step 1, are called **eigenspaces** and those of Theorem 4.3.3 are called **generalized eigenspaces**. Every eigenspace is also a generalized eigenspace, but the converse is not true.

Theorem 4.3.4 *Let* M *be a square matrix of size* q *and assume that all eigenvalues* $r_1,...,r_k$ *of* M *are real . To each eigenvalue* r_i *there corresponds a generalized eigenspace* $\mathbf{X}_i$ *such that*

(i) *For every* $\mathbf{x}\in\mathbf{X}_i$, *we have* $M\mathbf{x}\in\mathbf{X}_i$.

(ii) $\mathbf{X}_1\oplus\mathbf{X}_2\oplus...\oplus\mathbf{X}_k=\mathbf{R}^q$.

Theorem 4.3.4 is not as helpful as Theorem 4.3.2 for writing the orbits of the dynamical system governed by a matrix M whose eigenvalues are real but not necessarily semisimple. The following example illustrates the situation.

Example 4.3.10 Let

$$M = \begin{pmatrix} 3 & 0 & 1 \\ 0 & 4 & 1 \\ 0 & 0 & 4 \end{pmatrix}.$$

We have $\det(rI-M)=(r-3)(r-4)^2$. The eigenvalues of M are $r_1=3$ and $r_2=4$. Their algebraic multiplicities are 1 and 2, respectively. The geometric multiplicity of 4 is 1 and 4 is not semisimple.

From $(4I-M)\mathbf{x}=\mathbf{0}$, $\mathbf{x}=(x,y,z)$, we obtain $x-z=0$ and $z=0$. Hence, $\mathbf{e}_2=(0,1,0)$ is an eigenvector. Since $(4I-M)^2\mathbf{x}=\mathbf{0}$ requires only $x-z=0$, we obtain that $\mathbf{x}_3=(1,0,1)$ is a solution of $(4I-M)^2\mathbf{x}=\mathbf{0}$ and it is not an eigenvector. Thus $\mathbf{X}_2=\text{Span}\{\mathbf{e}_2,\mathbf{x}_3\}$. Since $M\mathbf{e}_2=4\mathbf{e}_2$ and $M\mathbf{x}_3=\mathbf{e}_2+4\mathbf{x}_3$, we see that $\mathbf{X}_2$ is invariant under the action of M. An eigenvector corresponding to $r_1=3$ is $\mathbf{e}_1=(1,0,0)$. Hence, $\mathbf{X}_1=\text{Span}\{\mathbf{e}_1\}$ and $\mathbf{R}^3=\mathbf{X}_1\oplus\mathbf{X}_2$.

We can now write the matrix T and its inverse T^{-1}. Hence, given $\mathbf{y}_0\in\mathbf{R}^3$ we can find b_1,b_2, and b_3 such that $\mathbf{y}_0=b_1\mathbf{e}_1+b_2\mathbf{e}_2+b_3\mathbf{x}_3$. Applying the matrix M, we obtain

$$\mathbf{y}_1 = M\mathbf{y}_0 = b_1 3\mathbf{e}_1 + b_2 4\mathbf{e}_2 + b_3 M\mathbf{x}_3 = b_1 3\mathbf{e}_1 + b_2 4\mathbf{e}_2 + b_3(\mathbf{e}_2 + 4\mathbf{x}_3).$$

We see that the lack of a third eigenvector complicates the form of $\mathbf{y}_1$. A further step gives

$$\mathbf{y}_2 = b_1 3^2\mathbf{e}_1 + b_2 4^2\mathbf{e}_2 + b_3 4\mathbf{e}_2 + 4b_3(\mathbf{e}_2 + 4\mathbf{x}_3)$$

$$= b_1 3^2\mathbf{e}_1 + b_2 4^2\mathbf{e}_2 + b_3 2 \times 4\mathbf{e}_2 + b_3 4^2\mathbf{x}_3.$$

In general, we obtain

$$\mathbf{y}_n = b_1 3^n\mathbf{e}_1 + b_2 4^n\mathbf{e}_2 + n \times 4^{n-1} b_3\mathbf{e}_2 + b_3 4^n\mathbf{x}_3$$

$$= b_1 3^n\mathbf{e}_1 + 4^{n-1}(4b_2 + nb_3)\mathbf{e}_2 + b_3 4^n\mathbf{x}_3.$$

The final form is not as neat as in the case when the eigenvalues are semisimple, although we must recognize that it is of some help in studying the orbits of M.

Here is another example.

Example 4.3.11 Let

$$M = \begin{pmatrix} 1 & 0 & 1 \\ 0 & 1 & 0 \\ 0 & 1 & 2 \end{pmatrix}.$$

The eigenvalues of M are $r_1=1$ and $r_2=2$ with algebraic multiplicities 2 and 1, respectively. The eigenvalue 1 is not semisimple. Two linearly independent solutions of $(I-M)^2\mathbf{x}=\mathbf{0}$ are $\mathbf{e}_1=(1,0,0)$, which is an eigenvector, and $\mathbf{v}=(0,1,-1)$, which is not an eigenvector. An eigenvector corresponding to 2 is $\mathbf{w}=(1,0,1)$. Given the vector $\mathbf{x}_0=(3,1,0)=2\mathbf{e}_1+\mathbf{v}+\mathbf{w}$, we have $\mathbf{x}_n=2\mathbf{e}_1+M^n\mathbf{v}+2^n\mathbf{w}$ and our problem is to determine $M^n\mathbf{v}$. Since $M\mathbf{v}=(-1,1,-1)=-\mathbf{e}_1+\mathbf{v}$, we obtain $M^2\mathbf{v}=M(-\mathbf{e}_1+\mathbf{v})=-M\mathbf{e}_1+M\mathbf{v}=-\mathbf{e}_1+M\mathbf{v}=-2\mathbf{e}_1+\mathbf{v}$. In general, we find that $M^n(\mathbf{v})=-n\mathbf{e}_1+\mathbf{v}$. Hence,

$$\mathbf{x}_n = 2\mathbf{e}_1 - n\mathbf{e}_1 + \mathbf{v} + 2^n\mathbf{w} = (2 - n)\mathbf{e}_1 + \mathbf{v} + 2^n\mathbf{w}.$$

In fact, for n=4 we have

$$\mathbf{x}_4 = M^4\mathbf{x}_0 = (14,1,15) \quad \text{and}$$

$$-2\mathbf{e}_1 + \mathbf{v} + 16\mathbf{w} = (-2,0,0) + (0,1,-1) + (16,0,16) = (14,1,15).$$

We see that Theorem 4.3.4 still provides a quite simple way to find the orbit of $\mathbf{x}_0$.

SDT*: *when* $\sigma(M)$ *has complex elements (*)

Before stating the spectral decomposition theorem in its full generality we briefly examine the situation when M has a pair of simple complex conjugate eigenvalues. Assume that $s_1=s+it$ and $s_2=s-it$ are simple eigenvalues of M. We consider $s_1=s+it$ and we show that there is a two-dimensional invariant subspace **X** corresponding to it. The reader can easily verify that the analysis and the conclusions do not change if we start from s_2.

There exists a complex vector $\mathbf{z}=\mathbf{u}+i\mathbf{v}$, with $\mathbf{u},\mathbf{v}\in\mathbf{R}^q$ such that

$$M(\mathbf{u} + i\mathbf{v}) = s_1(\mathbf{u} + i\mathbf{v}), \tag{4.3.10}$$

i.e.,

$$M\mathbf{u} + iM\mathbf{v} = s\mathbf{u} - t\mathbf{v} + i(t\mathbf{u} + s\mathbf{v}). \tag{4.3.11}$$

In fact, since $\det(s_1I-M)=0$, we know that the homogeneous system $(s_1I-M)\mathbf{z}=\mathbf{0}$ has a nonzero solution. In other words, there exists $\mathbf{z}\neq\mathbf{0}$ such that $M\mathbf{z}=(s+it)\mathbf{z}$. The vector $\mathbf{z}$ must have complex entries, for otherwise the left-hand side of (4.3.10) would be a vector with real entries, while the right-hand side would be a vector with complex entries. Hence,

$$\mathbf{z} = (a_1 + ib_1,\dots,a_q + ib_q) = (a_1,\dots,a_q) + i(b_1,\dots,b_q) = \mathbf{u} + i\mathbf{v}.$$

The vectors $\mathbf{u}$ and $\mathbf{v}$ are linearly independent. In fact, assume that $a\mathbf{u}+b\mathbf{v}=\mathbf{0}$. Multiply (4.3.11) by b and use $b\mathbf{v}=-a\mathbf{u}$ to obtain

$$(b - ia)M\mathbf{u} = (bs + at)\mathbf{u} + i(bt - as)\mathbf{u}.$$

Therefore,

$$bM\mathbf{u} = (bs + at)\mathbf{u} \quad \text{and} \quad aM\mathbf{u} = (as - bt)\mathbf{u}.$$

Multiply the first equality by a and the second by b, and subtract to arrive at $\mathbf{0}=(a^2+b^2)t\mathbf{u}$. Since $\mathbf{u}\neq\mathbf{0}$ and $t\neq0$ (otherwise, the eigenvalue is real) we obtain $a^2+b^2=0$. Hence, a=b=0 and the two vectors are linearly independent.
From (4.3.11) we derive

$$M\mathbf{u} = s\mathbf{u} - t\mathbf{v} \quad \text{and} \quad M\mathbf{v} = t\mathbf{u} + s\mathbf{v}. \tag{4.3.12}$$

Therefore, the two-dimensional subspace $\mathbf{X}$ spanned by the linearly independent vectors $\mathbf{u}$ and $\mathbf{v}$ is invariant under the action of M. The reader can easily verify that $\mathbf{X}$ is also the subspace corresponding to $s_2=s-it$.

Thus, to the pair of simple complex conjugate eigenvalues s+it, s–it, there corresponds a two-dimensional invariant subspace $\mathbf{X}$ of $\mathbf{R}^q$. The simplest way to find $\mathbf{u}$ and $\mathbf{v}$ is to solve (4.3.10), using the guidelines of the following example.

Example 4.3.12 The characteristic polynomial of

$$M = \begin{pmatrix} 1 & .3 & -.5 \\ .4 & 1 & -.4 \\ .4 & .5 & .1 \end{pmatrix}$$

is $(r-.5)(.68-1.6r+r^2)$ and the eigenvalues of M are $r_1=1/2$, $s_1=.8+.2i$, $s_2=.8-.2i$. An eigenvector corresponding to 1 is $\mathbf{x}_1=(1,0,1)$. Let $\mathbf{X}_1=\mathrm{Span}\{\mathbf{x}_1\}$. To find $\mathbf{u}$ and $\mathbf{v}$

corresponding to the two complex conjugate eigenvalues we set $\mathbf{w}=\mathbf{u}+i\mathbf{v}=(a,b,c)$ and rewrite equation (4.3.10) in the form $(s_1I-M)\mathbf{w} = \mathbf{0}$, i.e.,

$$(s_1I - M)\mathbf{w} = \begin{pmatrix} -.2+.2i & -.3 & .5 \\ -.4 & -.2+.2i & .4 \\ -.4 & -.5 & .7+.2i \end{pmatrix}(a, b, c) = (0, 0, 0).$$

The coordinates (a,b,c) are complex numbers. We obtain

$$\begin{aligned} &(-2 + 2i)a - 3b + 5c = 0, \\ &4a + (2 - 2i)b - 4c = 0, \\ &4a + 5b - (7 + 2i)c = 0. \end{aligned}$$

Subtracting the second equation from the third gives $(3+2i)b-(3+2i)c=0$. Hence, $c=b$. Substituting into the first equation, we arrive at $(1-i)a-b=0$. Setting $b=1-i$, we obtain $a=1$. Hence,

$$\mathbf{w} = (1,1 - i,1 - i) = (1,1,1) + i(0,-1,-1) = \mathbf{u} + i\mathbf{v}.$$

The equality (see 4.3.12)

$$M\mathbf{u} = .8\mathbf{u} - .2\mathbf{v}, \qquad M\mathbf{v} = .2\mathbf{u} + .8\mathbf{v},$$

implies

$$M(\alpha\mathbf{u} + \beta\mathbf{v}) = (.8\alpha + .2\beta)\mathbf{u} + (.8\beta - .2\alpha)$$

The subspace $\mathbf{X}=\mathrm{Span}\{\mathbf{u},\mathbf{v}\}$ is invariant under the action of M. The three vectors $\mathbf{e}_1$, $\mathbf{u}$, and $\mathbf{v}$ are linearly independent. Thus $\mathbf{R}^3=\mathrm{Span}\{\mathbf{e}_1,\mathbf{u},\mathbf{v}\}$ and $\mathbf{R}^3=\mathbf{X}_1\oplus\mathbf{X}$.

Although we did not illustrate it in our examples to avoid messy computations, it can be shown that when a complex eigenvalue a+ib has multiplicity k (i.e., it appears with exponent k in the factorization of the characteristic polynomial of M), then the invariant subspace corresponding to the pair a±ib has dimension 2k. As a consequence of this observation and of the comments made before Theorem 4.3.4, we obtain that $\mathbf{R}^q$ is the direct sum of the generalized eigenspaces corresponding to the real eigenvalues and of the invariant subspaces corresponding to the complex conjugate pairs. This is exactly the content of the spectral decomposition theorem.

Theorem 4.3.5 (Spectral Decomposition Theorem) *Let* $M : \mathbf{R}^q \to \mathbf{R}^q$. *Denote by* $\mathbf{X}_1, \ldots, \mathbf{X}_k$ *the generalized eigenspaces corresponding to the real eigenvalues* $r_1, \ldots, r_k$ *of* M, *and by* $\mathbf{Y}_1, \ldots, \mathbf{Y}_m$ *the invariant subspaces corresponding to the pairs of complex conjugate eigenvalues* $s_1=a_1\pm ib_1, \ldots, s_m=a_m\pm ib_m$ *of* M, *with* $\dim(\mathbf{Y}_j)= 2\,\mathrm{am}(s_j)$ [*where* $\mathrm{am}(s_j)$ *is the algebraic multiplicity of* s_j]. *Then*

$$\mathbf{R}^q = \mathbf{X}_1 \oplus \ldots \oplus \mathbf{X}_k \oplus \mathbf{Y}_1 \oplus \ldots \oplus \mathbf{Y}_m.$$

The next example illustrates how Theorem 4.3.5 can be used to study linear dynamical systems. Before presenting the example we explain how the action of M in the invariant subspaces $\mathbf{Y}_j$, j=1,2,...,m can be written so that it is more suitable for computational purposes. Assume that s±it is a pair of simple complex conjugate eigenvalues and **u** and **v** are two linearly independent vectors such that [see (4.3.12)] $M\mathbf{u}=s\mathbf{u}-t\mathbf{v}$ and $M\mathbf{v}=t\mathbf{u}+s\mathbf{v}$. Set $r^2=s^2+t^2$. Since $(s/r)^2+(t/r)^2=1$, we can find $\theta\in[0,2\pi)$ such that $s=r\cos\theta$ and $t=r\sin\theta$. Thus, (4.3.12) can be rewritten in the form

$$M\mathbf{u} = r(\cos\theta\,\mathbf{u} - \sin\theta\,\mathbf{v}),\ \ M\mathbf{v} = r(\sin\theta\,\mathbf{u} + \cos\theta\,\mathbf{v}). \tag{4.3.13}$$

Since for every vector $\mathbf{y}\in\mathbf{Y}=\text{Span}\{\mathbf{u},\mathbf{v}\}$ we have $\mathbf{y}=c_1\mathbf{u}+c_2\mathbf{v}=(c_1,c_2)_B$, we obtain

$$M\mathbf{y} = c_1M\mathbf{u} + c_2M\mathbf{v} = r(c_1\cos\theta + c_2\sin\theta, -c_1\sin\theta + c_2\cos\theta)_B.$$

It follows that the action of M on **y** can be expressed in the following way:

$$M\mathbf{y} = r\begin{pmatrix} \cos\theta & \sin\theta \\ -\sin\theta & \cos\theta \end{pmatrix}(c_1,c_2)_B.$$

This implies that (see Example 1.2.4)

$$M^n\mathbf{y} = r^n\begin{pmatrix} \cos n\theta & \sin n\theta \\ -\sin n\theta & \cos n\theta \end{pmatrix}(c_1,c_2)_B. \tag{4.3.14}$$

In particular,

$$M^n\mathbf{u} = r^n(\cos n\theta\,\mathbf{u} - \sin n\theta\,\mathbf{v}),\ M^n\mathbf{v} = r^n(\sin n\theta\,\mathbf{u} + \cos n\theta\,\mathbf{v}). \tag{4.3.15}$$

In the case when $\theta/(2\pi)=m/p$, where m and p are relatively prime, and r=1, all orbits of M on **Y** are periodic of period p. In fact,

$$M^p\mathbf{y} = r^p\begin{pmatrix} \cos 2m\pi & \sin 2m\pi \\ -\sin 2m\pi & \cos 2m\pi \end{pmatrix}(c_1,c_2)_B = (c_1,c_2)_B = \mathbf{y}.$$

Example 4.3.13 Let M be the matrix of Example 4.3.12. Since $\rho(M)<1$ we know that every orbit of M goes to **0**. We are interested in finding the form of an orbit. We have $M^n\mathbf{x}_1=(1/2)^n\mathbf{x}_1=(1/2)^n(1,0,1)$. We also have $\mathbf{u}=(1,1,1)$ and $\mathbf{v}=(0,-1,-1)$. The modulus and the argument of the complex number .8+.2i are

$$r = \sqrt{17}/5,\quad \theta = \arctan .25\ (\approx .245).$$

Let $\mathbf{x}_0=(x_0,y_0,z_0)$. We need to express $\mathbf{x}_0$ in the basis $B=\{\mathbf{x}_1,\mathbf{u},\mathbf{v}\}$. This is a change of basis problem. The matrix T and its inverse T^{-1} are

$$T = \begin{pmatrix} 1 & 1 & 0 \\ 0 & 1 & -1 \\ 1 & 1 & -1 \end{pmatrix},\qquad T^{-1} = \begin{pmatrix} 0 & -1 & 1 \\ 1 & 1 & -1 \\ 1 & 0 & -1 \end{pmatrix}.$$

Then $T^{-1}(x_0,y_0,z_0)=(-y_0+z_0,x_0+y_0-z_0,x_0-z_0)_B$. Hence,

$$\mathbf{x}_0 = (-y_0 + z_0)\,\mathbf{x}_1 + (x_0 + y_0 - z_0)\mathbf{u} + (x_0 - z_0)\mathbf{v}$$

and

$$M^n\mathbf{x}_0 = (-y_0 + z_0)M^n\mathbf{x}_1 + (x_0 + y_0 - z_0)M^n\mathbf{u} + (x_0 - z_0)M^n\mathbf{v}.$$

The appropriate substitutions give

$$M^n\mathbf{x}_0 = (-y_0 + z_0)2^{-n}\,\mathbf{x}_1 + (\sqrt{17}/5)^n[(x_0 + y_0 - z_0)(\cos n\theta\,\mathbf{u} - \sin n\theta\,\mathbf{v}) + (x_0 - z_0)(\sin n\theta\mathbf{u} + \cos n\theta\mathbf{v})].$$

In the one-dimensional subspace **X** spanned by **w**, the action of M^n is a compression of size 2^{-n}. In the subspace **Y**=Span{**u**,**v**} with the coordinates of every vector written in the basis B={**u**,**v**}, the action of M^n can be regarded as a compression of size $(\sqrt{17}/5)^n$ together with a rotation of an angle $n\theta$. The graph of Fig. 4.3.1 shows the behavior of the orbit starting at (3,3,–1).

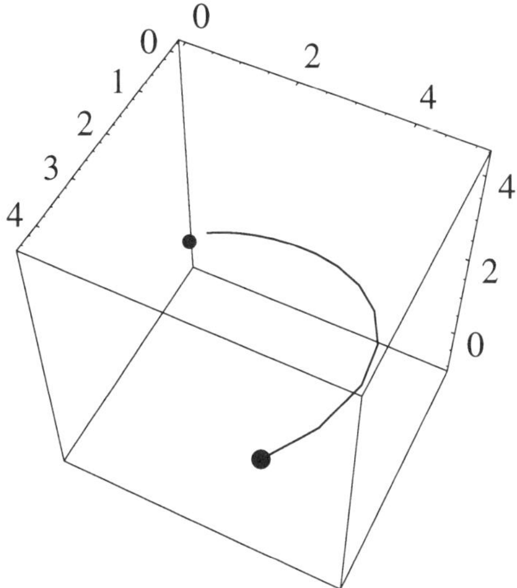

Fig. 4.3.1 The orbit starting at (3,3,-1) (larger dot) spirals toward the origin under the combined effects of the compression and the rotation. Rather than plotting the various states of the orbit, we have visualized it by connecting them with line segments. The changes in direction at the beginning of the orbit detect where the states are. Later, as the origin is approached, this feature is lost.

A similar analysis can be carried out when the eigenvalues s±it are semisimple, with multiplicity k. The invariant subspace **Y** has dimension 2k, but it is spanned by k pairs of vectors such that the span of each pair is invariant under the action of M. Hence, **Y** is the direct sum of these two-dimensional invariant subspaces. The action of M in each of these subspaces is exactly like the action illustrated in Example 4.3.13.

We conclude this section with a remark, which will be useful later.

Remark 4.3.2 Assume that **X** is a subspace of $\mathbf{R}^q$ which is invariant under the action of M. Define

$$\|M\|_{\mathbf{X}} = \max\{\|M\mathbf{x}\| : \|\mathbf{x}\| = 1, \mathbf{x} \in \mathbf{X}\}.$$

$\|M\|_{\mathbf{X}}$ is the operator norm of M, restricted to the invariant subspace **X**. Let $\sigma(M)_{\mathbf{X}} \subset \sigma(M)$ be the subset of eigenvalues of M such that

- The corresponding eigenvectors are in **X** if the eigenvalue r is real.
- The pairs of vectors **u**, **v** solution of (4.3.12) are in **X** if r is complex.

Then

$$\rho(M)_{\mathbf{X}} = \lim_{n\to\infty} \sqrt[n]{\|M^n\|_{\mathbf{X}}} \tag{4.3.16}$$

where $\rho(M)_{\mathbf{X}}=\max\{|r| : r \in \sigma(M)_{\mathbf{X}}\}$. The result remains valid if a different norm is used in **X** and the operator norm of M in **X** is computed using this different norm.

Example 4.3.14 Let

$$M = \begin{pmatrix} 2 & 1 \\ 0 & 1/2 \end{pmatrix}.$$

Then **X**=Span{(1,0)} and **Y**=Span{(2,–3)} are invariant subspaces of M. We have $\|M\|_{\mathbf{X}}=2$ and $\|M\|_{\mathbf{Y}}=1/2$. In fact,

$$M(1,0) = (2,0) \quad \text{and} \quad M\frac{1}{\sqrt{13}}(2,-3) = \frac{1}{2\sqrt{13}}(2,-3).$$

Moreover, $\sigma(M)_{\mathbf{X}}=\{2\}$ and $\sigma(M)_{\mathbf{Y}}=\{1/2\}$.

Problems

1. * Let

$$M = \begin{pmatrix} 0 & 2 \\ 1 & 1 \end{pmatrix}.$$

Find the eigenvalues and eigenvectors of M. Find the matrices T and T^{-1}. Use Theorem 4.3.2 to write the dynamical system $\mathbf{x}_{n+1}=M\mathbf{x}_n$ in the form (4.3.7).

2. * Let

$$M = \begin{pmatrix} 0 & 2 & 1 \\ -1/2 & 5/2 & 1 \\ 0 & 0 & 1/2 \end{pmatrix}.$$

Find the eigenvalues and eigenvectors of M, the matrices T and T^{-1}. Use Theorem 4.3.2 to write the dynamical system $\mathbf{x}_{n+1}=M\mathbf{x}_n$ in the form (4.3.7).

3. * Let

$$M = \begin{pmatrix} 5/4 & 3/4 & 0 \\ 3/4 & 5/4 & 0 \\ 0 & 0 & 1/2 \end{pmatrix}.$$

Find the eigenvalues and eigenvectors of M. Use Theorem 4.3.2 to write the dynamical system $\mathbf{x}_{n+1}=M\mathbf{x}_n$ in the form (4.3.7).

4. * Let M

$$M = \begin{pmatrix} 3/8 & -1/8 & 0 & 0 \\ -1/8 & 3/8 & 0 & 0 \\ 0 & 0 & 2 & 0 \\ -3/4 & 3/4 & 0 & 2 \end{pmatrix}.$$

Find the eigenvalues and eigenvectors of M. Use Theorem 4.3.2 to write the dynamical system $\mathbf{x}_{n+1}=M\mathbf{x}_n$ in the form (4.3.7).

5. * Let

$$M = \begin{pmatrix} 2 & 1 \\ 0 & 2 \end{pmatrix}.$$

The only eigenvalue of M is 2. Find an eigenvector corresponding to it. Find the form of the orbit of any point $\mathbf{x}_0=(x_0,y_0)$.

6. * Let

$$M = \begin{pmatrix} 5/2 & -2 & -1 \\ -1/6 & 2/3 & -2/3 \\ 4/3 & -4/3 & 4/3 \end{pmatrix}.$$

Show that the eigenvalues of M are 1/2 (simple) and 2 (not semisimple). Find the eigenspace corresponding to 1/2 and the generalized eigenspace corresponding to 2. Use Theorem 4.3.4 to write the orbit of (1,2,–3) in a form similar to the ones found in Examples 4.3.10 and 4.3.11.

7. * Let

$$M = \begin{pmatrix} 1/2 & 0 & 0 \\ 0 & 1/2 & 0 \\ 0 & 1 & 1/2 \end{pmatrix}.$$

Show that 1/2 is the only eigenvalue of M with geometric multiplicity 2. Find two eigenvectors corresponding to it. Use Theorem 4.3.4 to find the form of the orbit of (0,1,1).

8. * Let

$$M = \begin{pmatrix} 2 & 0 & 0 \\ 0 & 2 & 0 \\ 1 & 1 & 2 \end{pmatrix}.$$

Show that 2 is the only eigenvalue of M and that its geometric multiplicity is 2. Use Theorem 4.3.4 to find the form of the orbit of (1,0,1).

9. * Let

$$M = \begin{pmatrix} 9/2 & -4 & -5 \\ -3/2 & 2 & 0 \\ 5 & -5 & -2 \end{pmatrix}.$$

Show that the eigenvalues of M are 1/2, 2+3i and 2–3i. Find the eigenspace corresponding to 1/2 and the invariant subspace corresponding to the complex conjugate pair. Use Theorem 4.3.5 to find the form of the orbit of (1,1,1).

10. * Let

$$M = \begin{pmatrix} -7 & 4 & 5 & 5 \\ -10 & 6 & 6 & 7 \\ 11 & -9/2 & -9/2 & -17/2 \\ -16 & 7 & 9 & 12 \end{pmatrix}.$$

Show that the eigenvalues of M are 2, 1/2, 2+i and 2–i. Find the eigenspaces corresponding to 2 and 1/2. Find the invariant subspace corresponding to the pair of complex conjugate eigenvalues 2±i.

Section 4. THE ORIGIN AS A SADDLE POINT

Stable and unstable subspaces

The two cases analyzed in Section 2 are somehow extreme. In the first, every eigenvalue of M has modulus smaller than 1 and every orbit goes to **0**. In the second every eigenvalue of M has modulus larger than 1 and every orbit with initial state $\mathbf{x}_0 \neq \mathbf{0}$ goes to infinity. In this section we analyze the case when some eigenvalues of M have modulus smaller than 1 and the others have modulus larger than 1.

The stationary state **0** is now called a **saddle point**. To understand the behavior of a dynamical system governed by a matrix having this property, we start with a simple example.

Example 4.4.1 Let

$$M = \begin{pmatrix} 2^{-1} & 2 \\ 0 & 3 \end{pmatrix}.$$

The eigenvalues of M are 2^{-1} and 3. An eigenvector of M corresponding to 2^{-1} is $\mathbf{e}_1=(1,0)$; while $\mathbf{u}=(4,5)$ corresponds to 3. The orbit starting at any vector of the form (t,0) will go to zero, since $M^n(t,0)=(2^{-n}\,t,0)$. For this reason the one-dimensional subspace (the *x*-axis) spanned by the vector $\mathbf{e}_1$ is called the **stable subspace** of M and is denoted by **S**.

Similarly, the orbit starting at any vector of the form (4t,5t), with $t\neq0$, goes to infinity, since $M^n(4t,5t) = 3^n(4t,5t)$. The span of the vector **u**, which is also a one-dimensional subspace of $\mathbf{R}^2$, is called the **unstable subspace** of M and denoted by **U**. Since $\mathbf{e}_1=(1,0)$ and $\mathbf{u}=(4,5)$ are linearly independent, they are a basis of $\mathbf{R}^2$. Hence $\mathbf{R}^2=\mathbf{S}\oplus\mathbf{U}$ and every vector $\mathbf{x}\in\mathbf{R}^2$ can be written in a unique manner as a linear combination of $\mathbf{e}_1$ and **u**, $\mathbf{x}=a\mathbf{e}_1+b\mathbf{u}$, for some real numbers a and b. This implies that

$$M^n\mathbf{x} = 2^{-n}a\mathbf{e}_1 + 3^n b\mathbf{u}.$$

As $n\to\infty$ the part $2^{-n}a\mathbf{e}_1$ goes to zero, while $3^n b\mathbf{u}$ goes to infinity. The orbit starting at any point $\mathbf{x}=a\mathbf{e}_1+b\mathbf{u}$ with $b\neq0$ goes to infinity while approaching the unstable subspace of M.

In summary, every vector $\mathbf{x}\in\mathbf{R}^2$ can be written as a sum of two vectors, one (proportional to $\mathbf{e}_1$) belonging to the stable subspace **S** and the other (proportional to **u**) belonging to the unstable subspace **U**. Each state of the orbit of **x** is the sum of the corresponding states of the orbits of the $\mathbf{e}_1$ component and of the **u** component of **x**. The orbit of the $\mathbf{e}_1$ component goes to **0**, and the orbit of the **u** component goes to infinity. Consequently, the orbit of **x** goes to infinity while approaching the unstable subspace.

The situation encountered in Example 4.4.1 is typical of this type of problems. The behavior of the system is based on the following consequence of the spectral decomposition theorem.

Theorem 4.4.1 *Let* M *be a matrix of size* q, *with eigenvalues* $r_1,\ldots,r_p$. *Assume that* $|r_i|<1$, $i=1,2,\ldots,k$ *and* $|r_i|>1$, $i=k+1,\ldots,p$. *Denote by* **S** *the direct sum of all generalized eigenspaces and invariant subspaces corresponding to the eigenvalues* r_i, $i=1,2,\ldots,k$ *and by* **U** *the direct sum of the generalized eigenspaces and invariant subspaces corresponding to the remaining eigenvalues. Then the two subspaces* **S** *and* **U** *are invariant and such that*

i. $\mathbf{R}^q = \mathbf{S} \oplus \mathbf{U}$.

ii. *For every* $s_0\in\mathbf{S}$ *we have* $\|s_n\|\to0$ *as* $n\to\infty$.

iii. *For every* $\mathbf{u}_0\in\mathbf{U}$, $\mathbf{u}_0\neq\mathbf{0}$, *we have* $\|\mathbf{u}_n\|\to\infty$ *as* $n\to\infty$.

Proof. (i) By the spectral decomposition theorem we have $\mathbf{R}^q=\mathbf{S}\oplus\mathbf{U}$ where **S** and **U** are invariant under the action of M.

(ii) Since **S** is invariant we can study the action of M on the elements of **S** by considering (see Remark 4.3.2)

$$\|M\|_S = \max\{\|Ms\| : s\in \mathbf{S}, \|s\| = 1\}, \qquad \rho(M)_S = \max\{|r_i|: i = 1,2,\dots,k\}<1.$$

Therefore, we can apply Theorem 4.2.1 to obtain that the orbit of every point $s_0\in \mathbf{S}$ goes to $\mathbf{0}$ as $n\to\infty$.

(iii) Similar considerations hold for **U** with Theorem 4.2.2 replacing Theorem 4.2.1 to obtain that for every $\mathbf{u}_0\in \mathbf{U}$, $\mathbf{u}_0\neq\mathbf{0}$, the orbit goes to infinity as $n\to\infty$. QED

The subspaces **S** and **U** are called **stable** and **unstable** subspaces, respectively. Every vector $\mathbf{x}_0\in \mathbf{R}^q$ can be written in a unique manner as $\mathbf{x}_0=\mathbf{s}_0+\mathbf{u}_0$ with $s_0\in \mathbf{S}$ and $\mathbf{u}_0\in \mathbf{U}$. Thus, $\mathbf{x}_n=\mathbf{s}_n+\mathbf{u}_n$. Whenever $\mathbf{u}_0\neq\mathbf{0}$ the orbit of $\mathbf{x}_0$ goes to infinity while approaching the unstable subspace **U**.

Example 4.4.2 Let L : $\mathbf{R}^3\to\mathbf{R}^3$ be a linear operator represented in the standard basis by the matrix

$$M = \begin{pmatrix} 1 & 2 & 0 \\ -2 & 1 & 1 \\ 0 & 0 & .5 \end{pmatrix}.$$

The eigenvalues of M are $r_1=1/2$, $r_2=1+2i$, and $r_3=1-2i$. Notice that $|r_2|=\sqrt{5}$, $\theta=\arctan 2$. An eigenvector corresponding to r_1 is $\mathbf{w}=(8,-2,17)$. A pair of vectors **u** and **v** such that $M(\mathbf{u}+i\mathbf{v})=r_2(\mathbf{u}+i\mathbf{v})$ is $\mathbf{u}=(1,0,0)=\mathbf{e}_1$, $\mathbf{v}=(0,1,0)=\mathbf{e}_2$. Consequently, the stable subspace of M is $\mathbf{S}=\mathrm{Span}\{(8,-2,17)\}$, and the unstable subspace of M is $\mathbf{U}=\mathrm{Span}\{\mathbf{e}_1,\mathbf{e}_2\}$. The matrices T and T^{-1} are

$$T = \begin{pmatrix} 8 & 1 & 0 \\ -2 & 0 & 1 \\ 17 & 0 & 0 \end{pmatrix}, \qquad T^{-1} = \begin{pmatrix} 0 & 0 & 1/17 \\ 1 & 0 & -8/17 \\ 0 & 1 & 2/17 \end{pmatrix}.$$

The vector $\mathbf{x}_0=(x_0,y_0,z_0)$, written in the basis B is $(z_0/17, x_0-8z_0/17, y_0+2z_0/17)_B$, i.e.,

$$\mathbf{x}_0 = \frac{z_0}{17}\mathbf{w} + \left(x_0 - \frac{8z_0}{17}\right)\mathbf{e}_1 + \left(y_0 + \frac{2z_0}{17}\right)\mathbf{e}_2.$$

It follows that

$$s_0 = \frac{z_0}{17}\mathbf{w} \quad \text{and} \quad \mathbf{u}_0 = \left(x_0 - \frac{8z_0}{17}\right)\mathbf{e}_1 + \left(y_0 + \frac{2z_0}{17}\right)\mathbf{e}_2.$$

We find

$$\mathbf{x}_n = \frac{z_0}{17} 2^{-n}(8,-2,17) + (x_0 - 8z_0/17) \sqrt{5^n} (\cos n\theta\, \mathbf{e}_1 - \sin n\theta\, \mathbf{e}_2)$$

$$+ (y_0 + 2z_0/17) \sqrt{5^n} (\sin n\theta\, \mathbf{e}_1 + \cos n\theta\, \mathbf{e}_2) = \mathbf{s}_n + \mathbf{u}_n,$$

with

$$\mathbf{s}_n = \frac{z_0}{17} 2^{-n}(8,-2,17)$$

and

$$\mathbf{u}_n = (x_0 - 8z_0/17) \sqrt{5^n} (\cos n\theta\, \mathbf{e}_1 - \sin n\theta\, \mathbf{e}_2)$$

$$+ (y_0 + 2z_0/17) \sqrt{5^n} (\sin n\theta\, \mathbf{e}_1 + \cos n\theta\, \mathbf{e}_2).$$

Thus, $\|\mathbf{s}_n\| \to 0$ and $\|\mathbf{u}_n\| \to \infty$ as $n \to \infty$.

It is important to underline the fact that the orbit of every initial state which does not belong to the stable subspace will go to infinity. It may take many iterations to detect this behavior, particularly if the $\mathbf{s}_0$ component of the starting point is very large and the $\mathbf{u}_0$ component is very small. In fact, in this case the trajectory will move towards the origin during the initial iterations. Eventually the orbit will get closer to the unstable subspace, and from then on its states will move away from **0** at an exponential rate.

Comparing trajectories

A remark may be helpful. We have analyzed dynamical systems in $\mathbf{R}^q$ governed by matrices with some eigenvalues in modulus smaller than 1, and the others with modulus larger than 1. In these cases we can express every vector $\mathbf{x}$ of $\mathbf{R}^q$ as a sum $\mathbf{x}=\mathbf{s}+\mathbf{u}$, where $\mathbf{s}$ is a vector of the stable subspace **S** while $\mathbf{u}$ is in the unstable subspace **U**. The nth state of the orbit of $\mathbf{x}$ is the sum of the nth states of the orbits of $\mathbf{s}$ and $\mathbf{u}$. Given two initial conditions $\mathbf{x}_0$ and $\mathbf{y}_0$, the distance $\|\mathbf{x}_n - \mathbf{y}_n\|$ between the corresponding states of $O(\mathbf{x}_0)$ and $O(\mathbf{y}_0)$ goes to infinity unless the two initial states differ by a vector belonging to the stable subspace, i.e., $\mathbf{x}_0 - \mathbf{y}_0 \in \mathbf{S}$. In this case, the distance goes to 0.

In fact, let $\mathbf{x}_0 - \mathbf{y}_0 = \mathbf{z}_0$. Then $\mathbf{x}_n - \mathbf{y}_n = \mathbf{z}_n$. Since $\|\mathbf{z}_n\| \to \infty$ if $\mathbf{z}_0 \notin \mathbf{S}$ and $\|\mathbf{z}_n\| \to 0$ if $\mathbf{z}_0 \in \mathbf{S}$, we obtain that

$\|\mathbf{x}_n - \mathbf{y}_n\| \to \infty$ in the first case, and $\|\mathbf{x}_n - \mathbf{y}_n\| \to 0$ in the second case.

Example 4.4.3 Let M be as in Example 4.4.1 and consider the initial states $\mathbf{x}_0=(1,2)$ and $\mathbf{y}_0=(2,1)$. Their distance is $2^{.5}$. Recall that (1,0), (4,5) are the eigenvectors of M and $\mathbf{S}=\text{Span}\{(1,0)\}$. Since $(1,2)-(2,1)=(-1,1) \notin \mathbf{S}$, $\|\mathbf{x}_n - \mathbf{y}_n\|$ should go to ∞. In fact,

$$(1,2) = -\frac{3}{5}(1,0) + \frac{2}{5}(4,5), \qquad (2,1) = \frac{6}{5}(1,0) + \frac{1}{5}(4,5).$$

Therefore,

$$\mathbf{x}_n = M^n(1,2) = M^n\left(-\frac{3}{5}(1,0) + \frac{2}{5}(4,5)\right) = -\frac{3}{5}2^{-n}(1,0) + \frac{2}{5}3^n(4,5)$$

and

$$\mathbf{y}_n = M^n(2,1) = M^n\left(\frac{6}{5}(1,0) + \frac{1}{5}(4,5)\right) = \frac{6}{5}2^{-n}(1,0) + \frac{1}{5}3^n(4,5).$$

For n very large we can neglect the contribution to the distance between $\mathbf{x}_n$ and $\mathbf{y}_n$ coming from the terms containing 2^{-n}. Therefore, $\|\mathbf{x}_n - \mathbf{y}_n\|$ is given approximately by the norm of the vector $(3^n/5)(4,5)$, which is $(3^n/5)41^{.5}$. The distance goes to infinity. Let us now change $\mathbf{y}_0$ and take the point $\mathbf{z}_0=(3,2)$. Since $(3,2)-(1,2)=(2,0)\in \mathbf{S}$, the distance between the orbits of (3,2) and (1,2) should go to 0 as $n\to\infty$. In fact,

$$(3,2) = \frac{7}{5}(1,0) + \frac{2}{5}(4,5)$$

and we derive

$$M^n(3,2) = \frac{7}{5}2^{-n}(1,0) + \frac{2}{5}3^n(4,5).$$

The distance between corresponding states of the orbits O((1,2)) and O((3,2)) is given by the norm of the vector $2^{-n}(2,0)$, which obviously goes to 0 as $n\to\infty$.

Problems

1. Analyze the dynamical system defined in $\mathbf{R}^2$ by the matrix

$$M = \begin{pmatrix} 2^{-1} & 0 \\ 2 & 3 \end{pmatrix}.$$

In particular, find the eigenvalues and eigenvectors of M, and the stable and unstable subspaces of M.

2. Let M be the matrix of Problem 1 and consider the vector (1,2). Notice that the orbit starting from (1,2) goes to infinity. Compare this orbit with the one starting at (1.1,2), and with the one starting at (6,–2). Make some comments on the dynamics of these three orbits.

3. * Let

$$M = \begin{pmatrix} 2^{-1} & 0 & 2 \\ 0 & 2 & 0 \\ 0 & 1 & 3 \end{pmatrix}.$$

Study the dynamical system in $\mathbf{R}^3$ governed by the matrix M. Determine the stable and unstable subspaces of M.

4. Let M be as in Problem 1. Consider the dynamical system in $\mathbf{R}^2$ governed by the map $A\mathbf{x}=(I-M)\mathbf{x}$. Find the eigenvalues and eigenvectors of A and the stable and unstable subspaces of A.

5. * Let

$$M = \begin{pmatrix} 2 & 0 & 1 \\ 0 & 2^{-1} & 4^{-1} \\ 0 & -1 & 2^{-1} \end{pmatrix}.$$

Verify that M has one real and two complex conjugate eigenvalues. Find an eigenvector corresponding to the real eigenvalue and the two-dimensional invariant subspace corresponding to the pair of complex conjugate eigenvalues. Find the stable and unstable subspaces of M.

6. * Let

$$M = \begin{pmatrix} 9/8 & -3 & -7/8 \\ 3/2 & 2 & -3/2 \\ -7/8 & 3 & 9/8 \end{pmatrix}.$$

Verify that the eigenvalues of M are 1/4 and $2\pm3i$. Find an eigenvector corresponding to the real eigenvalue and the two-dimensional invariant subspace corresponding to the pair of complex conjugate eigenvalues. Find the stable and unstable subspaces of M. Find the form of the orbit of (1,1,1).

Section 5. **EIGENVALUES WITH MODULUS 1**(*)

We now extend the analysis of the previous sections to include those cases when some of the eigenvalues have modulus 1. For organizational purposes we shall investigate separately the following situations.

- There is an eigenvalue equal to 1.
- There is an eigenvalue equal to –1.
- There is a pair of complex conjugate eigenvalues with modulus 1.

Before studying each case we observe that by the SDT we can focus our attention on the action of M in the generalized eigenspaces of (I–M), (I+M) and in the invariant subspaces corresponding to each pair of complex conjugate eigenvalues with modulus 1. We shall always denote by

- $\mathbf{X}_1$, the (generalized) eigenspace corresponding to the eigenvalue 1.
- $\mathbf{X}_{-1}$, the (generalized) eigenspace corresponding to the eigenvalue –1.
- $\mathbf{X}_C$, the 2k-dimensional invariant subspace corresponding to $s\pm it$, with k denoting the algebraic multiplicity of s+it.

***The action of* M *on* $\mathbf{X}_1$**

Assume first that $r_1=1$ is semisimple with multiplicity k and let $\mathbf{u}_1,\mathbf{u}_2,...,\mathbf{u}_k$ be k linearly independent eigenvectors such that Ker(I–M)=Span$\{\mathbf{u}_1, \mathbf{u}_2,...,\mathbf{u}_k\}=\mathbf{X}_1$. Every vector $\mathbf{u}\in\mathbf{X}_1$ can be written as $\mathbf{u}=a_1\mathbf{u}_1+...+a_k\mathbf{u}_k$ for some constants $a_1,...,a_k$. Hence,

$$M\mathbf{u} = a_1M\mathbf{u}_1 + ... + a_kM\mathbf{u}_k = a_1\mathbf{u}_1 + ... + a_k\mathbf{u}_k = \mathbf{u}.$$

Consequently, M acts like the identity on $\mathbf{X}_1$. Every vector of $\mathbf{X}_1$ is a stationary state of the dynamical system defined by M.

Example 4.5.1 Let

$$M = \begin{pmatrix} 2^{-1} & 2 \\ 0 & 1 \end{pmatrix}.$$

The eigenvalues of M are 2^{-1} and 1. An eigenvector corresponding to 2^{-1} is $\mathbf{e}_1=(1,0)$. The vector $\mathbf{v}=(4,1)$ is an eigenvector corresponding to 1. Every vector of $\mathbf{R}^2$ can be written as a linear combination of $\mathbf{e}_1$ and $\mathbf{v}$, $\mathbf{x}=a\mathbf{e}_1+b\mathbf{v}$. Hence,

$$M^n\mathbf{x} = aM^n\mathbf{e}_1 + bM^n\mathbf{v} = a2^{-n}\mathbf{e}_1 + b\mathbf{v}.$$

We see that as $n\to\infty$ the trajectory starting from $\mathbf{x}$ converges to $b\mathbf{v}$. For example, given

$$\mathbf{x}_0 = (1,1) = -3(1,0) + (4,1),$$

we have

$$\mathbf{x}_n = -\frac{3}{2^n}(1,0) + (4,1) \to (4,1) \text{ as } n\to\infty.$$

Example 4.5.2 Let

$$M = \begin{pmatrix} 1 & 0 & 1 \\ 0 & 1 & 1 \\ 0 & 0 & .5 \end{pmatrix}.$$

The eigenvalues of M are $r_1=r_2=1$, and $r_3=.5$. This last eigenvalue is simple, while 1 is semisimple with algebraic multiplicity 2. Two linearly independent eigenvectors corresponding to r=1 are $\mathbf{e}_1$ and $\mathbf{e}_2$. An eigenvector corresponding to $r_3=.5$ is $\mathbf{s}=(2,2,-1)$. Therefore, M is the identity in the two-dimensional subspace $\mathbf{X}_1$ spanned by the two vectors $\mathbf{e}_1$ and $\mathbf{e}_2$. Every vector $\mathbf{x}_0\in\mathbf{R}^3$ can be written in a unique manner as $\mathbf{x}_0=a\mathbf{e}_1+ b\mathbf{e}_2+c\mathbf{s}$. Consequently,

$$\mathbf{x}_n = M^n\mathbf{x}_0 = a\mathbf{e}_1 + b\mathbf{e}_2 + 2^{-n}c\mathbf{s} \to a\mathbf{e}_1 + b\mathbf{e}_2 \text{ as } n \to \infty.$$

To see the importance of the assumption that the algebraic and geometric multiplicity be the same we now examine an example in which this condition is violated.

Example 4.5.3 Let

$$M = \begin{pmatrix} 1 & 1 & 0 \\ 0 & 1 & 1 \\ 0 & 0 & .5 \end{pmatrix}.$$

The eigenvalues of M are 1 and .5. The second is simple, the first has algebraic multiplicity 2 and geometric multiplicity 1. The generalized kernel $\mathrm{Ker}(I-M)^2$ is spanned by the vectors $\mathbf{e}_1=(1,0,0)$, which is also an eigenvector, and $\mathbf{e}_2=(0,1,0)$. Hence, $\mathbf{X}_1=\mathrm{Span}\{\mathbf{e}_1,\mathbf{e}_2\}$. Consider the vector $\mathbf{x}_0=(1,1,0)$, which belongs to $\mathbf{X}_1$ since $\mathbf{x}_0=\mathbf{e}_1+\mathbf{e}_2$, but is not in Ker(I–M) (not an eigenvector). Then $M\mathbf{x}_0=(2,1,0)$, and in general, $M^n(1,1,0)= (n+1,1,0)$. The orbit $O(\mathbf{x}_0)$ goes to infinity.

Let us discuss the situation theoretically. Assume that 1 is not semisimple. Then $\mathbf{X}_1=\mathrm{Ker}(I-M)^p$, where p is the algebraic multiplicity of 1. There exists at least one vector $\mathbf{v}\in \mathrm{Ker}(I-M)^2$, $\mathbf{v}\notin \mathrm{Ker}(I-M)$. Then $\mathbf{v}-M\mathbf{v}\neq\mathbf{0}$ and $(I-M)^2\mathbf{v}=\mathbf{0}$. Thus,

$$M^2\mathbf{v} - 2M\mathbf{v} + \mathbf{v} = \mathbf{0}, \quad \text{i.e.} \quad M^2\mathbf{v} = 2M\mathbf{v} - \mathbf{v} = 2(M\mathbf{v} - \mathbf{v}) + \mathbf{v}.$$

Therefore,

$$M^3\mathbf{v} = 2M^2\mathbf{v} - M\mathbf{v} = 4M\mathbf{v} - 2\mathbf{v} - M\mathbf{v} = 3(M\mathbf{v} - \mathbf{v}) + \mathbf{v}.$$

In general, we obtain

$$M^n\mathbf{v} = n(M\mathbf{v} - \mathbf{v}) + \mathbf{v}.$$

Since $M\mathbf{v}-\mathbf{v}\neq\mathbf{0}$, we see that $\|M^n\mathbf{v}\|\to\infty$ as $n\to\infty$.

This result suggests the following conclusion that we state without proof. When 1 is not semisimple, we consider the vectors that span the generalized kernel, $\mathrm{Ker}(I-M)^p=\mathbf{X}_1$, and divide them into two groups: those that are eigenvectors, and those that are not eigenvectors. The first group spans Ker(I–M) and the second spans another subspace Z of $\mathbf{X}_1$. We now write $\mathbf{X}_1=\mathrm{Ker}(I-M)\oplus Z$. Every vector in Ker(I–M) is a stationary state of our system, while the orbit of every vector in **Z** goes to infinity. Consequently, the orbit of every vector in $\mathbf{X}_1$ which is not in Ker(I–M) will go to infinity. We offer here another example to corroborate this conclusion.

Example 4.5.4 Let

$$M = \begin{pmatrix} .25 & 1 & .25 & 0 \\ -.5 & 2 & .5 & 0 \\ .25 & -1 & .25 & 0 \\ -1 & 1 & 1 & 1 \end{pmatrix}.$$

The eigenvalues of M are 1 and .5. The first has algebraic multiplicity 3 and geometric multiplicity 1.

An eigenvector corresponding to 1 is obviously $\mathbf{u}=(0,0,0,1)=\mathbf{e}_4$. Since

$$(I - M)^3 = \begin{pmatrix} .0625 & 0 & .0625 & 0 \\ 0 & 0 & 0 & 0 \\ .0625 & 0 & .0625 & 0 \\ 0 & 0 & 0 & 0 \end{pmatrix},$$

the generalized eigenspace is spanned by $\mathbf{e}_4$, $\mathbf{e}_2=(0,1,0,0)$ and $\mathbf{u}_3=(1,0,-1,0)$. Let $\mathbf{x}_0=\mathbf{e}_4+2\mathbf{e}_2-\mathbf{u}_3=(-1,2,1,1)$. Figure 4.5.1 shows how the logarithm of the norm of $\mathbf{x}_n$ grows for n=0,...,60. We see that the orbit of $\mathbf{x}_0$ goes to infinity.

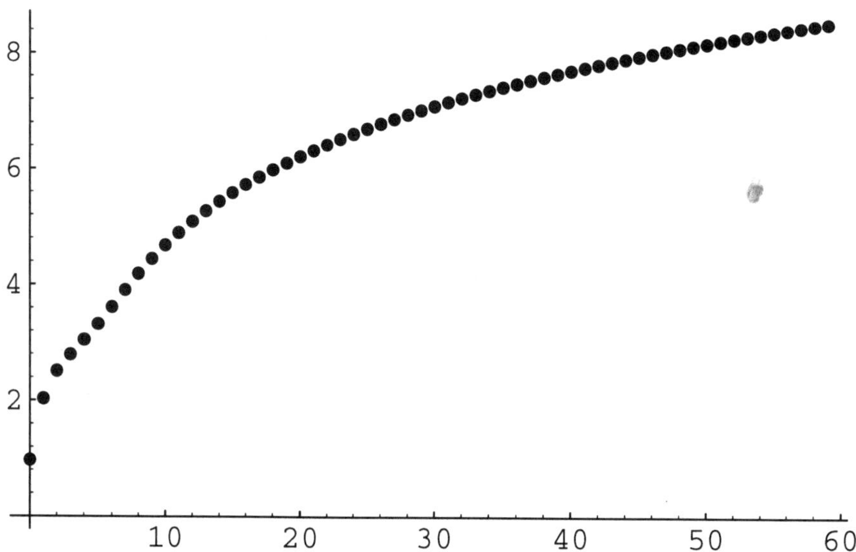

Fig. 4.5.1 On the horizontal axis we have plotted the iteration number n and on the vertical axis the natural logarithm of $\|\mathbf{x}_n\|$, with $\mathbf{x}_0=(-1,2,1,1)$.

The action of M *on* $\mathbf{X}_{-1}$

Assume that $r_1=-1$ is semisimple with multiplicity k and let $\mathbf{u}_1,\mathbf{u}_2,...,\mathbf{u}_k$ be k linearly independent eigenvectors such that $\mathrm{Ker}(I+M)=\mathrm{Span}\{\mathbf{u}_1,\mathbf{u}_2,...,\mathbf{u}_k\}=\mathbf{X}_{-1}$. Every vector $\mathbf{u}\in\mathbf{X}_{-1}$ can be written as $\mathbf{u}=a_1\mathbf{u}_1+...+a_k\mathbf{u}_k$ for some constants $a_1,...,a_k$. Hence,

$$\mathbf{Mu} = a_1 M\mathbf{u}_1 + \ldots + a_k M\mathbf{u}_k = -a_1\mathbf{u}_1 - \ldots - a_k\mathbf{u}_k = -\mathbf{u}$$

$$M(M(\mathbf{u})) = M(-\mathbf{u}) = \mathbf{u}.$$

Thus, the orbit starting from every element (except $\mathbf{0}$) of $\mathbf{X}_{-1}$ is periodic of period 2.

Example 4.5.5 Let

$$M = \begin{pmatrix} -1 & 1 \\ 0 & 2^{-1} \end{pmatrix}.$$

The eigenvalues of M are -1 and 2^{-1} with corresponding eigenvectors $\mathbf{e}_1=(1,0)$ and $\mathbf{v}=(2,3)$, respectively. Every vector $\mathbf{y} \in \mathbf{R}^2$ can be written as a linear combination of $\mathbf{e}_1$ and $\mathbf{v}$, $\mathbf{y}=a\mathbf{e}_1+b\mathbf{v}$. Hence,

$$M^n\mathbf{y} = aM^n\mathbf{e}_1 + bM^n\mathbf{v} = a(-1)^n\mathbf{e}_1 + b2^{-n}\mathbf{v}.$$

As $n \to \infty$ the trajectory starting from $\mathbf{y}$ converges to the periodic orbit of period 2,

$$\mathbf{x}_0 = a\mathbf{e}_1, \quad \mathbf{x}_1 = -a\mathbf{e}_1.$$

For example, with

$$\mathbf{y} = (1,1) = 3^{-1}(1,0) + 3^{-1}(2,3),$$

we have $\mathbf{y}_{2n} \to 3^{-1}(1,0)$ and $\mathbf{y}_{2n+1} \to -3^{-1}(1,0)$ as $n \to \infty$.

When $r=-1$ is not semisimple we encounter the same situation found previously for the eigenvalue 1. The following example illustrates the behavior of M on $\mathbf{X}_{-1}$ in this case.

Example 4.5.6 Let

$$M = \begin{pmatrix} -.75 & 1 & 1.25 & 0 \\ -.5 & 0 & .5 & 0 \\ 1.25 & -1 & -.75 & 0 \\ -1 & 1 & 1 & -1 \end{pmatrix}.$$

The eigenvalues of M are -1 and .5. The first has algebraic multiplicity 3 and geometric multiplicity 1. An eigenvector corresponding to -1 is obviously $\mathbf{u}_1=(0,0,0,1)=\mathbf{e}_4$.
The matrix $-(I+M)^3$ is

$$-\begin{pmatrix} 1.6875 & 0 & 1.6875 & 0 \\ 0 & 0 & 0 & 0 \\ 1.6875 & 0 & 1.6875 & 0 \\ 0 & 0 & 0 & 0 \end{pmatrix}$$

whose kernel is spanned by the vectors $\mathbf{e}_4$, $\mathbf{e}_2$=(0,1,0,0) and $\mathbf{u}_3$=(1,0,–1,0).

Lct $\mathbf{x}_0=\mathbf{e}_4-2\mathbf{e}_2+\mathbf{u}_3$=(1,–2,–1,1). The graph (see Fig. 4.5.2) shows how the logarithm of the norm of $\mathbf{x}_n$ grows for n=0,...,60. The orbit goes to infinity. For comparison purposes look at Example 4.5.4 and notice that the behavior is qualitatively the same.

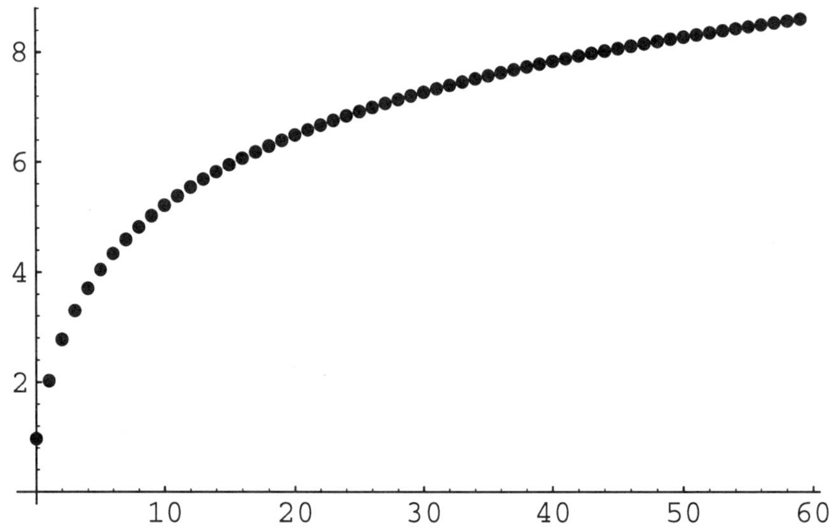

Fig. 4.5.2 On the horizontal axis we have plotted the iteration number n and on the vertical axis the natural logarithm of $\|\mathbf{x}_n\|$, with $\mathbf{x}_0$=(1,–2,–1,1).

***The action of* M *on* $\mathbf{X}_C$**

We now consider the case when M has a pair of complex conjugate eigenvalues with modulus 1, say r_1=s+it and r_2=s–it with $s^2+t^2=1$. Let us first assume that the two eigenvalues are simple. There is an invariant subspace $\mathbf{X}_C$ of dimension 2 spanned by two linearly independent vectors **u** and **v** found by solving the equations

$$M\mathbf{w} = (s + it)\mathbf{w} \quad \text{with} \quad \mathbf{w} = \mathbf{u} + i\mathbf{v}.$$

We also have [see (4.3.12)]

$$M\mathbf{u} = s\mathbf{u} - t\mathbf{v},\ M\mathbf{v} = t\mathbf{u} + s\mathbf{v}.$$

Since $1=s^2+t^2$ we can find $\theta\in[0,2\pi)$ such that s=cosθ and t=sinθ. Consequently, we can write

$$M\mathbf{u} = \cos\theta\ \mathbf{u} - \sin\theta\ \mathbf{v}, \qquad M\mathbf{v} = \sin\theta\ \mathbf{u} + \cos\theta\ \mathbf{v}$$

and [see (4.3.14)]

$$M^n\mathbf{u} = \cos n\theta\ \mathbf{u} - \sin n\theta\ \mathbf{v}, \quad M^n\mathbf{v} = \sin n\theta\ \mathbf{u} + \cos n\theta\ \mathbf{v}.$$

Given a vector $\mathbf{y} \in \mathbf{Y}$, $\mathbf{y}=c_1\mathbf{u}+c_2\mathbf{v}$ we have

$$M^n\mathbf{y} = r^n(c_1\cos n\theta + c_2\sin n\theta, -c_1\sin n\theta + c_2\cos n\theta)_B.$$

As mentioned in Section 4, when $\theta/(2\pi)=m/p$, where m and p are relatively prime integers, all orbits of M in $\mathbf{X}_C$ are periodic of period p.

Example 4.5.7 Let

$$M = \begin{pmatrix} \sqrt{2} & -\sqrt{2} \\ 1/\sqrt{2} & 0 \end{pmatrix}.$$

The eigenvalues of M are $2^{-.5}(1\pm i)$. A pair $\mathbf{u}$ and $\mathbf{v}$ such that $M\mathbf{w}=2^{-.5}(1+i)\mathbf{w}$ with $\mathbf{w}=\mathbf{u}+i\mathbf{v}$ is given by $\mathbf{u}=(1,1)$ and $\mathbf{e}_1=(1,0)$. The angle θ is $\pi/4$. Therefore, all non-stationary orbits of the dynamical system governed by M are periodic of period 8. For example, let $\mathbf{x}_0=(0,1)$. Then $\mathbf{x}_0=\mathbf{u}-\mathbf{e}_1$ and

$$M^n\mathbf{x}_0 = (\cos n\theta - \sin n\theta)\mathbf{u} - (\sin n\theta + \cos n\theta)\mathbf{e}_1 = -(-\cos n\theta + \sin n\theta, \sin n\theta + \cos n\theta)_B.$$

Substituting θ with $\pi/4$, $\mathbf{u}$ with (1,1) and $\mathbf{e}_1$ with (1,0), we arrive at

$$M^n\mathbf{x}_0 = \left(-2\sin\frac{n\pi}{4}, \cos\frac{n\pi}{4} - \sin\frac{n\pi}{4}\right).$$

A graphical representation of the orbit (see Fig. 4.5.3) clearly shows its periodicity.

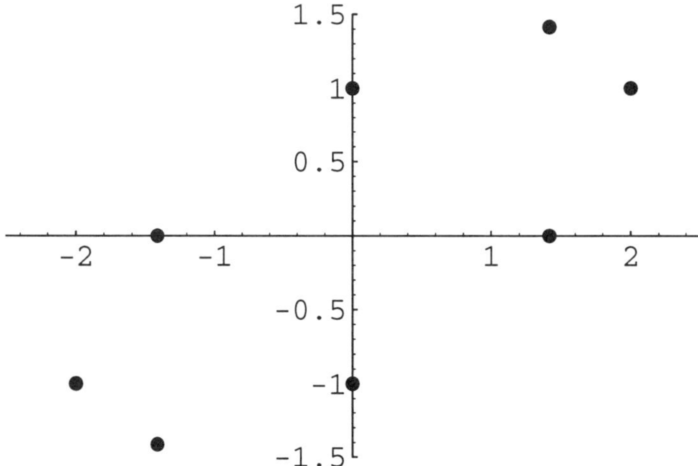

Fig. 4.5.3 Orbit of (0,1) of the dynamical system governed by M.

Notice that M is not a rotation of the plane since the angle between $\mathbf{x}_0=(0,1)$ and $\mathbf{x}_1=(-\sqrt{2},0)$ is $\pi/2$, while the angle between $\mathbf{x}_3=(-\sqrt{2},-\sqrt{2})$ and $\mathbf{x}_4=(0,-1)$ is $\pi/4$.

The previous analysis can be extended to the case when r_1=s+it and r_2=s–it are semisimple, with multiplicity k. In this case $\mathbf{X}_C$ can be expressed as a direct sum of k two-dimensional invariant subspaces $\mathbf{X}_{C1},...,\mathbf{X}_{Ck}$ spanned by $\{\mathbf{u}_1,\mathbf{v}_1\},...,$ $\{\mathbf{u}_k,\mathbf{v}_k\}$. In $\mathbf{X}_{Ci}$ the action of M is the same as the action analyzed in the previous case of simple eigenvalues. Just replace **u** and **v** with $\mathbf{u}_i$ and $\mathbf{v}_i$. When the two eigenvalues are not semisimple we have the same problems found before for the action on $\mathbf{X}_1$ and $\mathbf{X}_{-1}$. We illustrate both situations in the following two examples.

Example 4.5.8 Let

$$M = \begin{pmatrix} r & r & -r & 0 \\ -r & 0 & 0 & r \\ 0 & -r & r & r \\ -r & -r & -r & 1/r \end{pmatrix}$$

with $r=1/\sqrt{2}$. The eigenvalues of M are r(1±i). The eigenvalues are semisimple with multiplicity 2. The angle θ is π/4.

A pair $\mathbf{u}_1,\mathbf{v}_1$ (with r=1+i) is given by $\mathbf{u}_1$=(0,1,0,1) and $\mathbf{v}_1$=(–1,0,0,0}. The two-dimensional invariant subspace is $\mathbf{X}_{C1}$={(–b,a,0,a) : a,b∈**R**}. A second pair is given by $\mathbf{u}_2$=(–1,0,1,0) and $\mathbf{v}_2$=(1,–1,0,0) with associated invariant subspace $\mathbf{X}_{C2}$ ={(–c+d,–d,c,0): c,d∈**R**}. We obviously have $\mathbf{X}_C=\mathbf{R}^4$. Every vector $\mathbf{x}_0$ can be written as $\mathbf{x}_0=\mathbf{y}_0+\mathbf{z}_0$ with $\mathbf{y}_0\in\mathbf{X}_{C1}$ and $\mathbf{z}_0\in\mathbf{X}_{C2}$. Hence, $\mathbf{x}_n=\mathbf{y}_n+\mathbf{z}_n$, with $\mathbf{y}_n$ and $\mathbf{z}_n$ computed as in Example 4.4.7. We cannot provide a graphical representation of the orbits, which are in $\mathbf{R}^4$. Therefore, we produce a numerical output. Assume that the initial condition is

$$\mathbf{x}_0 = \mathbf{u}_1 + \mathbf{v}_1 + \mathbf{u}_2 + \mathbf{v}_2 = (-1,0,1,1) = \mathbf{y}_0 + \mathbf{z}_0$$

with $\mathbf{y}_0$=(–1,1,0,1) and $\mathbf{z}_0$=(0,–1,1,0). We have

$$\mathbf{y}_n = M^n\mathbf{u}_1 + M^n\mathbf{v}_1 = \cos n\theta\,\mathbf{u}_1 - \sin n\theta\,\mathbf{v}_1 + \sin n\theta\,\mathbf{u}_1 + \cos n\theta\,\mathbf{v}_1$$

$$= (\cos n\theta + \sin n\theta)\mathbf{u}_1 + (\cos n\theta - \sin n\theta)\mathbf{v}_1$$

$$= (-\cos n\theta + \sin n\theta,\ \cos n\theta + \sin n\theta,\ 0,\ \cos n\theta + \sin n\theta);$$

$$\mathbf{z}_n = M^n\mathbf{u}_2 + M^n\mathbf{v}_2 = (\cos n\theta + \sin n\theta)\mathbf{u}_2 + (\cos n\theta - \sin n\theta)\mathbf{v}_2$$

$$= (-2\sin n\theta,\ -\cos n\theta + \sin n\theta,\ \cos n\theta + \sin n\theta,\ 0).$$

Hence, $\mathbf{x}_n$=(–cosnθ–sinnθ, 2sinnθ, cosnθ+sinnθ, cosnθ+sinnθ). Since θ=π/4, we obtain

$$\begin{array}{llll} (-1,0,1,1) & = \mathbf{x}_0 & (1,0,-1,-1) & = \mathbf{x}_4 \\ (1/r)(-1,1,1,1) & = \mathbf{x}_1 & (1/r)(1,-1,-1,-1) & = \mathbf{x}_5 \\ (-1,2,1,1) & = \mathbf{x}_2 & (1,-2,-1,-1) & = \mathbf{x}_6 \\ (0,1/r,0,0) & = \mathbf{x}_3 & (0,-1/r,0,0) & = \mathbf{x}_7 \\ (-1,0,1,1) & = \mathbf{x}_8 = \mathbf{x}_0. & & \end{array}$$

Example 4.5.9 Let

$$M = \begin{pmatrix} -1 & 2+1/r & 1 & 2+r \\ -r & 2r & r & 2r \\ -1+r & 2-r & 1 & 2-r \\ 0 & -1/r & -1/r & 0 \end{pmatrix},$$

with $r=(2)^{.5}$. The eigenvalues of M are $(1/(2)^{.5})(1\pm i)$. They are not semisimple, since their geometric multiplicity is 1. In Fig. 4.5.4 we have plotted the iteration number n on the horizontal axis for n=140,...,180 and the norm of $\mathbf{x}_n$ on the vertical axis with $\mathbf{x}_0=(1,-1,1,1)$. We clearly see that the orbit is going to infinity.

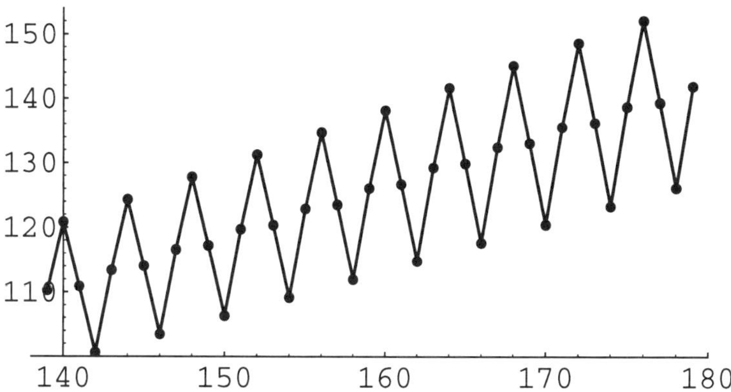

Fig. 4.5.4 Growth of the norm of $\mathbf{x}_n$ for n=120,...,182. See Example 4.5.9.

Problems

1. Let

$$M = \begin{pmatrix} 1 & 0 \\ 2 & 2^{-1} \end{pmatrix}.$$

Find the eigenvalues and eigenvectors of M and study the dynamical system governed by M.

2. * Let

$$M = \begin{pmatrix} 1 & 0 & 1 \\ 0 & 2^{-1} & 4^{-1} \\ 0 & -1 & 2^{-1} \end{pmatrix}.$$

Find the eigenvalues and eigenvectors of M and study the dynamical system governed by M.

3. * Let

$$M = \begin{pmatrix} 1 & 0 & 1 \\ 0 & 1 & 0 \\ 0 & 0 & 2^{-1} \end{pmatrix}.$$

Find the eigenvalues and eigenvectors of M and study the dynamical system governed by M.

4. * Let

$$M = \begin{pmatrix} 2 & 0 & 1 \\ 0 & 1 & 0 \\ 0 & 0 & 1 \end{pmatrix}.$$

Find the eigenvalues and eigenvectors of M and study the dynamical system governed by M.

5. Let

$$M = \begin{pmatrix} -1 & 0 \\ 1 & 1 \end{pmatrix}.$$

Find the eigenvalues and eigenvectors of M and study the dynamical system governed by M.

6. Let

$$M = \begin{pmatrix} -1 & 0 \\ 1 & -1 \end{pmatrix}.$$

Verify that -1 is the only eigenvalue of M and that its algebraic and geometric multiplicity are not the same. Study the dynamical system generated by the matrix M. Which orbits of the system are unbounded?

7. * Let

$$M = \begin{pmatrix} 3/8 & 11/8 & 5/4 \\ -3/8 & 5/8 & 3/4 \\ -1/8 & 7/8 & 5/4 \end{pmatrix}.$$

Verify that the eigenvalues of M are 1/4 and 1. Determine that 1 is not semisimple. Find an eigenvector corresponding to 1/4 and the generalized kernel of (I–M). Which orbits of the system governed by M go to infinity?

8. * Let

$$M = \begin{pmatrix} 1 & .5+r & -.5(1+r) & -.5r \\ 0 & 2+.5r & -.5(3+r) & -1.5 \\ 0 & 1.5(1+r) & -1-r & -.5(3+r) \\ 0 & 1.5-r & .5(-3+r) & -1+.5r \end{pmatrix}$$

where $r=\sqrt{3}$. Verify that the eigenvalues of M are $1, -1, .5(1\pm i\sqrt{3})$. Verify that $\mathbf{e}_1$ is an eigenvector corresponding to 1 and $\mathbf{w}=(0,1,1,1)$ is an eigenvector corresponding to -1. Verify also that a pair of vectors corresponding to the complex conjugate eigenvalues are $\mathbf{u}=(1,1,2,-1)$ and $\mathbf{v}=(-1,1,0,1)$. Write the dynamical system $\mathbf{x}_{n+1}=M\mathbf{x}_n$ using the eigenspaces and the invariant subspace of M.

Section 6. **AFFINE SYSTEMS**

When 1 is not an eigenvalue

In all those cases when 1 is not an eigenvalue of M, the results obtained in the previous sections for dynamical systems of the form

$$\mathbf{x}_{n+1} = M\mathbf{x}_n \tag{4.6.1}$$

can be extended to affine systems

$$\mathbf{x}_{n+1} = M\mathbf{x}_n + \mathbf{b} = N(\mathbf{x}_n). \tag{4.6.2}$$

In fact, when $\det(I-M)\neq 0$, the equation $\mathbf{x}-M\mathbf{x}=\mathbf{b}$ has the only solution $\mathbf{x}_s=(I-M)^{-1}\mathbf{b}$. Given a point $\mathbf{x}_0\in\mathbf{R}^q$, we now have from (4.6.2)

$$\mathbf{x}_{n+1} = M\mathbf{x}_n + \mathbf{b} = M\mathbf{x}_n - M\mathbf{x}_s + M\mathbf{x}_s + \mathbf{b} = M(\mathbf{x}_n - \mathbf{x}_s) + \mathbf{x}_s$$

since $M\mathbf{x}_s+\mathbf{b}=\mathbf{x}_s$. Therefore,

$$\mathbf{x}_{n+1} - \mathbf{x}_s = M(\mathbf{x}_n - \mathbf{x}_s). \tag{4.6.3}$$

Consequently, with the change of variable $\mathbf{y}=\phi(\mathbf{x})=\mathbf{x}-\mathbf{x}_s$, we obtain

$$\mathbf{y}_{n+1} = M\mathbf{y}_n. \tag{4.6.4}$$

We recognize that ϕ is a conjugacy between (4.6.2) and (4.6.4), i.e., $\phi N=M\phi$. Hence, by applying to (4.6.4) the theory developed in the previous sections we can derive the information we need in (4.6.2).

- In particular, if $\rho(M)<1$, then all orbits of (4.6.4) converge to $\mathbf{0}$. Consequently, all orbits of (4.6.2) converge to $\mathbf{x}_s$.
- When all eigenvalues of M have modulus larger than 1, all nonstationary orbits of (4.6.4) go to infinity. Consequently, all orbits of N, except the one starting at $\mathbf{x}_s$, go to infinity.
- In the case when some eigenvalues of M have modulus smaller than 1 and the others have modulus larger than 1, the stable and unstable subspaces **S** and **U** are replaced by $\mathbf{T}=\mathbf{x}_s+\mathbf{S}$ and $\mathbf{V}=\mathbf{x}_s+\mathbf{U}$. In other words, we translate **S** and **U** by adding to each of their vectors the fixed vector $\mathbf{x}_s$. For every element in **T** the orbit goes to $\mathbf{x}_s$, while for every other element of $\mathbf{R}^q$ the orbit goes to infinity. **T** and **V** are called **stable and unstable linear varieties** of the system.

The following examples illustrate the various situations.

Example 4.6.1 Consider the system $\mathbf{x}_{n+1}=M\mathbf{x}_n+(1,2)=N(\mathbf{x}_n)$, where

$$M = \begin{pmatrix} 1 & -1 \\ .5 & -1 \end{pmatrix}.$$

The eigenvalues of M are $(.5)^{.5}$ and $-(.5)^{.5}$. The only solution of the system $\mathbf{x}-M\mathbf{x}=(1,2)$ is $\mathbf{x}_s=(0,1)$. Hence, every orbit of the dynamical system $\mathbf{x}_{n+1}=N(\mathbf{x}_n)$ converges to (0,1).

Example 4.6.2 Let

$$M = \begin{pmatrix} .5 & 0 \\ 3 & 2 \end{pmatrix}, \qquad \mathbf{b} = (1,-8).$$

For the dynamical system $\mathbf{x}_{n+1}=N(\mathbf{x}_n)=M\mathbf{x}_n+(1,-8)$ we obtain $\mathbf{x}_s=(2,2)$. The eigenvalues of M are .5 and 2 with corresponding eigenvectors $\mathbf{s}=(1,-2)$ and $\mathbf{e}_2=(0,1)$. Thus $\mathbf{S}=\text{Span}\{(1,-2)\}$ is the stable subspace of M and $\mathbf{U}=\text{Span}\{(0,1)\}$ is the unstable subspace. Accordingly,

$$\mathbf{T} = \{(2 + u, 2 - 2u) : u\in \mathbf{R}\}, \qquad \mathbf{V} = \{(2, v + 2): v\in \mathbf{R}\}.$$

All orbits starting at points of $\mathbf{T}$ converge to (2,2), while orbits starting at points of $\mathbf{V}$ diverge to infinity. For example, let us take u=–2 and consider the initial condition $\mathbf{x}_0=(0,6)\in \mathbf{T}$. With the change of variable $\mathbf{y}=\mathbf{x}-(2,2)$ we obtain $\mathbf{y}_0=(-2,4)=-2\mathbf{s}$. Hence, $\mathbf{y}_n= M^n\mathbf{y}_0=-M^n2\mathbf{s}=-2M^n\mathbf{s}=-2(.5)^n\mathbf{s}=(.5)^n(-2,4)$. The orbit of $\mathbf{x}_0$ is

$$\mathbf{x}_n = \mathbf{y}_n+\mathbf{x}_s = (.5)^n(-2,4) + (2,2)\rightarrow(2,2) \text{ as } n\rightarrow\infty.$$

Example 4.6.3 Let (see Example 4.4.2)

$$M = \begin{pmatrix} 1 & 2 & 0 \\ -2 & 1 & 1 \\ 0 & 0 & .5 \end{pmatrix}.$$

The eigenvalues of M are $r_1=1/2$, $r_2=1+2i$, and $r_3=1-2i$. An eigenvector corresponding to 1/2 is $\mathbf{s}=(8,-2,17)$. A pair $\mathbf{u}$, $\mathbf{v}$ corresponding to the complex conjugate eigenvalues $1\pm2i$ is $\mathbf{u}=(1,0,0)=\mathbf{e}_1$ and $\mathbf{v}=(0,1,0)=\mathbf{e}_2$. Hence, $\mathbf{S}=\text{Span}\{(8,-2,17)\}$, $\mathbf{U}=\text{Span}\{\mathbf{e}_1,\mathbf{e}_2\}$. Consider the dynamical system

$$\mathbf{x}_{n+1} = M\mathbf{x}_n + (2,-2,2) = N\mathbf{x}_n.$$

We find $\mathbf{x}_s=(1,-1,4)$. Consequently,

$$\mathbf{T} = \mathbf{x}_s+\mathbf{S} = \{(1 + 8a, -1 - 2a, 4 + 17a) : a\in \mathbf{R}\}$$

and

$$\mathbf{V} = \mathbf{x}_s+\mathbf{U} = \{(1 + b, -1 + c, 4) : b,c\in \mathbf{R}\}.$$

Let us take an initial condition in **T**. For example, with a=1 we get $\mathbf{x}_0$=(9,–3,21). With the change of variable $\mathbf{y}=\mathbf{x}-\mathbf{x}_s$ we obtain $\mathbf{y}_0$=(9,–3,21)–(1,–1,4)=(8,–2,17)=**s**. Thus $\mathbf{y}_n= M^n\mathbf{y}_0=M^n\mathbf{s}=2^{-n}(8,-2,17)$. Therefore, the orbit of $\mathbf{x}_0$ is

$$\mathbf{x}_n = \mathbf{y}_n + \mathbf{x}_s = (1,-1,4) + 2^{-n}\,(8,-2,17)\rightarrow\mathbf{x}_s \text{ as } n\rightarrow\infty.$$

Example 4.6.4 Let

$$M = \begin{pmatrix} -2.5 & 1.5 \\ -3 & 2 \end{pmatrix}.$$

The eigenvalues of M are –1 and .5 with corresponding eigenvectors **u**=(1,1) and **v**=(1,2). Consider the system $\mathbf{x}_{n+1}=M\mathbf{x}_n+(-2,2)$. Solving **x**=M**x**+(–2,2) we obtain $\mathbf{x}_s=(5,13)=-3\mathbf{u}+8\mathbf{v}$. The given system is conjugate to the system $\mathbf{y}_{n+1}=M\mathbf{y}_n$ and the conjugacy is $\mathbf{y}=\phi(\mathbf{x})=\mathbf{x}-\mathbf{x}_s$. More precisely, $\phi N=M\phi$. Assume that $\mathbf{y}_0=a\mathbf{u}+b\mathbf{v}$. Then $O(\mathbf{y}_0)$ converges to the periodic orbit of period 2 {a**u**,–a**u**}. Consequently, the orbit $\mathbf{x}_{n+1}=M\mathbf{x}_n+(-2,2)$ converges to the periodic orbit of period 2 $\{\mathbf{x}_s+a\mathbf{u},\mathbf{x}_s-a\mathbf{u}\}$. For example, choose $\mathbf{x}_0=(2,-1)=5\mathbf{u}-3\mathbf{v}$. Then

$$\mathbf{y}_0 = \mathbf{x}_0 - \mathbf{x}_s = 5\mathbf{u} - 3\mathbf{v} - (-3\mathbf{u} + 8\mathbf{v}) = 8\mathbf{u} - 11\mathbf{v}.$$

Hence, the orbit of $\mathbf{y}_0$ converges to {8**u**,–8**u**} and the orbit of $\mathbf{x}_0$ converges to $\{\mathbf{x}_s+8\mathbf{u},\mathbf{x}_s-8\mathbf{u}=\{(13,21),(-3,5)\}$. In fact, using Mathematica, after 20 iterates we find that $\mathbf{x}_{18}$=(13, 20.9999), $\mathbf{x}_{19}$=(–3.00002,4.99996), and $\mathbf{x}_{20}$=(13,21).

When* 1 *is an eigenvalue (*)

When r=1 is an eigenvalue of M, the problem of finding a conjugacy between (4.6.2) and (4.6.1) is more complicated since the equation

$$\mathbf{x} - M\mathbf{x} = \mathbf{b} \tag{4.6.5}$$

may have infinitely many solutions or it may have none. From dimKer(I – M)≥1 we have

$$\dim\mathrm{Im}(I - M) \leq q - 1. \tag{4.6.6}$$

Consequently, (4.6.5) has infinitely many solutions when $\mathbf{b}\in \mathrm{Im}(I-M)$ and no solutions otherwise. We study both cases.

Assume that the eigenvalue r=1 is **semisimple**. Denote by $\mathbf{X}_1$, $\mathbf{X}_R$ the **invariant** subspaces corresponding, respectively, to r=1 and to the remaining eigenvalues of M. Every $\mathbf{x}\in \mathbf{R}^q$ can be written in a unique way as

$$\mathbf{x} = \mathbf{x}_1 + \mathbf{x}_R, \quad \mathbf{x}_1\in \mathbf{X}_1, \quad \mathbf{x}_R\in \mathbf{X}_R. \tag{4.6.7}$$

The equation **x**–M**x**=**b** takes the form

$$\mathbf{x}_1 + \mathbf{x}_R - M\mathbf{x}_1 - M\mathbf{x}_R = \mathbf{b}_1 + \mathbf{b}_R, \quad \mathbf{b}_1\in \mathbf{X}_1, \quad \mathbf{b}_R\in \mathbf{X}_R. \tag{4.6.8}$$

Since $M\mathbf{x}_1=\mathbf{x}_1$ (recall that M is the identity on $\mathbf{X}_1$) we obtain

$$\mathbf{x}_R - M\mathbf{x}_R = \mathbf{b}_1 + \mathbf{b}_R. \tag{4.6.9}$$

There are two possibilities: $\mathbf{b}_1 \neq \mathbf{0}$, $\mathbf{b}_1=\mathbf{0}$.

I. $\mathbf{b}_1 \neq \mathbf{0}$.

Equation (4.6.9) does not have any solution since $\mathbf{x}_R - M\mathbf{x}_R \in \mathbf{X}_R$, while $\mathbf{b}_1 \in \mathbf{X}_1$. No reduction of (4.6.2) to (4.6.1) is possible. We need to study (4.6.2) directly. Let $\mathbf{x}_0 = \mathbf{x}_{10} + \mathbf{x}_{R0}$ with $\mathbf{x}_{10} \in \mathbf{X}_1$ and $\mathbf{x}_{R0} \in \mathbf{X}_R$. Let us see what is happening in $\mathbf{X}_1$. Since $M\mathbf{x}_{10} = \mathbf{x}_{10}$ and $M\mathbf{b}_1=\mathbf{b}_1$ ($\mathbf{b}_1$ is an eigenvector) we obtain

$$\mathbf{x}_{11} = M\mathbf{x}_{10} + \mathbf{b}_1 = \mathbf{x}_{10} + \mathbf{b}_1, \mathbf{x}_{12} = M\mathbf{x}_{11} + \mathbf{b}_1 = \mathbf{x}_{10} + 2\mathbf{b}_1$$

and, in general,

$$\mathbf{x}_{1n} = \mathbf{x}_{10} + n\mathbf{b}_1.$$

The component $\mathbf{x}_{1n}$ of $\mathbf{x}_n$ goes to infinity and the other component, $\mathbf{x}_{Rn}$, cannot balance this trend since the two subspaces have only $\mathbf{0}$ in common. Consequently, every orbit of the system goes to infinity. The same conclusion holds when 1 is not semisimple and $\mathbf{b}_1$ is not an eigenvector.

Example 4.6.5 Let

$$M = \begin{pmatrix} 1 & 1 \\ 0 & .5 \end{pmatrix}, \quad \mathbf{b} = (3,1).$$

The eigenvalues of M are 1 and .5 with corresponding eigenvectors (1,0) and (2,–1). Since $\mathbf{b}$=5(1,0)–(2,–1) we have $\mathbf{b}_1 \neq \mathbf{0}$. Hence, every orbit of the system

$$\mathbf{x}_{n+1} = M\mathbf{x}_n + \mathbf{b}$$

goes to infinity.

II. $\mathbf{b}_1 = \mathbf{0}$ ($\mathbf{b} = \mathbf{b}_R$).

Equation (4.6.9) becomes

$$\mathbf{x}_R - M\mathbf{x}_R = \mathbf{b} = \mathbf{b}_R. \tag{4.6.10}$$

Since in $\mathbf{X}_R$ there are no eigenvalues of M equal to 1, this equation has a unique solution $\mathbf{z}_R$. All solutions of (4.6.10) are obtained by adding to $\mathbf{z}_R$ any vector of $\mathbf{X}_1$. In particular, it is convenient to choose $\mathbf{0}$, just to simplify the computational effort. We now use the procedure illustrated before when $1 \notin \mathrm{Spec}(M)$ to find the orbits of the system governed by N. Given an initial condition $\mathbf{x}_0 = \mathbf{x}_{10} + \mathbf{x}_{R0}$, we have (recall that $M\mathbf{z}_R + \mathbf{b} = \mathbf{z}_R$)

$$\mathbf{x}_1 = N(\mathbf{x}_0) = M\mathbf{x}_{10} + M\mathbf{x}_{R0} + \mathbf{b} = \mathbf{x}_{10} + M(\mathbf{x}_{R0} - \mathbf{z}_R) + M\mathbf{z}_R + \mathbf{b}$$

$$\mathbf{x}_2 = N(\mathbf{x}_1) = M\mathbf{x}_1+\mathbf{b} = M[\mathbf{x}_{10} + M(\mathbf{x}_{R0} - \mathbf{z}_R) + \mathbf{z}_R)] + \mathbf{b}$$

$$= \mathbf{x}_{10} + M^2(\mathbf{x}_{R0} - \mathbf{z}_R) + M\mathbf{z}_R+\mathbf{b} = \mathbf{x}_{10} + M^2(\mathbf{x}_{R0} - \mathbf{z}_R) + \mathbf{z}_R.$$

In general, we have

$$\mathbf{x}_n = N(\mathbf{x}_{n-1}) = \mathbf{x}_{10} + M^n(\mathbf{x}_{R0} - \mathbf{z}_R) + \mathbf{z}_R.$$

We see that the behavior of the system is reduced to its evolution in $\mathbf{X}_R$. For example, if all remaining eigenvalues of M have modulus smaller than 1, then $M^n(\mathbf{x}_{R0}-\mathbf{z}_R)\to\mathbf{0}$ and we obtain $\mathbf{x}_n\to\mathbf{x}_{10}+\mathbf{z}_R$. Other cases are analyzed in a similar manner.

The reader may wonder what would have happened if we would have selected an element different from $\mathbf{0}$ in $\mathbf{X}_1$ to produce a solution of (4.6.10). Going through the previous steps it is easy to find out that the final outcome is independent of this choice. For this reason we selected the easiest choice: $\mathbf{0}$.

Example 4.6.6 Let

$$M = \begin{pmatrix} 1 & 0 \\ 1 & -.5 \end{pmatrix}, \quad \mathbf{b} = (0,2).$$

The eigenvalues of M are 1 and –.5. An eigenvector corresponding to 1 is $\mathbf{u}=(3,2)$ and an eigenvector corresponding to –.5 is $\mathbf{e}_2$. Consequently, $\mathbf{b}=\mathbf{b}_R=2\mathbf{e}_2$ and $\mathbf{b}_1=\mathbf{0}$. The unique solution of (4.6.10) is found by determining a such that

$$(0,a) - M(0,a) = (0,2), \quad \text{i.e., } (0,a) - (0,-.5a) = (0,2)$$

or a=4/3. Thus $\mathbf{z}_R=(0,4/3)$. Notice that the origin is a sink for the action of M on $\mathbf{X}_R$. Consequently, given the initial condition

$$\mathbf{x}_0 = (2,3) = (2/3)(3,2) + (5/3)(0,1) = \mathbf{x}_{10} + \mathbf{x}_R, \quad \mathbf{x}_{10} = (2,4/3), \quad \mathbf{x}_R = (0,5/3),$$

we obtain that the orbit $O(\mathbf{x}_0)$ converges to $\mathbf{x}_{10}+\mathbf{z}_R=(2,4/3)+(0,4/3)=(2,8/3)$.

Problems

1. * Let

$$M = \begin{pmatrix} 2^{-1} & 0 \\ 2 & 3 \end{pmatrix}$$

and let $\mathbf{b}=(1,-2)$. Consider the dynamical system in $\mathbf{R}^2$ governed by the map $N(\mathbf{x})=M\mathbf{x}+\mathbf{b}$. Find the stationary state of this system and the stable and unstable linear varieties of N.

2. * Let

$$M = \begin{pmatrix} 2 & 0 & 1 \\ 0 & 2^{-1} & 4^{-1} \\ 0 & -1 & 2^{-1} \end{pmatrix}$$

and let $\mathbf{b}=(1,1,-1)$. Consider the dynamical system in $\mathbf{R}^3$ governed by the function $N(\mathbf{x})=M\mathbf{x}+\mathbf{b}$. Find the stationary state of the system and the stable and unstable linear varieties of N.

3. * Study the system $\mathbf{x}_{n+1}=M\mathbf{x}_n+(1, 2)$ where

$$M = \begin{pmatrix} 2^{-1} & 2 \\ 0 & 2 \end{pmatrix}.$$

Find the stationary state and the stable and unstable linear varieties of the system.

4. * Let $\mathbf{x}_{n+1}=M\mathbf{x}_n+(2,3)$,

$$M = \begin{pmatrix} 2^{-1} & 2 \\ 0 & 2 \end{pmatrix}.$$

Find the coordinates of $\mathbf{x}_{10}$ if $\mathbf{x}_0=(3,5)$.

5. * Let

$$M = \begin{pmatrix} -2 & -3 \\ 1.5 & 2.5 \end{pmatrix}.$$

Find the eigenvalues and eigenvectors of M. Let $\mathbf{b}=(2,-2)$. Study the dynamical system governed by the map $N(\mathbf{x})=M\mathbf{x}+\mathbf{b}$. In particular, determine the orbit of the point $\mathbf{x}_0=(1,3)$.

6. * Let

$$M = \begin{pmatrix} 3/4 & 1/4 \\ 1/4 & 3/4 \end{pmatrix}.$$

Find the eigenvalues and eigenvectors of M. Let $\mathbf{b}=(2,1)$. Study the dynamical system $\mathbf{x}_{n+1}=M\mathbf{x}_n+\mathbf{b}$.

7. Let A and B be two similar matrices, i.e., $B=PAP^{-1}$ for some invertible matrix P. Assume that every orbit of the system $\mathbf{x}_{n+1}=A\mathbf{x}_n+\mathbf{b}$ goes to infinity. Consider the dynamical systems $\mathbf{y}_{n+1}=B\mathbf{y}_n+\mathbf{c}$, $\mathbf{c}=P\mathbf{b}$. Is it true that every orbit of this system goes to infinity?

8. Assume that $1 \notin \sigma(M)$. Let $\mathbf{x}_0$, $\mathbf{x}_1$, and $\mathbf{x}_2$ be a periodic orbit of period 3 of $\mathbf{x}_{n+1}=M\mathbf{x}_n$. Find the corresponding periodic orbit of period 3 of $\mathbf{y}_{n+1}=M\mathbf{y}_n+\mathbf{b}$.

9. * Let M be the matrix of Problem 8 of Section 5. Consider the dynamical system $\mathbf{x}_{n+1}=M\mathbf{x}_n+(2,0,0,0)$. Study the orbits of this system and explain their behavior according to the initial state $\mathbf{x}_0$. Check at least one orbit numerically.

10. * Let M be the matrix of Problem 7 of Section 5. Consider the dynamical system $\mathbf{x}_{n+1}=M\mathbf{x}_n+(1,0,1/2)$. Study the orbits of this system and explain their behavior according to the initial state $\mathbf{x}_0$. Check at least one orbit numerically.

CHAPTER 5

NONLINEAR DYNAMICAL SYSTEMS

SUMMARY

The purpose of this chapter is to provide a qualitative analysis of stationary states and periodic orbits of nonlinear discrete dynamical systems in a dimension higher than 1. We first investigate the presence of bounded invariant sets (Section 1). Then we study the global stability of fixed points (Section 2), followed by the local stability of fixed points and periodic orbits (Section 3). Next we turn our attention to repelling stationary states and periodic orbits (Section 4). In the last section we discuss bifurcation.

Section 1. BOUNDED INVARIANT SETS

Let $\mathbf{x}_{n+1}=F(\mathbf{x}_n)$ be a discrete dynamical system defined in a (possibly) unbounded subset X of $\mathbf{R}^q$. It is frequently useful to find bounded regions $Q\subset X$ that are **invariant** under the action of F, i.e., $F(Q)\subset Q$. In this case every orbit starting from a point $\mathbf{x}_0\in Q$ will remain in Q. Mathematical models of real dynamical processes are supposed to have this property with respect to suitable bounded regions of X, since it is very unlikely that in a real situation the orbits will "escape to infinity." The most obvious examples of bounded invariant sets are fixed points and periodic orbits.

In Chapter 1 we mentioned that the set $L(\mathbf{x}_0)$ of limit points of a bounded orbit is bounded and invariant under the action of F. More precisely,

$$F(L(\mathbf{x}_0) = L(\mathbf{x}_0) \qquad (5.1.1)$$

provided that F is continuous. In this section we assume some additional properties besides continuity. The focus of our attention are three classes of maps for which the existence of bounded invariant sets can easily be established: contractions, dissipative maps, and quasi-bounded maps.

Contractions

Let X be a subset of $\mathbf{R}^q$. A map $F : X\rightarrow\mathbf{R}^q$ is said to be a **contraction** if there exists $k\in[0,1)$ such that

$$\|F(\mathbf{x}) - F(\mathbf{x})\| \leq k\, \|\mathbf{x} - \mathbf{y}\| \qquad (5.1.2)$$

for every $\mathbf{x}, \mathbf{y} \in X$.

Theorem 5.1.1 *Let* $F : \mathbf{R}^q\rightarrow\mathbf{R}^q$ *be a contraction with constant* $k\in[0,1)$. *Then there exists* $r>0$ *such that* $D(R)=\{\mathbf{x}: \|\mathbf{x}\|\leq R\}$ *is invariant for every* $R\geq r$.

Proof. From $F(\mathbf{x})=F(\mathbf{0})+F(\mathbf{x})-F(\mathbf{0})$ we derive

$$\|F(\mathbf{x})\| \leq \|F(\mathbf{0})\| + \|F(\mathbf{x}) - F(\mathbf{0})\| \leq \|F(\mathbf{0})\| + k\, \|\mathbf{x}\|.$$

Thus

$$\|F(\mathbf{x})\| \leq k\, \|\mathbf{x}\| + \|F(\mathbf{0})\|.$$

We are looking for the smallest $r>0$ such that $\|\mathbf{x}\|\leq r$ implies that $\|F(\mathbf{x})\|\leq r$. We need $r\leq kr+\|F(\mathbf{0})\|$. Hence, the smallest r for which the inequality is verified is $r=\|F(\mathbf{0})\|/(1-k)$. In fact, for every $R\geq r$ we have $R\geq\|F(\mathbf{0})\|/(1-k)$, i.e., $R\geq kR+\|F(\mathbf{0})\|$. Then, $\|\mathbf{x}\|\leq R$ implies that

$$\|F(\mathbf{x})\| \leq k\, \|\mathbf{x}\| + \|F(\mathbf{0})\| \leq kR + \|F(\mathbf{0})\| \leq R.$$

Hence, D(R) is invariant under the action of F. QED

Example 5.1.1 Let $F(x)=.8x+4$. Since $|F(x)-F(y)|\leq.8|x-y|$ the map F is a contraction. From $|F(0)|=4$ we obtain $r=20$. Thus every interval $[-R,R]$ with $R\geq20$ is invariant.

Example 5.1.2 Let $F(x)=\sin(x/2)+5$. By the mean value theorem we have $F(x)-F(y)=F'(c)(x-y)$. Since $|F'(x)|\leq1/2$ we obtain $|F(x)-F(y)|\leq.5|x-y|$. Hence, F is a contraction with constant .5. From $F(0)=5$ we obtain $r=10$. Thus $[-R,R]$ with $R\geq10$ is invariant under F.

Remark 5.1.1 It may happen that the map F does not satisfy inequality (5.1.2) when the Euclidean norm is used, but it does with a different norm. Theorem 5.1.1 can be adjusted to this situation.

Dissipative maps

The one-parameter family of maps of the form

$$F(h,\mathbf{x}) = \mathbf{x} + hG(\mathbf{x}) \tag{5.1.3}$$

can be obtained with the application of the one-step Euler's method to the continuous system

$$\mathbf{x}' = G(\mathbf{x}), \tag{5.1.4}$$

with $h=\Delta t$ (see Chapter 1, Section 4). We assume that G is continuous. We say that G is of **dissipative** or **point-dissipative** type if there exist two positive numbers r and k such that

$$G(\mathbf{x})\cdot\mathbf{x} \leq -k \tag{5.1.5}$$

for every vector $\|\mathbf{x}\|\geq r$.

For dynamical systems governed by maps of the form (5.1.3) and in which the G component is of dissipative type, we have the following result.

Theorem 5.1.2 *Let* F *be of the form* (5.1.3) *and assume that* G *satisfies* (5.1.5) *for some* $r,k>0$. *Then, given any real number* $R\geq2r$ *there exists* $d>0$ *such that for all* $h\leq d$ *the set* $D=D(\mathbf{0},R)$ *is invariant under the action of* $F(h,.)$.

Proof. Let $R\geq2r$ and $C=\max\{\|G(\mathbf{x})\| : \|\mathbf{x}\|\leq R\}$. Such a maximum value exists. In fact, G is continuous and the norm function is continuous. Hence, we can apply to the function $g(\mathbf{x})=\|G(\mathbf{x})\|$ the extreme value theorem (see Theorem 3.1.1) to obtain C. We now check the size of $F(h,\mathbf{x})$ for all vectors $\mathbf{x}$ such that $r\leq\|\mathbf{x}\|\leq R$. We have

$$\|\mathbf{x} + hG(\mathbf{x})\|^2 = \|\mathbf{x}\|^2 + 2hG(\mathbf{x})\cdot\mathbf{x} + h^2\|G(\mathbf{x})\|^2 \leq \|\mathbf{x}\|^2 - 2hk + h^2C^2.$$

Since $\|\mathbf{x}\|\leq R$ we obtain

$$\|\mathbf{x} + hG(\mathbf{x})\|^2 \leq R^2 + h(-2k + hC^2).$$

We want $||F(h,\mathbf{x})||\le R$. Hence, we choose d so that $-2k+dC^2\le 0$. This is our first condition on d.

Now we check the size of $F(h,\mathbf{x})$ when $||\mathbf{x}||<r$. We have

$$||F(h,\mathbf{x})|| \le ||\mathbf{x}|| + h||G(\mathbf{x})|| \le r + hC.$$

We want $||F(h,\mathbf{x})||\le R$. Hence, d ($h\le d$) must have the property $r+dC\le 2r\le R$, i.e., $dC\le r$. In conclusion, by selecting d so that both inequalities $dC\le r$ and $dC^2\le 2k$ are satisfied, we obtain the desired result. QED

Example 5.1.3 The Lorenz model for the study of weather patterns is

$$\begin{cases} x' = -ax + ay \\ y' = -xz + rx - y \\ z' = xy - bz. \end{cases}$$

Use the substitution $u=z-a-r$. The qualitative behavior of the system is not affected by this change since $u'=z'$. We obtain

$$\begin{cases} x' = -ax + ay \\ y' = -xu - ax - y \\ u' = xy - bu - ab - br. \end{cases}$$

Euler's one-step method gives (see Chapter 1, Section 4)

$$\begin{cases} x_{n+1} = x_n + ah(y_n - x_n) \\ y_{n+1} = y_n + h(-x_n u_n - ax_n - y_n) \\ u_{n+1} = u_n + h(x_n y_n - bu_n - b(a + r)). \end{cases}$$

We have obtained the discrete dynamical system

$$\mathbf{x}_{n+1} = F(h,\mathbf{x}_n) = \mathbf{x}_n + hG(\mathbf{x}_n), \tag{5.1.6}$$

where

$$G(\mathbf{x}) = G(x,y,u) = (a(y - x), -xu - ax - y, xy - bu - b(a + r)).$$

Notice that

$$G(\mathbf{x})\bullet\mathbf{x} = -ax^2 - y^2 - bu^2 - b(a + r)u.$$

This can be rewritten in the following way:

$$G(\mathbf{x})\bullet\mathbf{x} = -ax^2 - y^2 - b(u + (a + r)/2)^2 + b(a + r)^2/4.$$

Set $q^2 = b(a+r)^2/4$ and consider the ellipsoid $G(\mathbf{x})\bullet\mathbf{x}=0$, (see Fig. 5.1.1) i.e.,

$$ax^2 + y^2 + b(u + (a + r)/2)^2 = q^2.$$

The ellipsoid separates the remaining space into two open regions, the interior IN and the exterior EX. For every vector $\mathbf{x} \in$ IN we have $G(\mathbf{x}) \bullet \mathbf{x} > 0$, while for every $\mathbf{x} \in$ EX we have $G(\mathbf{x}) \bullet \mathbf{x} < 0$, and the quantity becomes more and more negative as we move away from the ellipsoid. Hence, the condition of Theorem 5.1.2 is satisfied by the discrete dynamical system obtained with the application of the one-step Euler's method to the Lorenz model.

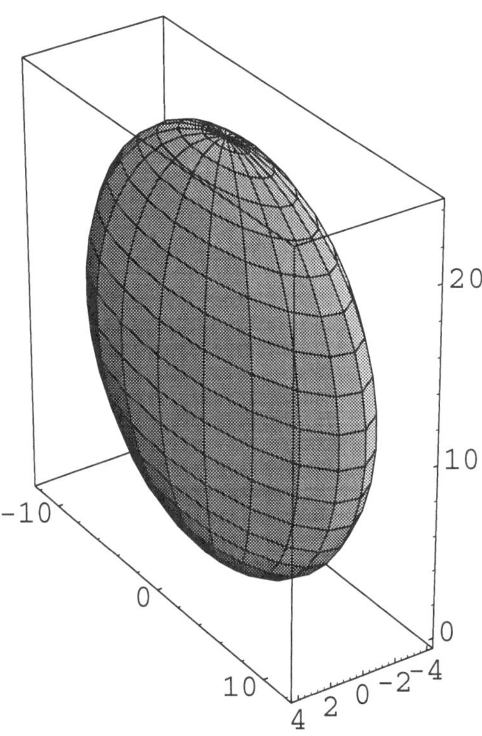

Fig. 5.1.1 Ellipsoid of the Lorenz map for a=8, b=1, and r=16 in the space *xyz*. Its sections with planes perpendicular to the *x*-axis (x=k, $-3\sqrt{2} \leq k \leq 3\sqrt{2}$) are circles centered at the point (k,0,12). Its sections with planes perpendicular to the *y*-axis (y=k, $-12 \leq k \leq 12$), are ellipses centered at (0,k,12) with major axis along z and minor along x. The ellipsoid contains (0,0,0) and (0,0,24). Its center is (0,0,12).

Quasi-bounded maps

Let $F : \mathbf{R}^q \rightarrow \mathbf{R}^q$ be continuous and r>0. Denote by $D^c(r)$ the set $D^c(r) = \{\mathbf{x} \in \mathbf{R}^q : \|\mathbf{x}\| \geq r\}$. Define

$$N(r) = \sup\{\|F(\mathbf{x})\|/\|\mathbf{x}\| : \mathbf{x} \in D^c(r)\}.$$

Notice that $r_1 < r_2$ implies that $D^c(r_2) \subset D^c(r_1)$ and $N(r_2) \leq N(r_1)$.

Hence, N(r) is not increasing as $r \rightarrow \infty$. We may have $N(r) = \infty$ for all r>0. In the case when $N(1) < \infty$, we say that F is **quasi-bounded**.

Example 5.1.4 Let $F(x)=x^2$, and $G(x)=\ln(1+x^2)$. F is not quasi-bounded. In fact $N(1)=\sup\{x^2/|x| : |x|\geq 1\}=\sup\{|x|, |x|\geq 1\}=\infty$. For G we have $N(1)= \sup\{\ln(1+x^2)/|x| : |x|\geq 1\}<\infty$ since the ratio goes to 0 as $|x|$ goes to ∞. Hence G is quasi-bounded.

Theorem 5.1.3 *Let* $F : \mathbf{R}^q \to \mathbf{R}^q$ *be quasi-bounded and assume that there exists* $r_0>0$ *such that* $N(r_0)\leq 1$. *Then there exists a real number* d *such that for all* $R\geq d$ *the set* $D(R)=\{\mathbf{x} : \|\mathbf{x}\|\leq R\}$ *is invariant.*

Proof. Since N(r) is not increasing we have $N(r)\leq 1$ for all $r\geq r_0$. Moreover the continuity of the function $g(\mathbf{x})=\|F(\mathbf{x})\|$ implies, by the extreme value theorem (see Theorem 3.1.1), that there exists a constant C such that $C=\max\{\|F(\mathbf{x})\|: \|\mathbf{x}\|\leq r_0\}$.

Let $d=\max\{r_0,C\}$. Then, for every $R\geq d$ we have

- $\|F(\mathbf{x})\| \leq C \leq R$ whenever $\|\mathbf{x}\| < r_0$.
- $\|F(\mathbf{x})\| \leq \|\mathbf{x}\| \leq R$ whenever $r_0 \leq \|\mathbf{x}\| \leq R$.

Hence, D(R) is invariant. QED

Example 5.1.5 Let $F(x,y)=\left(\frac{ay}{1+x^2}, \frac{bx}{1+y^2}\right)$. We find that

$$\frac{\|F(\mathbf{x})\|^2}{\|\mathbf{x}\|^2} = \frac{(ay)^2}{x^2 + y^2}\frac{1}{(1 + x^2)^2} + \frac{(bx)^2}{x^2 + y^2}\frac{1}{(1 + y^2)^2}.$$

If $|a|\leq 1$ and $|b|\leq 1$, the condition of Theorem 5.1.3 is satisfied by the discrete dynamical system governed by F.

Example 5.1.6 As mentioned in Chapter 1, Section 4, the dynamic evolution of a neural net following an initial input $\mathbf{x}_I$, which remains active during the entire process, can be modeled by the dynamical system

$$\mathbf{x}_{n+1} = (I - C)\mathbf{x}_n + T\Phi(\mathbf{x}_n) + \mathbf{x}_I. \tag{5.1.7}$$

I is the identity, C is the diagonal matrix

$$C = \begin{pmatrix} c_1 & 0 & \dots & 0 \\ 0 & c_2 & \dots & 0 \\ \dots & \dots & \dots & \dots \\ \dots & \dots & 0 & c_N \end{pmatrix},$$

T is the connectivity matrix, and $\Phi(\mathbf{x})=(\phi(x_1),\dots,\phi(x_N))$ is a nonlinear map with the property $\|\Phi(\mathbf{x})\|^2\leq N$ for all $\mathbf{x}\in \mathbf{R}^N$. Let

$$K(\mathbf{x}) = (I - C)\mathbf{x} + T\Phi(\mathbf{x})+ \mathbf{x}_I.$$

Then,

$$\frac{\|K(\mathbf{x})\|}{\|\mathbf{x}\|} \leq \|I - C\| + \|T\|\frac{\|\Phi(\mathbf{x})\|}{\|\mathbf{x}\|} + \frac{\|\mathbf{x}_I\|}{\|\mathbf{x}\|}.$$

As $||\mathbf{x}||\to\infty$, the second and third terms on the right-hand side go to 0. Hence, the map K is quasi-bounded. Moreover, there exists r_0 such that $N(r_0)\leq 1$ provided that $c_i\in(0,1]$ for all i=1,2,...,N.

Problems

1. Let $F(x)=.6x-5$. Prove that F is a contraction and find r>0 such that [–R,R] is invariant for every $R\geq r$.

2. Let $F(x)=x+1/(1+|x|)$. Prove that F is not a contraction.

3. * Let

$$M=\begin{pmatrix} .5 & 1 \\ 0 & .4 \end{pmatrix}.$$

Show that M is not a contraction (in the Euclidean norm). Find an integer $p\geq 1$ such that M^p is a contraction.

4. Let $F:\mathbf{R}^q\to\mathbf{R}^q$ be a contraction with constant k. Assume that $F(\mathbf{x}_s)=\mathbf{x}_s$. Verify that every ball centered at $\mathbf{x}_s$ is invariant under F.

5. * Find the ellipsoid of the Lorenz map for

$$a=10,\ b=\frac{8}{3},\ \text{and } r=\frac{470}{19}.$$

Check if the fixed points of F (see Chapter 1) belong to the ellipsoid. Verify that for every $\mathbf{x}=(x,y,u)$ inside the ellipsoid, we have $G(\mathbf{x})\bullet\mathbf{x}>0$.

6. Let $F(\mathbf{x})=f(||\mathbf{x}||)G(\mathbf{x})$, where $f:\mathbf{R}\to\mathbf{R}$ is such that

$$f(t)=\ln(1+t^2),\ \text{and } ||G(\mathbf{x})||\leq 0.5\ ||\mathbf{x}||.$$

Find a ball that is mapped into itself by F.

7. Verify that the map of Example 5.1.5 satisfies the condition of Theorem 5.1.3, provided that $|a|\leq 1$ and $|b|\leq 1$.

8. Verify that the map defined by a matrix M is always quasi-bounded, and that M^p satisfies the condition of Theorem 5.1.3 for some $p\geq 1$, provided that the spectral radius of M is less than 1.

9. Assume that the matrix M is symmetric and that its spectral radius is 1. Using the Euclidean norm, prove that every closed ball D centered at the origin is invariant. [Hint: For a symmetric matrix M we have $||M||=\rho(M)$].

10. Let $\Phi:\mathbf{R}^q\to\mathbf{R}^q$ be bounded, i.e., there exists a constant k>0 such that

$$||\Phi(\mathbf{x})||\leq k \text{ for all } \mathbf{x}\in\mathbf{R}^q.$$

Let T be a matrix of size q and C be a matrix such that $||C||<1$. Prove that the composite map $C+T\Phi$ is quasi-bounded and $N(r_0)<1$ for some $r_0>0$ (this idea is useful in the study of neural networks).

Section 2. **GLOBAL STABILITY OF FIXED POINTS**

In this section we study three different classes of maps with the property of admitting a unique fixed point which is a global sink. We start with the most famous class: contractions.

Banach contraction principle

We prove the celebrated Theorem of **Stephan Banach** (1892-1945) on contractions.

Theorem 5.2.1 (Banach) *Let* $F : \mathbf{R}^q \to \mathbf{R}^q$ *be a* **contraction**. *Then* F *has one and only one fixed point* $\mathbf{x}_s$ *and for every* $\mathbf{x}_0 \in \mathbf{R}^q$ *we have*

$$\|\mathbf{x}_n - \mathbf{x}_s\| \le k^n \|\mathbf{x}_0 - \mathbf{x}_s\|. \tag{5.2.1}$$

Proof. *Uniqueness*. The uniqueness of $\mathbf{x}_s$ is easy to establish, and it is left to the reader (Problem 1).

Convergence. Inequality (5.2.1) can be proved as follows. Let $\mathbf{x}_0$ be an arbitrary point of $\mathbf{R}^q$. Then

$$\|\mathbf{x}_1 - \mathbf{x}_s\| = \|F(\mathbf{x}_0) - F(\mathbf{x}_s)\| \le k\, \|\mathbf{x}_0 - \mathbf{x}_s\|,$$

and in general,

$$\|\mathbf{x}_n - \mathbf{x}_s\| = \|F(\mathbf{x}_{n-1}) - F(\mathbf{x}_s)\| \le k\|\mathbf{x}_{n-1} - \mathbf{x}_s\| \le \ldots \le k^n \|\mathbf{x}_0 - \mathbf{x}_s\|. \tag{5.2.2}$$

Since $0 \le k < 1$ we see that $\mathbf{x}_n = F(\mathbf{x}_{n-1}) \to \mathbf{x}_s$ as $n \to \infty$. More precisely, we have

$$\|\mathbf{x}_n - \mathbf{x}_0\| \downarrow 0 \text{ as } n \to \infty,$$

where the symbol $\downarrow 0$ means that the sequence is "decreasing" and it converges to 0.

Existence. Let $d = \|F(\mathbf{0})\|$ and choose R such that $d/(1-k) \le R$. In Theorem 5.1.1 we established that $D(\mathbf{0},R)$ is invariant under the action of F. Let $\mathbf{x}_0$ be any point of D. With $\mathbf{x}_1 = F(\mathbf{x}_0)$, $\mathbf{x}_2 = F(\mathbf{x}_1)$, ..., we have

$$\|\mathbf{x}_2 - \mathbf{x}_1\| = \|F(\mathbf{x}_1) - F(\mathbf{x}_0)\| \le k\, \|\mathbf{x}_1 - \mathbf{x}_0\| \le k2r,$$

$$\|\mathbf{x}_3 - \mathbf{x}_2\| = \|F(\mathbf{x}_2) - F(\mathbf{x}_1)\| \le k\, \|\mathbf{x}_2 - \mathbf{x}_1\| \le k^2 2r.$$

In general,

$$\|\mathbf{x}_{n+1} - \mathbf{x}_n\| = \|F(\mathbf{x}_n) - F(\mathbf{x}_{n-1})\| \le k\|\mathbf{x}_n - \mathbf{x}_{n-1}\| \le k^n 2r.$$

The above inequalities and the extreme value theorem (Theorem 3.1.1) imply that the minimum value of the continuous function $g : D \to \mathbf{R}$ defined by $g(\mathbf{x}) = \|F(\mathbf{x}) - \mathbf{x}\|$ is 0. Hence, there exists $\mathbf{x}_s \in D$, such that $g(\mathbf{x}_s) = 0$, i.e., $F(\mathbf{x}_s) = \mathbf{x}_s$. QED

Theorem 1 is known as the **Banach Contraction Principle** (BCP).

Remark 5.2.1 We mention a variant of the BCP, due to the Italian mathematician **Renato Caccioppoli** (1904-1959), who proved that the principle still holds if F is not a contraction, but one of its iterates is. There are important cases in which this happens. For example, a matrix M such that $\rho(M)<1$ may not be a contraction, but one of its iterates is as a consequence of the property

$$\lim_{n\to\infty} \sqrt[n]{\|M^n\|} = \rho(M) < 1.$$

Remark 5.2.2 The BCP remains true if $\mathbf{R}^q$ is replaced by any **closed set** $X\subset\mathbf{R}^q$, F maps X into itself, and either F or one of its iterates is a contraction in X.

Remark 5.2.3 The Banach Contraction Principle and Caccioppoli's extension remain valid when F or one of its iterates are contractions in a norm other than the Euclidean.

Example 5.2.1 A more elaborate version of this example will be analyzed in Chapter 7 when we study neural networks. Let $h : \mathbf{R}\to\mathbf{R}$ be defined as follows

$$h(x) = \begin{cases} 0 & x<0 \\ x & 0\le x\le 1 \\ 1 & x>1 \end{cases}$$

and let M be the matrix

$$M = \begin{pmatrix} .2 & .5 \\ 0 & .4 \end{pmatrix}.$$

The operator norm of M, using the maximum norm, is .7 since

$$\|M\|_\infty = \max\{\|M\mathbf{x}\|_\infty : \|\mathbf{x}\|_\infty=1\} = \|M(1,1)\|_\infty = .7.$$

Consequently, the function $F : \mathbf{R}^2\to\mathbf{R}^2$ defined by

$$F(x,y) = M(h(x),h(y))$$

is a contraction. In fact, from $\|M\|_\infty\le .7$ and $|h(x_1)-h(x_2)|\le|x_1-x_2|$, we derive

$$\|F(x,y) - F(u,v)\|_\infty \le .7\ \|(h(x) - h(u),\ h(y) - h(v))\|_\infty \le .7\ \|(x,y) - (u,v)\|_\infty.$$

Triangular maps

A map $F : \mathbf{R}^q\to\mathbf{R}^q$ is said to be **upper triangular** if

$$F(\mathbf{x}) = F(x_1,x_2,\ldots,x_q) = (F_1(x_1,\ldots,x_q),\ F_2(x_2,\ldots,x_q),\ldots,\ F_q(x_q)),$$

and **lower triangular** if

$$F(\mathbf{x}) = F(x_1, x_2,...,x_q) = (F_1(x_1), F_2(x_1,x_2),..., F_q(x_1,x_2,...,x_q)).$$

The derivative of an upper triangular or a lower triangular map is, respectively, an upper or a lower triangular matrix.

Example 5.2.2 Let $F(x,y,z)=(x^2+y+xz^2,yz,\sin z)$. Then F is upper triangular. In fact

$$F_1(x,y,z) = x^2 + y + xz^2,\ F_2(y,z) = yz,\ F_3(z) = \sin z.$$

Therefore,

$$F'(\mathbf{x}) = \begin{pmatrix} 2x+z^2 & 1 & 2xz \\ 0 & z & y \\ 0 & 0 & \cos z \end{pmatrix},$$

and for $\mathbf{x}_0=(1,0,0)$ we have

$$F'(1,0,0) = \begin{pmatrix} 2 & 1 & 0 \\ 0 & 0 & 0 \\ 0 & 0 & 1 \end{pmatrix}.$$

Example 5.2.3 Let $F(x,y,z)= (x^3, xy+\sin y, x+y+z^2)$. Then F is lower triangular.

For dynamical systems governed by such mappings we have the following result which we present without proof.

Theorem 5.2.2 *Let* $F : \mathbf{R}^q \to \mathbf{R}^q$ *be an upper (or lower) triangular map. Assume that* F' *is continuous and there is a constant* $k<1$ *such that*

$$\rho(F'(\mathbf{x})) \le k \quad \textit{for every } \mathbf{x}\in \mathbf{R}^q. \tag{5.2.3}$$

Then F *has one and only one fixed point* $\mathbf{x}_s$ *and for every* $\mathbf{x}_0\in \mathbf{R}^q$ *we have*

$$\mathbf{x}_n = F(\mathbf{x}_{n-1}) \to \mathbf{x}_s \quad \textit{as} \quad n \to \infty. \tag{5.2.4}$$

Assumption (5.2.3) is equivalent to

$$\left|\frac{\partial F_i}{\partial x_i}(\mathbf{x})\right| \le k \quad \text{for all } \mathbf{x}\in \mathbf{R}^q \text{ and } i = 1, 2,..., q. \tag{5.2.5}$$

In fact, since the matrix $F'(\mathbf{x})$ is a triangular matrix its eigenvalues are the entries of the main diagonal.

Example 5.2.4 Let $F(\mathbf{x})=F(x,y)=(y^2+x/2,y/3)$. Every orbit of the dynamical system converges to $\mathbf{0}$ since $F(\mathbf{0})=\mathbf{0}$ and $\rho(F'(\mathbf{x}))=.5$ for all $\mathbf{x}\in \mathbf{R}^2$. Therefore, we can apply Theorem 5.2.2.

Example 5.2.5 Recall that in a neural network the various neurons can be hierarchically arranged into three layers: input, hidden, and output layer. Accordingly, the entries of the N-dimensional vector **x**, which represents the energy levels of the neurons, can be organized so that the input neurons come first, followed by the hidden and the output neurons. The neural net is said to be **forward** if each neuron can affect only the neurons of the following level in the hierarchical order just established. Consequently, $t_{ij}=0$ for all $j\geq i$ and the dynamical system

$$\mathbf{x}_{n+1} = (I - C)\mathbf{x}_n + T\Phi(\mathbf{x}_n) + \mathbf{x}_I$$

is lower triangular. The conditions of Theorem 5.2.2 are not satisfied since the neuron response function ϕ, which enters in the definition of Φ, is not differentiable at x=0 and x=1. However, the Theorem can be generalized so that this occurrence can be included in its formulation. Condition (5.2.5) is verified provided that $c_i\in(0,1]$, i.e., no leakage coefficient is 0. In this setting, given an input vector $\mathbf{x}_I$, the net converges to the same fixed point regardless of its initial state $\mathbf{x}_0$.

Gradient maps (*)

There is another case in which the dynamical situation is similar to the one envisioned by Theorems 5.2.1 and 5.2.2. Assume that F is a gradient, i.e., there exists a differentiable function $f : \mathbf{R}^q \to \mathbf{R}$ such that

$$F(\mathbf{x}) = \nabla f(\mathbf{x}) = \left(\frac{\partial f}{\partial x_1},\ldots,\frac{\partial f}{\partial x_q}\right). \tag{5.2.6}$$

Also assume that f is of class C^2, i.e. continuous together with its first and second partial derivatives. It is well known that the mixed partial derivatives are equal

$$\frac{\partial^2 f}{\partial x_i \partial x_j} = \frac{\partial^2 f}{\partial x_j \partial x_i} \tag{5.2.7}$$

for every $1\leq i, j\leq n$. Hence, the derivative F′ of F is a symmetric matrix at every point **x** of $\mathbf{R}^q$.

Example 5.2.6 Let $F(\mathbf{x})=F(x,y)=(2x+y^3, 3xy^2-4y^3)$. Then $F(x,y)=\nabla f(x,y)$ with $f(x,y) = x^2+xy^3-y^4$. Notice that

$$F'(\mathbf{x}) = \begin{pmatrix} 2 & 3y^2 \\ 3y^2 & 6xy-12y^2 \end{pmatrix}$$

is symmetric for every **x**.

Given a symmetric matrix M, the operator norm $\|M\|$ as defined in Chapter 3, and using the Euclidean norm in $\mathbf{R}^q$, satisfies the equality

$$\rho(M) = \|M\|, \tag{5.2.8}$$

i.e., the spectral radius of the matrix equals its operator norm. With this machinery set up we now obtain for discrete dynamical systems in $\mathbf{R}^q$ a result which is similar to the one established in Chapter 2 for q=1.

Theorem 5.2.3 *Let* $F : \mathbf{R}^q \to \mathbf{R}^q$ *be a gradient. Assume that* F *has a fixed point* $\mathbf{x}_s$, *and the derivative* F′ *satisfies the following conditions*:

(i) $\rho(F'(\mathbf{x})) \le 1$ *for every* $\mathbf{x} \in \mathbf{R}^q$.
(ii) *There exists* $d > 0$ *such that* $\rho(F'(\mathbf{x})) < 1$ *for every* $\mathbf{x} \in B(\mathbf{x}_s, d)$.

Then $\mathbf{x}_s$ *is the only fixed point of* F *and for every* $\mathbf{x}_0 \in \mathbf{R}^q$ *we have*

$$\mathbf{x}_n = F(\mathbf{x}_{n-1}) \to \mathbf{x}_s \text{ as } n \to \infty. \tag{5.2.9}$$

Proof. We first establish two inequalities which will be used in the proof. By assumption (i), property (5.2.8), and the mean value inequality we obtain

$$\|F(\mathbf{x}) - F(\mathbf{y})\| \le \|\mathbf{x} - \mathbf{y}\| \tag{5.2.10}$$

for all $\mathbf{x},\mathbf{y} \in \mathbf{R}^q$. Moreover, if $\mathbf{x}=\mathbf{x}_s$ then we have ($F(\mathbf{x}_s)=\mathbf{x}_s$)

$$\|\mathbf{x}_s - F(\mathbf{y})\| < \|\mathbf{x}_s - \mathbf{y}\|. \tag{5.2.11}$$

This last inequality can be established as follows. In the line segment joining $\mathbf{x}_s$ with $\mathbf{y}$, select a point $\mathbf{z} \ne \mathbf{x}_s$ such that $\mathbf{z} \in B(\mathbf{x}_s, d)$. From the mean value inequality it follows that

$$\|\mathbf{x}_s - F(\mathbf{y})\| = \|F(\mathbf{x}_s) - F(\mathbf{y})\| \le \|F(\mathbf{x}_s) - F(\mathbf{z})\| + \|F(\mathbf{z}) - F(\mathbf{y})\|$$

$$\le \|F'(\mathbf{c}_1)\| \, \|\mathbf{x}_s - \mathbf{z}\| + \|F'(\mathbf{c}_2)\| \, \|\mathbf{z} - \mathbf{y}\|,$$

where $\mathbf{c}_1$ belongs to the open line segment joining $\mathbf{x}_s$ with $\mathbf{z}$ and $\mathbf{c}_2$ belongs to the open line segment joining $\mathbf{z}$ with $\mathbf{y}$. Hence, $\|F'(\mathbf{c}_1)\|<1$. Since the element $\mathbf{z}$ belongs to the line segment joining $\mathbf{x}_s$ with $\mathbf{y}$ we have $\|\mathbf{x}_s-\mathbf{y}\|=\|\mathbf{x}-\mathbf{z}\|+\|\mathbf{z}-\mathbf{y}\|$. Therefore, using equality (5.2.8), and assumptions (i) and (ii) of Theorem 5.2.3 we obtain (5.2.11).

Uniqueness of the fixed point. Assume that $\mathbf{y}_s$ is a fixed point. Then, from (5.2.11) we have

$$\|\mathbf{x}_s - \mathbf{y}_s\| = \|\mathbf{x}_s - F(\mathbf{y}_s)\| < \|\mathbf{x}_s - \mathbf{y}_s\|.$$

Hence, $\mathbf{y}_s=\mathbf{x}_s$, and the fixed point is unique.

Every orbit converges to $\mathbf{x}_s$. Let $\mathbf{x}_0$ be any point of $\mathbf{R}^q$. Then, again by (5.2.11)

$$\|\mathbf{x}_s - \mathbf{x}_1\| = \|\mathbf{x}_s - F(\mathbf{x}_0)\| < \|\mathbf{x}_s - \mathbf{x}_0\|.$$

It follows that the sequence $\|\mathbf{x}_s-\mathbf{x}_n\|$ is strictly decreasing. Hence, it converges to a value $c \ge 0$. We must show that c=0.

Assume c>0 and consider the set $S(\mathbf{x}_s,c)$, which contains $L(\mathbf{x}_0)$, the set of limit points of $O(\mathbf{x}_0)$. Define $g : S(\mathbf{x}_s,c)\rightarrow\mathbf{R}$ by $g(\mathbf{x})=\|\mathbf{x}_s-F(\mathbf{x})\|$. By the extreme value theorem, g has a maximum value in $S(\mathbf{x}_s,c)$. Such value has to be smaller than c, since from (5.2.11) we have $g(\mathbf{x})<c$ for all $\mathbf{x}\in S(\mathbf{x}_s,c)$. This however contradicts $F(L(\mathbf{x}_0))=L(\mathbf{x}_0)$. Thus, c=0. QED

The reader has certainly recognized that Theorem 5.2.3 is almost as sharp as Theorem 2.3.1, its one-dimensional counterpart. The hypothesis that F is a gradient is automatically satisfied in that case. Hence, it was not included explicitly. It would be nice to have a result in which the conclusion of Theorem 5.2.3 still holds, but the restriction that F be a gradient is removed. Unfortunately, counterexamples can be given to show that such a result does not hold in the stated generality.

The reader may have noticed certain similarities between Theorems 5.2.2 and 5.2.3. In **R** Theorem 5.2.2 is a particular case of Theorem 5.2.3. In a higher dimension the two results are different, since in Theorem 5.2.2 the derivative of F need not be a symmetric matrix.

Remark 5.2.4 The three previous results guarantee that all orbits converge to the unique stationary state of the system. Admittedly, their practical value is limited by the restrictions imposed on the function F. However, Theorem 5.2.1 holds true when F has the given properties in an invariant and closed set D of $\mathbf{R}^q$ (see Remark 5.2.2). Similarly, an invariant disk $D(\mathbf{x}_0,r)$, r>0, can replace $\mathbf{R}^q$ in the statements of Theorems 5.2.2 and 5.2.3. Continuity of F is required in $D(\mathbf{x}_0,r)$ and differentiability in $B(\mathbf{x}_0,r)$.

Problems

1. Let $F : \mathbf{R}^q\rightarrow\mathbf{R}^q$ be a contraction. Prove that the dynamical system generated by F has at most one fixed point.

2. Prove that if F^p is a contraction and $F^p(\mathbf{x}_0)=\mathbf{x}_0$, then $F(\mathbf{x}_0)=\mathbf{x}_0$. Notice that this result may not hold when F is not a contraction.

3. Prove that Theorem 5.2.1 still holds if F is not a contraction, but one of its iterates is.

4. Let $F : \mathbf{R}^q\rightarrow\mathbf{R}^q$ be a contraction with constant k<1 and let $\mathbf{x}_s$ be the only fixed point of F. Consider the orbit starting from a point $\mathbf{x}_0$. Prove that

$$\|\mathbf{x}_{n+1} - \mathbf{x}_n\| \rightarrow 0 \text{ as } n \rightarrow \infty.$$

5. Let $F : \mathbf{R}^q\rightarrow\mathbf{R}^q$ be a contraction with constant k<1. Let $\mathbf{x}_0$ be any initial state. Prove the following inequality:

$$\|\mathbf{x}_n - \mathbf{x}_s\| \leq \frac{k^n}{1-k}\|\mathbf{x}_1 - \mathbf{x}_0\|.$$

(Hint: We have $\|\mathbf{x}_0-\mathbf{x}_s\|\leq\|\mathbf{x}_0-\mathbf{x}_1\|+\|\mathbf{x}_1-\mathbf{x}_s\|\leq\|\mathbf{x}_0-\mathbf{x}_1\|+k\|\mathbf{x}_0-\mathbf{x}_s\|$.)

This result allows us to estimate how close we are to the fixed point without knowing it.

6. Let M be a square matrix of size q. Assume that the spectral radius of M is smaller than 1. Using the equality

$$\rho(M) = \lim_{n\to\infty} \sqrt[n]{\|M^n\|},$$

prove that for some integer n, the operator norm of M^n, using the Euclidean norm in $\mathbf{R}^q$, is smaller than 1. Derive from this result that the action of M^n in $\mathbf{R}^q$ is a contraction.

7. Let M be as in the previous problem. Using Caccioppoli's version of the Banach Contraction Principle, prove that every orbit of the dynamical system governed by M converges to the origin. This provides another way of establishing the result obtained in the previous chapter about the convergence to **0** of every orbit of the dynamical system defined by M.

8. Let $F(x,y)=(.5x+y^2,.3y)$. Show that every orbit goes to zero.

9. * Let $F(x,y)=(y^2+\cos x,\cos y)$. Show that the dynamical system governed by the map F has only one stationary state (x_s,y_s). Numerically explore the behavior of several different orbits and verify that they all converge to the unique fixed point (x_s,y_s).

10. In the proof of Theorem 5.2.3 we stated that $g(\mathbf{x})<c$ for all $\mathbf{x}\in S(\mathbf{x}_s,c)$ contradicts the property $F(L(\mathbf{x}_0))=L(\mathbf{x}_0)$. Explain why.

11. * Let $F(x,y,z)=(\sin(.5x)+y^2-z^2,\cos(.7y)+e^z,\arctan z)$. Does the function F satisfy the conditions of Theorem 5.2.2? Explore the orbits of F numerically.

Section 3. **SINKS**

The situations encountered in the previous section are certainly ideal. More frequently, however, a fixed point attracts only orbits which start close to it. We are now interested in finding conditions under which this behavior can be guaranteed. The results we prove are very similar to the companion results obtained for the case of functions of one variable, and again they involve the derivative of the function at the fixed point. However, the presence of several variables makes the conditions more difficult to verify. The proofs are a bit more complex.

Theorem 5.3.1 *Let* $U\subset\mathbf{R}^q$ *be open and* $\mathbf{x}_s$ *be a fixed point of a continuous function* $F : U\to U$. *Assume that there exists* $r>0$ *such that* F *is differentiable on* $B(\mathbf{x}_s,r)$ *except possibly at* $\mathbf{x}_s$ *and* $\|F'(\mathbf{x})\|\leq 1$. *Then* $\mathbf{x}_s$ *is stable.*

Proof. Let $\mathbf{x}_0 \in B(\mathbf{x}_s, r)$. By the mean value inequality we have

$$\|\mathbf{x}_1 - \mathbf{x}_s\| = \|F(\mathbf{x}_0) - F(\mathbf{x}_s)\| \leq \|F'(\mathbf{c})\| \, \|\mathbf{x}_0 - \mathbf{x}_s\|$$

where $\mathbf{c}$ is a point in the open line segment joining $\mathbf{x}_0$ with $\mathbf{x}_s$. Since $\|F'(\mathbf{c})\| \leq 1$ we obtain that $\mathbf{x}_s$ is stable (select $\delta = r$ in the definition of stability). QED

Theorem 5.3.2 *Let* $U \subset \mathbf{R}^q$ *be open and* $\{\mathbf{x}_0, \mathbf{x}_1, ..., \mathbf{x}_{p-1}, ...\}$ *be a periodic orbit of period p of a continuous function* $F : U \to U$. *Assume that there exists* $r > 0$ *such that for every* $i = 0, 1, ..., p-1$, F *is differentiable on* $B(\mathbf{x}_i, r)$ *except possibly at* $\mathbf{x}_i$ *and* $\|(F^p)'(\mathbf{x})\| \leq 1$. *Then the periodic orbit is stable.*

Proof. By Theorem 5.3.1 each state $\mathbf{x}_i$, $i = 0, 1, ..., p-1$ is a stable fixed point of F^p. Hence, the periodic orbit is stable. QED

We now present definitions and theorems on sinks.

Definition 5.3.1

Let $U \subset \mathbf{R}^q$ be open, and $F : U \to U$ be continuous. A fixed point $\mathbf{x}_s$ of F is a **sink** (an **attractor**) if $\mathbf{x}_s$ is stable and there is $r > 0$ such that $\mathbf{x}_0 \in U$ and $\|\mathbf{x}_0 - \mathbf{x}_s\| \leq r$ imply $\|\mathbf{x}_n - \mathbf{x}_s\| \to 0$ as $n \to \infty$.

Let $U \subset \mathbf{R}^q$ be open, and $F : U \to U$ be continuous. A periodic orbit $\{\mathbf{x}_0, \mathbf{x}_1, ..., \mathbf{x}_{p-1}, ...\}$ of period p of F is a **sink** (an **attractor**) if each state $\mathbf{x}_i$ is a sink of F^p.

The readers have probably noticed the difference between the above definition of sink and the one used in Chapter 2 for one-dimensional systems. Here we have added the requirement "$\mathbf{x}_s$ is stable." In $\mathbf{R}$ this condition is not needed, since it can be shown that a fixed point cannot be unstable and have the property $|x_n - x_s| \to 0$ as $n \to \infty$ for all points $x_0 \in (x_s - r, x_s + r)$ (see [Martelli-Marshall, 1995]). In a higher dimension, however, we can have an unstable fixed point $\mathbf{x}_s$ such that $\|\mathbf{x}_n - \mathbf{x}_s\| \to 0$ as $n \to \infty$ for all $\mathbf{x}_0 \in B(\mathbf{x}_s, r)$. Example 6.1.3 of the next chapter provides an unstable fixed point $\mathbf{x}_s \in \mathbf{R}^2$ such that $\|\mathbf{x}_n - \mathbf{x}_s\| \to 0$ as $n \to \infty$ for all $\mathbf{x}_0 \in \mathbf{R}^2$!

The following lemma is needed in the proofs of the main results of this section and the next. In Chapter 2 we mentioned that given a continuous function g, defined on an open interval I, and such that $|g(x_0)| < 1$ for some $x_0 \in I$, and given any real number $k \in (|g(x_0)|, 1)$, there exists $r > 0$ such that $|g(x)| \leq k$ for all $x \in [x_0 - r, x_0 + r]$. We need an extension of this result with I replaced by an open set U and $[x_0 - r, x_0 + r]$ replaced by $D(\mathbf{x}_0, r)$. Here is the result, formally stated.

Lemma 5.3.1 *Let* $U \subset \mathbf{R}^q$ *be open and* $g : U \to \mathbf{R}$ *be continuous at* $\mathbf{x}_0$. *Assume that* $|g(\mathbf{x}_0)| < 1$. *Let* k *be such that* $|g(\mathbf{x}_0)| < k < 1$. *Then there exists* $r > 0$ *with the property that* $|g(\mathbf{x})| \leq k$ *for all* $\mathbf{x} \in D(\mathbf{x}_0, r)$.

Some of the inequalities of Lemma 5.3.1 can be reversed, namely $|g(\mathbf{x}_0)| > 1$ implies that given k such that $|g(\mathbf{x}_0)| > k > 1$, there exists $r > 0$ with the property that $|g(\mathbf{x})| \geq k > 1$ whenever $\mathbf{x} \in D(\mathbf{x}_0, r)$.

We are now ready to give a condition for $\mathbf{x}_s$ to be a sink.

Theorem 5.3.3 *Let* $U \subset \mathbf{R}^q$ *be open, and let* $\mathbf{x}_s$ *be a fixed point of a continuous function* $F: U \to U$. *Assume that* F *is differentiable in* $B(\mathbf{x}_s, d)$ *for some* $d>0$ *with continuous derivative at* $\mathbf{x}_s$. *Then* $\rho(F'(\mathbf{x}_s))<1$ *implies that* $\mathbf{x}_s$ *is a sink.*

Proof. Let $M=F'(\mathbf{x}_s)$. Since $\rho(M)<1$, there exists a norm $\|.\|_a$ such that $\|M\|_a<1$ (see Theorem 3.2.3). Let $k \in (\|M\|_a, 1)$. From Lemma 5.3.1 and the continuity of F' at $\mathbf{x}_s$ we derive the existence of $r<d$ such that $\|F'(\mathbf{x})\|_a \leq k$ for every $\mathbf{x} \in D(\mathbf{x}_s, r)$, where the closed ball $D(\mathbf{x}_s, r)$ has to be understood in the a-norm.

Let $\mathbf{x}_0 \in D(\mathbf{x}_s, r)$. By the mean value inequality we obtain

$$\|\mathbf{x}_1 - \mathbf{x}_s\|_a \leq \|F'(\mathbf{c})\|_a \|\mathbf{x}_0 - \mathbf{x}_s\|_a \leq k\|\mathbf{x}_0 - \mathbf{x}_s\|_a. \quad (5.3.1)$$

In general, after n steps, we arrive at

$$\|\mathbf{x}_n - \mathbf{x}_s\|_a \leq k^n \|\mathbf{x}_0 - \mathbf{x}_s\|_a. \quad (5.3.2)$$

Inequality (5.3.2) shows the stability of $\mathbf{x}_s$ (let $\delta=r$ in the definition of stability) and the convergence to $\mathbf{x}_s$ of every orbit $O(\mathbf{x}_0)$, with $\mathbf{x}_0 \in D(\mathbf{x}_s, r)$, in the a-norm. Since all norms are equivalent we obtain stability and convergence in the Euclidean norm. QED

The reader can see that Theorem 5.3.3 is the multidimensional version of the corresponding result for scalar maps (Theorem 2.2.6), with some minor adjustments in the proof's strategy.

Example 5.3.1 Let $F(\mathbf{x})=(x^2-y, -y^3+5z, z/2+x^2)$. Notice that $F(\mathbf{0})=\mathbf{0}$ and $\rho(F'(\mathbf{0}))$ $=.5$. Hence, $\mathbf{0}$ is an attractor for the dynamical system defined by F in $\mathbf{R}^3$. Not all trajectories of F converge to $\mathbf{0}$. For example, the orbit $O(\mathbf{x}_0)$, $\mathbf{x}_0=(1,1,1)$ is unbounded.

Recall that the origin is a sink for the linear dynamical system generated by a matrix M if $\rho(M)<1$, and that every orbit converges to $\mathbf{0}$. The addition of nonlinear terms usually destroys the property that every orbit converges to $\mathbf{0}$. The local behavior, however, is preserved if the nonlinear part is small near the origin. More precisely, the following result holds.

Corollary 5.3.1 *Let* $F: \mathbf{R}^q \to \mathbf{R}^q$ *be of the form* $F(\mathbf{x}) = M\mathbf{x} + h(\mathbf{x})$. *Assume that*

i. $\rho(M) < 1$.

ii. *There exist* $c > 0$, $a > 1$ *such that* $\|h(\mathbf{x})\| \leq c\|\mathbf{x}\|^a$.

Then the origin is a sink.

Proof. Notice that $F(\mathbf{0})=\mathbf{0}$ and $F'(\mathbf{0})=M$, since

$$\lim_{\|\mathbf{x}\|\to 0} \frac{h(\mathbf{x})}{\|\mathbf{x}\|} = \mathbf{0}.$$

Apply Theorem 5.3.3. QED

Example 5.3.2 Let

$$M=\begin{pmatrix} .5 & 3 \\ 0 & .25 \end{pmatrix},$$

and let $h(\mathbf{x})=(x^2,xy)$. The origin is a fixed point of the dynamical system governed by the function $F(\mathbf{x})=M\mathbf{x}+h(\mathbf{x})$. Moreover, since $h(\mathbf{x})$ contains only quadratic terms in the coordinates of $\mathbf{x}$, we obtain that $h(\mathbf{x})/\|\mathbf{x}\|\to\mathbf{0}$ as $\mathbf{x}\to\mathbf{0}$. Hence, Corollary 5.3.1 can be applied to the dynamical system governed by F. Since $\rho(M)=.5$, the origin is a sink.

We now establish for periodic orbits a result similar to Theorem 5.3.3. Its proof is based on a property of the spectrum for product of matrices. Given the integers k,...,1,0 write the string k,k–1,...,1,0,k,k–1,...,1,0. Any group of k+1 elements from this string is called a **cyclic permutation** of k,k–1,...,1,0 provided that the elements of the group are consecutive and the order in which they appear in the string is preserved. For example, for k=5, the group 2,1,0,5,4,3 is a cyclic permutation while 5,4,3,2,0,1 is not.

Proposition 5.3.1 *Let* $M_0,\dots,M_k$ *be* k+1 *matrices of size* q. *Then* $\sigma(M_k\dots M_0)$ *is independent of the cyclic permutation.*

Proof. We prove the result for k=2 and for the cyclic permutations 2,1,0 and 1,0,2. The reader is invited to adjust the proof to the general case.

Since $\det(M_2M_1M_0)=\det(M_1M_0M_2)$, $0\in\sigma(M_2M_1M_0)$ if and only if $0\in\sigma(M_1M_0M_2)$. Let $r\neq 0$, $r\in\sigma(M_2M_1M_0)$. For simplicity assume that r is real. There exists an eigenvector $\mathbf{x}\neq\mathbf{0}$, such that $M_2M_1M_0\mathbf{x}=r\mathbf{x}$. Notice that $M_1M_0\mathbf{x}\neq\mathbf{0}$. Apply M_1M_0 to both sides of $M_2M_1M_0\mathbf{x}=r\mathbf{x}$. We obtain

$$M_1M_0M_2M_1M_0\mathbf{x} = rM_1M_0\mathbf{x}.$$

Hence, $M_1M_0\mathbf{x}$ is an eigenvector of $M_1M_0M_2$ corresponding to the same eigenvalue r. QED

The following theorem gives a sufficient condition for a periodic orbit to be a sink. The result is the multi-dimensional version of Theorem 2.2.7.

Theorem 5.3.4 *Let* $U\subset\mathbf{R}^q$ *be open and let* $\{\mathbf{x}_0,\mathbf{x}_1,\dots,\mathbf{x}_{p-1},\dots\}$ *be a periodic orbit of period* p *of a continuous function* $F: U\to U$. *Assume that there exists* d>0 *such that* F *is differentiable in* $B(\mathbf{x}_i,d)$ *with continuous derivative at* $\mathbf{x}_i$ *for all* i=0,1, ..., p–1. *Then the periodic orbit is a sink provided that* $\rho(M_0)<1$ *where* $M_0=(F^p)'(\mathbf{x}_0)$.

Proof. By Theorem 5.3.3 the point $\mathbf{x}_0$ is a sink for the dynamical system defined by F^p. Let $M_i=(F^p)'(\mathbf{x}_i)$, i=1,2,...,p–1. By the chain rule we have

$$M_i = (F^p)'(\mathbf{x}_i) = F'(\mathbf{x}_{i-1})F'(\mathbf{x}_{i-2})...F'(\mathbf{x}_0)F'(\mathbf{x}_{p-1})...F'(\mathbf{x}_i).$$

The group i–1,...,1,0, p–1,...,i+1,i is a cyclic permutation of p–1,...,1,0, which is derived from the derivative of F^p at $\mathbf{x}_0$: $(F^p)'(\mathbf{x}_0)=F'(\mathbf{x}_{p-1})...F'(\mathbf{x}_0)$. Proposition 5.3.1 implies $\rho(M_i)<1$, i=0,1,...,p–1. It follows that each $\mathbf{x}_i$, i=0,1,...,p–1 is a sink for the dynamical system defined by F^p. Thus, the periodic orbit is a sink. QED

Example 5.3.3 Let $F(x,y)=(1-x^2+y,y^2)$. Then {(0,0), (1,0)} is a periodic orbit of period 2. We have

$$F_{\mathbf{x}}(\mathbf{0}) = \begin{pmatrix} 0 & 1 \\ 0 & 0 \end{pmatrix},\ F_{\mathbf{x}}(1,0) = \begin{pmatrix} -2 & 1 \\ 0 & 0 \end{pmatrix}.$$

Therefore,

$$F^2{}_{\mathbf{x}}(0,0) = \begin{pmatrix} 0 & -2 \\ 0 & 0 \end{pmatrix}$$

and $\rho(F^2{}_{\mathbf{x}})(0,0)=0$. The periodic orbit is a sink. Notice that $F^2{}_{\mathbf{x}}(1,0)$ is the zero matrix. Hence, $F^2{}_{\mathbf{x}}(0,0)\neq F^2{}_{\mathbf{x}}(1,0)$, but the two matrices have the same eigenvalues.

Remark 5.3.1 It is possible to obtain an extension of Theorem 2.2.8 to a higher dimensional setting, using again Lemma 2.2.3 and the property $F(L(\mathbf{x}_0))=L(\mathbf{x}_0)$. A similar extension is possible for periodic orbits.

Problems

1. Let $F(\mathbf{x})=(x^2-y,-y^3+5z,z/2+x^2)$. Find a fixed point of F different from **0** and analyze its stability.

2. Let $F(\mathbf{x})=(x^2+2y,y^2+3x^2)$. Find the eigenvalues of $F_{\mathbf{x}}(0,0)$ and verify that the point (0,0) is a sink.

3. * Let $F(\mathbf{x})=(ax+y,ay-xz+.01x,xy+.9z)$. For which values of the parameter a is the origin a sink?

4. * Let $F(\mathbf{x})=(.9x+.1y,.99y+.01rx-.01xz,.01xy+.97334z)$. For which values of r>0 is the origin a sink?

5. * Let $F(\mathbf{x})=(1.4x-.2xy,.96y+.2xy)$. Find the fixed points of this system. Verify that the origin is not a sink.

6. * Let $F(\mathbf{x})=(y^2-2x+1,2x-3y^2+1)$. Verify that the orbit starting at (0,0) is periodic. Is the periodic orbit a sink?

7. * Let $F(\mathbf{x})=(x\cos y - y\sin x, x\sin y - y\cos x)$. Verify that the orbit starting at the point (π, π) is periodic. Is the periodic orbit a sink?

Section 4. REPELLERS AND SADDLES

Repelling states

In the first part of this section we analyze those cases when all orbits which start close to a fixed point or to a periodic orbit will move away at some later time. The conditions which ensure this type of behavior are similar to the ones encountered in the preceding chapter for the linear case, although the results here do not have global validity.

Definition 5.4.1
Let $U\subset \mathbf{R}^q$ be open and $F : U\rightarrow U$ be continuous. A stationary state $\mathbf{x}_s\in U$ is said to be a **source** (a **repeller**) if there exists $r>0$ such that $0<\|\mathbf{x}_s-\mathbf{x}_0\|\leq r$ implies that $\|\mathbf{x}_s-\mathbf{x}_m\|>r$ for some $m\geq 1$.

Let $U\subset\mathbf{R}^q$ be open and $F : U\rightarrow U$ be continuous. A periodic orbit $\{\mathbf{x}_0,\ldots,\mathbf{x}_{p-1},\ldots\}$ (of period p) of F is a **source** (a **repeller**) if each state $\mathbf{x}_i$ is a source of F^p.

The following theorem gives a sufficient condition for a fixed point to be a source (see Theorem 2.2.9).

Theorem 5.4.1 *Let* $U\subset\mathbf{R}^q$ *be open and let* $\mathbf{x}_s$ *be a fixed point of a continuous function* $F : U\rightarrow U$. *Assume that* F *is differentiable in* $B(\mathbf{x}_s,d)$ *for some* $d>0$ *with continuous derivative at* $\mathbf{x}_s$. *Then* $|\lambda|>1$ *for all* $\lambda\in\sigma(F'(\mathbf{x}_s))$ *implies that* $\mathbf{x}_s$ *is a source.*

Proof. Since the eigenvalues of $M=F'(\mathbf{x}_s)$ have modulus larger than 1, M is invertible. The eigenvalues of the inverse are the reciprocal of the eigenvalues of M. Consequently the spectral radius of the inverse is smaller than 1, $\rho(M^{-1})<1$ and there is a norm $\|.\|_a$ such that $\|M^{-1}\|_a<1$. For every pair of vectors $\mathbf{x}$, $\mathbf{y}$ we have

$$\|\mathbf{x}-\mathbf{y}\|_a = \|M^{-1}M\mathbf{x} - M^{-1}M\mathbf{y}\|_a \leq \|M^{-1}\|_a\ \|M(\mathbf{x}-\mathbf{y})\|_a, \qquad (5.4.1)$$

which shows that

$$\frac{\|M(\mathbf{x}-\mathbf{y})\|_a}{\|\mathbf{x}-\mathbf{y}\|_a} \geq K \qquad (5.4.2)$$

with $K>1$ (K is the reciprocal of $\|M^{-1}\|_a$). Since F is differentiable in $B(\mathbf{x}_s,d)$ with continuous derivative at $\mathbf{x}_s$, we have

$$F(\mathbf{x}) - F(\mathbf{x}_s) = M(\mathbf{x}-\mathbf{x}_s) + o(\mathbf{x}-\mathbf{x}_s). \qquad (5.4.3)$$

The higher-order term $o(\mathbf{x}-\mathbf{x}_s)$ has the property

$$\lim_{\|\mathbf{x}-\mathbf{x}_s\|_a \to 0} \frac{o(\mathbf{x}-\mathbf{x}_s)}{\|\mathbf{x}-\mathbf{x}_s\|_a} = \mathbf{0}. \tag{5.4.4}$$

Thus, we can find $0<r_1\leq d$ such that $\|\mathbf{x}-\mathbf{x}_s\|_a\leq r_1$ implies that

$$\|o(\mathbf{x}-\mathbf{x}_s)\|_a/\|\mathbf{x}-\mathbf{x}_s\|_a \leq (K-1)/2.$$

From (5.4.3) we derive

$$\frac{\|F(\mathbf{x})-F(\mathbf{x}_s)\|_a}{\|\mathbf{x}-\mathbf{x}_s\|_a} \geq \frac{\|M(\mathbf{x}-\mathbf{x}_s)\|_a}{\|\mathbf{x}-\mathbf{x}_s\|_a} - \frac{\|o(\mathbf{x}-\mathbf{x}_s)\|_a}{\|\mathbf{x}-\mathbf{x}_s\|_a}$$

$$\geq K - \frac{\|o(\mathbf{x}-\mathbf{x}_s)\|_a}{\|\mathbf{x}-\mathbf{x}_s\|_a} \geq \frac{K+1}{2} = k > 1.$$

Let $0<\|\mathbf{x}_0-\mathbf{x}_s\|_a\leq r_1$. Then $\|\mathbf{x}_1-\mathbf{x}_s\|_a=\|F(\mathbf{x}_0)-F(\mathbf{x}_s)\|_a\geq k\|\mathbf{x}_0-\mathbf{x}_s\|_a$. Assume that $\|\mathbf{x}_1-\mathbf{x}_s\|_a\leq r_1$. Then, repeating the above argument we get $\|\mathbf{x}_2-\mathbf{x}_s\|_a\geq k^2\|\mathbf{x}_0-\mathbf{x}_s\|_a$. In general, as long as $\|\mathbf{x}_i-\mathbf{x}_s\|_a\leq r_1$ we can continue with the above inequalities and obtain

$$\|\mathbf{x}_{i+1}-\mathbf{x}_s\|_a \geq k^i\|\mathbf{x}_0-\mathbf{x}_s\|_a.$$

Since $k>1$ there exists a positive integer m such that $\|\mathbf{x}_m-\mathbf{x}_s\|_a>r_1$. The equivalence between the a-norm and the Euclidean norm ensures the existence of two constants, a,A, $0<a<A$ such that $a\|\mathbf{x}\|\leq\|\mathbf{x}\|_a\leq A\|\mathbf{x}\|$. Let $r=\min\{r_1/A,d\}$. Then $\|\mathbf{x}_0-\mathbf{x}_s\|\leq r$ implies that

$$\|\mathbf{x}_0-\mathbf{x}_s\|_a \leq A\|\mathbf{x}_0-\mathbf{x}_s\| \leq Ar \leq r_1.$$

Hence, according to the previous result, there exists $m\geq 1$ such that $\|\mathbf{x}_m-\mathbf{x}_s\|_a>r_1$. It follows that

$$A\|\mathbf{x}_m-\mathbf{x}_s\| \geq \|\mathbf{x}_m-\mathbf{x}_s\|_a > r_1, \quad \text{i.e.,} \quad \|\mathbf{x}_m-\mathbf{x}_s\| > r_1/A \geq r.$$

Thus, the orbit is a source with $r=r_1/A$. QED

Example 5.4.1 Let $F(x,y)=(x^2+2y,y^2+3x)$. The point (0,0) is a stationary state. Moreover, a simple computation shows that the two eigenvalues of the derivative of F at (0,0) have modulus larger than 1. Hence, (0,0) is a source.

We now establish for periodic orbits a companion result to Theorem 5.4.1.

Theorem 5.4.2 *Let* $U\subset\mathbf{R}^q$ *be open and let* $\{\mathbf{x}_0,\mathbf{x}_1,\ldots,\mathbf{x}_{p-1},\ldots\}$ *be a periodic orbit of period* p *of a continuous function* $F: U\to U$. *Assume that there exists* $d>0$ *such that* F *is differentiable in* $B(\mathbf{x}_i,d)$ *with continuous derivative at* $\mathbf{x}_i$ *for all* $i=0,1,\ldots,p-1$. *Then the periodic orbit is a source provided that* $M_0=(F^p)'(\mathbf{x}_0)$ *is invertible and* $\rho(M_0^{-1})<1$.

Proof. Since $\rho(M_0^{-1})<1$ we obtain that $|\lambda|>1$ for all $\lambda\in\sigma(M_0)$. Hence, by Theorem 5.4.1, $\mathbf{x}_0$ is a source for F^p. By Proposition 5.3.1 the eigenvalues of $M_i=(F^p)'(\mathbf{x}_i)$, $i=1,\ldots,p-1$ coincide with the eigenvalues of M_0. This implies that each $\mathbf{x}_i$ is a source for F^p. It follows that the periodic orbit is a source. QED.

Example 5.4.2 Let $F(x,y)=(2x+y^2-2,x^2-x-y^3)$. Then

$$\mathbf{x}_1 = F(1,1) = (1,-1) \quad \text{and} \quad F(1,-1) = (1,1) = \mathbf{x}_0.$$

Hence, $O(\mathbf{x}_0)$ is a periodic orbit of period 2. The two eigenvalues of $F^2{}_{\mathbf{x}}(1,1)$ have modulus larger than 1. The periodic orbit is a source.

Remark 5.4.1 (See also Remark 5.3.1.) It is possible to extend Theorem 2.2.11 to a higher-dimensional setting. A similar extension can be obtained for periodic orbits.

In the case of dynamical systems in several dimensions we find a situation similar to the one mentioned for dynamical systems in the real line. Given a repelling (or attracting) periodic orbit $O(\mathbf{x}_0)$ of period p, and an initial state $\mathbf{y}_0$ close to $\mathbf{x}_0$, $\|\mathbf{y}_n-\mathbf{x}_n\|$ may not grow (decrease) with n, but only with np, i.e., over an entire period. The following example illustrates this situation.

Example 5.4.3 Let $F(x,y)=(-(5/3)x^2+(2/3)y+1,-(5/3)y^2+(2/3)x+1)$. The orbit $\{\mathbf{x}_0=(0,0), \mathbf{x}_1=(1,1)\}$ is periodic of period 2. The eigenvalues of the derivative of F at the origin are $r_1=-r_2=-2/3$, while the eigenvalues at (1,1) are $s_1=-4$ and $s_2=-8/3$. The eigenvalues of the derivative of $F(F(\mathbf{x}))$ at the origin are $t_1=8/3$ and $t_2=-16/9$. Hence, the orbit is a source.

From $\mathbf{y}_0=(.01,.01)$ we get $\mathbf{y}_1=(1.0065,1.0065)$ and $\mathbf{y}_2\approx(-.0174,-.0174)$. We see that $\|\mathbf{y}_1-\mathbf{x}_1\|<\|\mathbf{y}_0-\mathbf{x}_0\|$, but $\|\mathbf{y}_2-\mathbf{x}_2\|=\|\mathbf{y}_2-\mathbf{x}_0\|\approx(16/9)\|\mathbf{y}_0-\mathbf{x}_0\|$. The separation constant is approximately equal to the absolute value of the smallest eigenvalue of the derivative of $F(F(\mathbf{x}))$ at the origin.

Saddles

So far we have examined the behavior of a dynamical system $F : \mathbf{R}^q\to\mathbf{R}^q$ around a fixed point or a periodic orbit in the case when all eigenvalues of the derivative have modulus smaller than 1 or larger than 1. Following the guidelines of the linear situation, we now analyze the **saddle case**. We shall confine ourselves to fixed points.

Definition 5.4.2
A stationary state $\mathbf{x}_s$ is said to be a **saddle** when some eigenvalues of $F'(\mathbf{x}_s)$ have modulus smaller than 1, and the others have modulus larger than 1.

In the linear case, we found an invariant subspace **S** corresponding to the group of eigenvalues with modulus smaller than 1, and another **U** corresponding to the eigenvalues with modulus larger than 1. They were called **stable** and **unstable subspaces** of the system. In the nonlinear case the situation is more complex.

Assume for simplicity that $\mathbf{x}_S=\mathbf{0}$ and F is invertible. Denote by G the inverse of F. Find **S** and **U** for the matrix $M=F'(\mathbf{0})$. These two subspaces constitute the first-order approximation of $\mathbf{M}_S$ and $\mathbf{M}_U$, which are now called the **stable** and **unstable manifolds** of F at **0**. The stable manifold $\mathbf{M}_S$ is defined as the set of those initial states $\mathbf{x}_0$ such that $F^n(\mathbf{x}_0)\to\mathbf{0}$ as $n\to\infty$, while the unstable manifold $\mathbf{M}_U$ is defined as the set of those initial states $\mathbf{x}_0$ such that $G^n(\mathbf{x}_0)\to\mathbf{0}$ as $n\to\infty$.

Example 5.4.4 Let $F(x,y)=(2x-y^3,.5y)$. Notice that $F(0,0)=(0,0)$ is the only fixed point of F. The inverse of F is the map

$$G(u,v) = .5(u + 8v^3, 4v).$$

The derivative of F at (0,0) is the matrix

$$F_{\mathbf{x}}(\mathbf{0}) = \begin{pmatrix} 2 & 0 \\ 0 & .5 \end{pmatrix}.$$

The stable subspace of this matrix is the one-dimensional subspace **S** spanned by $\mathbf{e}_2$, while the unstable subspace **U** is spanned by $\mathbf{e}_1$. Looking at G we see that the trajectory of any point of the form $(u,0)$ goes to the origin. Hence, the unstable manifold of F is $\mathbf{M}_U=\mathbf{U}$.

The stable manifold $\mathbf{M}_S$ of F is not the stable subspace **S** of $F'(\mathbf{0})$, because the orbit of an initial condition $\mathbf{x}_0=(0,y)$, with $y\neq 0$, goes to infinity. To find $\mathbf{M}_S$ let us select an initial condition $\mathbf{x}_0=(x,y)$ and find what relation between the two coordinates x and y will ensure that the orbit $O(\mathbf{x}_0)$ goes to **0**. After n+1 iterations we obtain

$$(2^n[2x - y^3 - 4^{-2}y^3 - \dots - 4^{-2n}y^3],\ 2^{-n-1}y).$$

The second component of this vector goes to zero, regardless of the value of y. The first one will go to zero if

$$2x = y^3 + 4^{-2}y^3 + \dots + 4^{-2n}y^3 + \dots = y^3(1 + 4^{-2} + 4^{-4} + \dots) = \frac{16}{15}y^3.$$

In fact, in this case

$$2x - (y^3 + 4^{-2}y^3 + \dots + 4^{-2n}y^3) = \frac{16}{15}y^3 4^{-2(n+1)},$$

and $2^n\,4^{-2(n+1)}\to 0$ as $n\to\infty$. Notice that 16/15 is the sum of the geometric series $1+4^{-2}+4^{-4}+\dots$. The stable manifold $\mathbf{M}_S$ of F at the origin is the curve $x=(8/15)y^3$ (see Fig. 5.4.1).

As we see, even in a simple case like Example 5.4.4, the problem of determining the stable and unstable manifold at a saddle point is not trivial. It is worth emphasizing once more the behavior of a saddle point (see a similar comment in Chapter 4), namely that every initial condition outside the stable manifold gives a trajectory that will move away from the saddle.

In Example 5.4.4 trajectories starting from every initial condition outside the curve $x=(8/15)y^3$ will go to infinity. The closer we are to the curve, the longer it takes to detect this behavior. At the beginning we may have the impression that the trajectory is moving toward the origin, but eventually the states will move closer to the unstable manifold and the distance between them and the origin will start increasing.

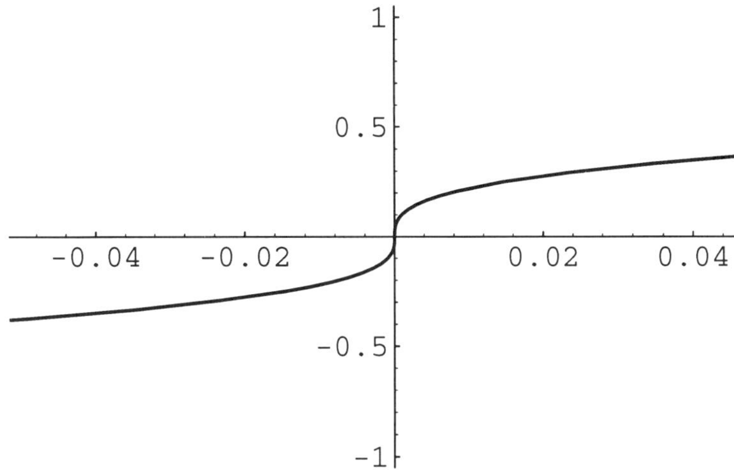

Fig. 5.4.1 The cubic $x=(8/15)y^3$ is the stable manifold of the dynamical system governed by the function $F(x,y)=(2x-y^3,.5y)$ at the origin.

The behavior of a dynamical system close to a saddle point is important in the study of the global behavior of the system. A trajectory leaving a saddle point along its unstable manifold and coming back to it along its stable manifold is called **homoclinic**. A trajectory connecting two saddle points is called **heteroclinic**. About one hundred years ago **Jules-Henri Poincaré** (1854-1912) discovered that the behavior of dynamical systems with heteroclinic or homoclinic trajectories is usually very complex. Poincaré can be rightly considered the father of the theory of dynamical systems.

Problems

1. Let $F(\mathbf{x})=(x^2+2y,y^2+3x)$. Show that (0,0) is a stationary state. Verify that (0,0) is a repeller.

2. * Let $F(\mathbf{x})=(2x+y^2-2,x^2-x-y^3)$. Show that the point (1,1) is periodic of period 2. Compute the eigenvalues of F^2 at the point (1,1). Conclude that the periodic orbit is a repeller.

3. * Let $F(r,\mathbf{x})=(.9x+.1y,\ .99y+.01rx-.01xz,.01xy+.97334z)$. Show that the origin is a saddle when $r>1$.

4. * The Hénon map is defined by $F(\mathbf{a},\mathbf{x})=(1-ax^2+y,bx)$, $a,b>0$. Find the fixed points of the Hénon map and check their stability in the case when $a=b=1$.

5. * Consider again the Hénon map with b=1. Check the stability of the fixed points.

6. Verify that the map $F(\mathbf{x})=(2x-y^3,.5y)$ is one-to-one and its inverse is the map $G(\mathbf{u})=.5(u+8v^3,4v)$.

7. Verify that after n+1 iterations the orbit starting from (x,y) and governed by the map F of Problem 6, reaches the state

$$(2^n[2x - y^3 - 4^{-2}y^3 - \ldots - 4^{-2n}y^3],\ 2^{-n-1}y).$$

8. Show that the orbit of Problem 7 will go to zero if the initial condition satisfies the equality $x=8y^3/15$. In particular, explain why the equality $x_0=(2y_0)^3/15$ implies that $x_{n+1}=1/(15\times 2^{4n})$.

9. * Verify that (0,0,–a–r) is a saddle point of the Lorenz map whenever r>1.

10. Verify that the origin is a saddle point of the predator-prey model for competition of species (1.4.10) without harvesting. Verify that in the presence of "heavy" fishing, the origin may become a sink.

11. Consider the dynamical system $\mathbf{x}_{n+1}=F(\mathbf{x}_n)$, where $F(\mathbf{x})=(2x,y/2+7x^2)$. Show that the origin is a saddle. Find the inverse of F and the stable and unstable manifolds of the origin.

Section 5. BIFURCATION

Dynamical systems in $\mathbf{R}^q$, q>1, offer bifurcation phenomena which are more complex than the ones encountered in one-dimensional systems. They can provide, although in a more complex setting, the same types of bifurcation points we found in the real line, and they have additional types that are not found in **R**. The following example gives an illustration of bifurcation phenomena similar to the ones occurring in **R**.

Example 5.5.1 Let

$$\begin{cases} x_{n+1} = ax_n(1 - x_n - y_n) \\ y_{n+1} = ax_ny_n. \end{cases}$$

We see that $(a,\mathbf{x}_1(a))$ with $\mathbf{x}_1(a)=(0,0)$ is a branch of fixed points regardless of the values of the parameter a. From

$$\begin{cases} x = ax(1 - x - y) \\ y = axy \end{cases}$$

we obtain, with $y=0$, the additional branch of fixed points

$$(a,\mathbf{x}_2(a))\ ,\ \mathbf{x}_2(a) = ((a-1)/a,\ 0).$$

For $y \neq 0$ we have

$$(a, \mathbf{x}_3(a)),\ \mathbf{x}_3(a) = (1/a,\ (a-2)/a).$$

It can be easily verified that $\mathbf{0}$ is a sink for $a \in (0,1)$ and is a saddle for $a>1$. From $F(a,\mathbf{x})=(ax(1-x-y), axy)$ we obtain

$$F_{\mathbf{x}}(a, \mathbf{x}_2(a)) = \begin{pmatrix} 2-a & 1-a \\ 0 & a-1 \end{pmatrix}.$$

The eigenvalues are $2-a$ and $a-1$. Hence, the fixed point $\mathbf{x}_2(a)$ is a sink for $a \in (1,2)$. At $a=2$ the stability is passed on to the fixed points of the branch $(a, \mathbf{x}_3(a))$. Notice that the two branches intersect each other at $a=2$, while at $a=1$ we have $\mathbf{x}_2(1)=\mathbf{0}$. Thus $a=1$ and $a=2$ are both bifurcation points.

However, even the simple map of Example 5.5.1 shows that the analysis of bifurcation points in $\mathbf{R}^q$, $q>1$, is more complex than the corresponding analysis in the one-dimensional case. We shall now mention two cases that are more frequently seen in applications and illustrate them with some examples. The theoretical justification of the results presented here will be omitted since it goes beyond the purpose of this book. We shall limit our study to one-parameter family of maps.

Bifurcation from the trivial branch: $\mathbf{x}(a)=\mathbf{0}$

We assume that the dynamical system has the form

$$\mathbf{x}_{n+1} = T(a, \mathbf{x}_n) = aM\mathbf{x}_n + F(a, \mathbf{x}_n), \tag{5.5.1}$$

where M is a matrix of size q. The function F and its derivative $F_{\mathbf{x}}$ are assumed to be continuous with $F(a,\mathbf{0})=\mathbf{0}$, and $F_{\mathbf{x}}(a,\mathbf{0})$ the zero matrix.

Thus, $\mathbf{x}_1(a)=\mathbf{0}$ is a fixed point of T for every value of a. We call $(a,\mathbf{x}_1(a))=(a,\mathbf{0})$ the *trivial* branch of fixed points of (5.5.1). We are interested in finding nonzero fixed points. More precisely, we are interested in finding those values a_0 that satisfy the conditions given in the following definition.

Definition 5.5.1

A parameter value a_0 is said to be a **bifurcation point** from the trivial branch if there exists a continuous function $\mathbf{x}_2 : J \to \mathbf{R}^q$, where J is an interval, $a_0 \in J$ and the interior of J is not empty, with the following properties

(i) $\mathbf{x}_2(a_0) = \mathbf{0}$.
(ii) $\mathbf{x}_2(a) \neq \mathbf{0}$ for all $a \in J$, $a \neq a_0$.
(iii) $\mathbf{x}_2(a)$ is a fixed point of $T(a,\mathbf{x})$ for all $a \in J$.

Notice that the situation presented in Example 5.5.1 can be reduced to the present setting with $M\mathbf{x}=(x,0)$ and $F(a,\mathbf{x})=a(-x^2-xy, xy)$. Here is another example.

Example 5.5.2 Let

$$T(a,\mathbf{x}) = T(a,(x,y)) = (ax(1+x),\ a(-x+2y+y^2)).$$

Then, $M\mathbf{x}=(x,-x+2y)$ and $F(a,\mathbf{x})=a(x^2,y^2)$. We are looking for the nontrivial solutions (x,y) of the system, depending on the parameter a (see Fig. 5.5.1),

$$\begin{cases} x = ax + ax^2 \\ y = a(-x + 2y + y^2). \end{cases}$$

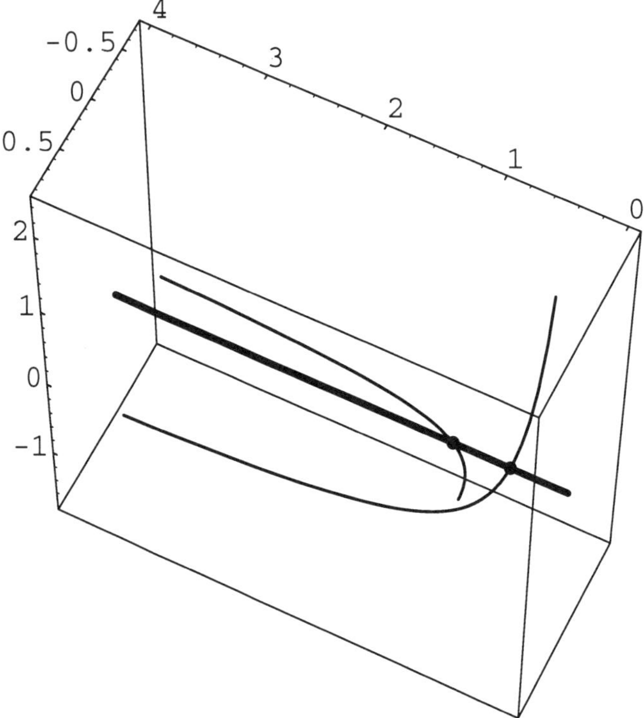

Fig. 5.5.1 Bifurcation diagram of the map $T(a,\mathbf{x})=T(a,(x,y))=(ax(1+x),\ a(-x+2y+y^2))$ from the branch of trivial solutions (the straight line through the two bifurcation points). The parameter a ranges from 0 to 4.

For $x \neq 0$, the first equation gives $x=(1-a)/a$. From the second we obtain $y=x$ and $y=-1$. The solution $y=-1$ is discarded since we are looking for bifurcation from the "trivial" branch $(a,\mathbf{0})$. We are left with the function

$$\mathbf{x}_2(a) = (a^{-1} - 1,\ a^{-1} - 1)\ .$$

Notice that $\mathbf{x}_2(1)=(0,0)=\mathbf{0}$. We can choose (0,2) as the interval J containing the point 1, such that for all $a \in J$ the other two properties are verified. Hence, $a_0=1$ is a bifurcation point.

For $x=0$, the second equation gives $y=0$ and $y=a^{-1}-2$. The first solution gives us again $\mathbf{x}_1(a)=\mathbf{0}$. From the other solution we derive

$$\mathbf{x}_3(a) = (0, a^{-1} - 2).$$

Since $\mathbf{x}_3(.5)=(0,0)=\mathbf{0}$, and in the interval J=(0,1) the function $\mathbf{x}_3(a)$ satisfies all the required properties, $a_0=.5$ is another bifurcation point.

Notice that the two bifurcation points 1 and .5 are the reciprocals of the eigenvalues of the matrix M since $\sigma(M)=\{1,2\}$. Hence, 1 is an eigenvalue of aM in both cases. Just as in the one-dimensional case, it can be shown that this condition is **necessary** for a parameter a_0 to be a bifurcation point from the branch of trivial solutions, i.e., 1 must be an eigenvalue of a_0M. Having such a property, however, does not guarantee that a_0 is in fact a bifurcation point as the following example shows.

Example 5.5.3 Let $T(a,\mathbf{x})=a(x,y)-(y^3,-x^3)$. Then $(a,\mathbf{0})$ is a branch of fixed points (the trivial branch). The only eigenvalue of M is 1. Hence, a=1 is the only possible bifurcation point from the trivial branch. However with $a\neq1$ and from

$$(x, y) = a(x, y) - (y^3, -x^3),$$

we get $(1-a)x=-y^3$ and $(1-a)y=x^3$. Multiplying the first equality by $-y$, the second by x and adding them together we obtain $x^4+y^4=0$, whose only solution is $x=y=0$. Thus the system does not have any branches of nontrivial fixed points. Notice that the eigenvalue 1 has algebraic multiplicity 2.

The necessary condition stated above becomes **sufficient** if the eigenvalue 1 of a_0M has **odd algebraic multiplicity**.

Example 5.5.4 Let $T(a,\mathbf{x})=a(x+z,2y+6z,y+z)+(ax^2-y^2,y^2-z^2,z^3)$. The matrix M is

$$M = \begin{pmatrix} 1 & 0 & 1 \\ 0 & 2 & 6 \\ 0 & 1 & 1 \end{pmatrix}$$

and $F(a,\mathbf{x})=(ax^2-y^2,y^2-z^2, z^3)$. Thus, $F(a,\mathbf{0})=\mathbf{0}$ and $F_{\mathbf{x}}(a,\mathbf{0})$ is the zero matrix. The eigenvalues of M are 1, –1, and 4. Since they are all simple we obtain that 1, –1, and 1/4 are bifurcation points. The reader can see that it is quite challenging to determine explicitly as functions of a the three different branches of nonzero fixed points.

Remark 5.5.1 In Example 5.5.2 we did not consider the fixed points $\mathbf{x}_4(a)=(a^{-1}-1,-1)$ since the branch $(a,\mathbf{x}_4(a))$ does not intersect the trivial branch. However, it intersects $(a,\mathbf{x}_3(a))$, $\mathbf{x}_3(a)=(0,a^{-1}-2)$ when a=1. We already know that $a_0=1$ is a bifurcation point. However, in the present situation, the bifurcation (of transcritical type) is taking place along the branch $(a,\mathbf{x}_3(a))$ [or $(a,\mathbf{x}_4(a))$] rather than along the trivial branch $(a,\mathbf{0})$. Could we have expected this occurrence? Since our previous discussion regarded only bifurcation from the trivial branch, no information was given regarding other possibilities. Hence, it seems appropriate to provide our readers with a few

comments on a feature of all potential bifurcation points at least when only branches of fixed points are involved.
Consider the function

$$H(a,\mathbf{x}) = \mathbf{x} - F(a,\mathbf{x}) \tag{5.5.2}$$

where F is at least C^1. Notice that H is a map from $\mathbf{R}\times\mathbf{R}^q$ into $\mathbf{R}^q$. Since $\dim\mathrm{Ker}H'(a,\mathbf{x})+\dim\mathrm{Im}H'(a,\mathbf{x})=q+1$ and $\dim\mathrm{Im}H'(a,\mathbf{x})\leq q$ we see that the kernel of $H'(a,\mathbf{x})$ is at least one-dimensional. Assume that there is a "known" branch $\alpha(a)=(a,\mathbf{x}(a))$ of fixed points of F, defined in some open interval J, and that $\dim\mathrm{Ker}H'(a,\mathbf{x}(a))=1$ for a generic value of $a\in J$. The possible bifurcation points are those values of a for which $\dim\mathrm{Ker}(H'(\alpha(a)))>1$.

Let us verify this feature in the case of Example 5.5.2. We first compute the derivative of H which is the map from $\mathbf{R}\times\mathbf{R}^2\to\mathbf{R}^2$ defined by

$$H(a,\mathbf{x}) = (x - ax(1 + x),\ y - a(-x + 2y + y^2)).$$

Its derivative is

$$H'(a,\mathbf{x}) = \begin{pmatrix} -x-x^2 & 1-a-2ax & 0 \\ x-2y-y^2 & a & 1-2a-2ay \end{pmatrix}.$$

Evaluating it along the curve $\alpha(a)=(a,\mathbf{x}_3(a))=(a,0,a^{-1}-2)$ we obtain

$$H'(\alpha(a)) = \begin{pmatrix} 0 & 1-a & 0 \\ -1/a^2+2/a & a & -1+2a \end{pmatrix}.$$

We see that the vector $\alpha'(a)=(1,0,-a^{-2})\in \mathrm{Ker}H'(\alpha(a))$ and the rank of $H'(\alpha(a))$ is always 2 except for a=1 and a=1/2. At a=1 or a=1/2 we have $\mathrm{rank}(H'(\alpha(a)))=1$. Hence, the $\dim\mathrm{Ker}(H'(\alpha(1)))=\dim\mathrm{Ker}(H'(\alpha(1/2)))=2$. It follows that a=1 and a=1/2 are the only possible bifurcation points. In this case both values are bifurcation points since at a=1 the branch $(a,\mathbf{x}_3(a))$ intersects $(a,\mathbf{x}_4(a))$ and at a=1/2 it intersects the trivial branch $(a,\mathbf{0})$.

Hopf bifurcation

We shall confine our attention to $\mathbf{R}^2$. Consider the one-parameter family of maps

$$\mathbf{x}_{n+1} = F(a,\mathbf{x}_n),\ \mathbf{x}\in\mathbf{R}^2. \tag{5.5.3}$$

We assume that F is differentiable at least twice, with continuous second derivative (F is of class C^2). Denote the eigenvalues of $F_\mathbf{x}(a,\mathbf{x})$ by $r_1(a)$ and $r_2(a)$.

Definition 5.5.2
Assume that there is a branch of fixed points $(a,\mathbf{x}(a))$ such that for a in an open interval J the eigenvalues of $F_\mathbf{x}(a,\mathbf{x}(a))$ are complex, i.e.,

$$r_1(a) = \rho(a)[\cos\theta(a) + i \sin\theta(a)];$$
$$r_2(a) = \rho(a)[\cos\theta(a) - i \sin\theta(a)]. \qquad (5.5.4)$$

Since F is of class C^2, the two functions $\rho(a)$ and $\theta(a)$ are C^1. Let $a_0 \in J$ be such that $\rho(a)<1$ for $a<a_0$ and $\rho(a)>1$ for $a>a_0$. Assume that

(j) $\frac{d}{da}\rho(a) > 0$ at $a = a_0$.

(jj) $\theta(a_0)/\pi$ is irrational or $\theta(a_0)/2\pi=m/n$, m and n relatively prime and $n \neq 3, 4$.

Then, at $a=a_0$ we have **Hopf bifurcation**. Orbits close to the branch of fixed points $(a,\mathbf{x}(a))$ (the fixed points are sources for $a>a_0$) are attracted by either an elliptical invariant curve (when $\theta(a_0)/\pi$ is irrational) or by a periodic orbit of period n when $\theta(a_0)/(2\pi)=m/n$, m and n relatively prime and $n\neq3$, 4. In the cases n=3 or n=4 the bifurcation picture is more complicated.

Example 5.5.5 Let $\mathbf{x}_{n+1}=F(a,\mathbf{x})$, where $F(a,\mathbf{x})=(ax(1-x-y),axy)$. There are three branches of stationary states:

$(a,\mathbf{0})$, $(a,\mathbf{x}_{s2}(a))$, with $\mathbf{x}_{s2}(a)=(1-a^{-1},0)$ and $(a,\mathbf{x}_{s3}(a))$ with $\mathbf{x}_{s3}(a)=(a^{-1},1-2a^{-1})$.

The matrix $F_\mathbf{x}(a,\mathbf{x})$ evaluated along the branch $(a,\mathbf{x}_3(a))$ is

$$F_\mathbf{x}(a,\mathbf{x}_{s3}(a)) = \begin{pmatrix} 0 & -1 \\ a-2 & 1 \end{pmatrix}.$$

The eigenvalues of this matrix are complex for $a>2.25$ and their modulus is smaller than 1 for $a\in(2.25,3)$. At $a_0=3$ we have a Hopf bifurcation. Moreover, $\theta(3)/(2\pi)= 1/6$. The orbits close to $(a^{-1},1-2a^{-1})$ for $a>3$ are attracted by a periodic orbit of period 6, that is itself a function of a (see Fig. 5.5.2). This example will be studied in Chapter 7. Notice that along the curve $\alpha(a)=(a,a^{-1},1-2a^{-1})$ the derivative of the function $H(a,\mathbf{x})=\mathbf{x}-F(a,\mathbf{x})$ is

$$H'(\alpha(a)) = \begin{pmatrix} -1/a^2 & 1 & 1 \\ -1/a+2/a^2 & 2-a & 0 \end{pmatrix}.$$

Hence, $H'(\alpha(a))\alpha'(a)=(0,0)=\mathbf{0}$, as expected (see Remark 5.5.1). The rank of $H'(\alpha(a))$ goes down to 1 only when a=2, for which $\mathbf{x}_{s3}(a)=\mathbf{x}_{s2}(a)$. Hopf's bifurcation takes place for a=3 and it does not fall into the scheme of Remark 5.5.1. This is expected since the periodic orbit is not a solution of $\mathbf{x}-F(a,\mathbf{x})=\mathbf{0}$.

There is no counterpart to Hopf bifurcation in the one-dimensional case. This topic, as well as the others presented before, require a mathematical background more sophisticated than the one on which this book is based. We have mentioned them here to make our readers aware that discrete dynamical systems have a lot more to say than what we can include in these notes. Upon acquiring greater mathematical skills, the reader will be able to gain further insight into this fascinating field.

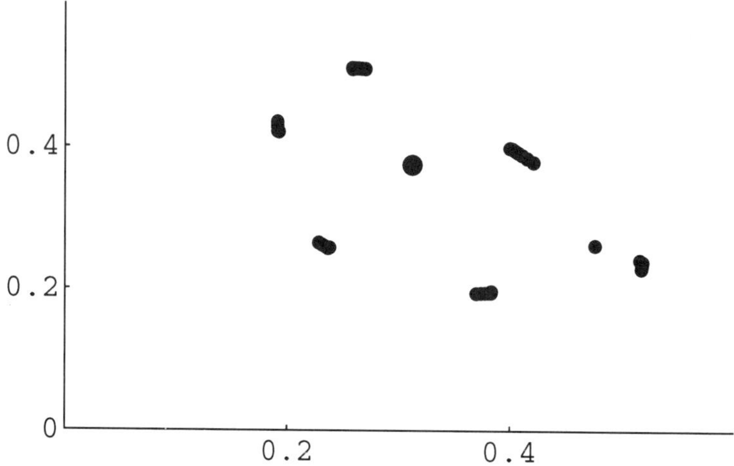

Fig. 5.5.2 The orbit starting at (.4,.1) for a=3.2 converges to a periodic orbit of period 6. The center point is the stationary state $\mathbf{x}_3(3.2)$.

Problems

1. Work out the details of Example 5.5.2. In particular verify that the function $\mathbf{x}_{s2}(a) = (a^{-1}-1, a^{-1}-1)$ satisfies the two properties (ii) $\mathbf{x}_{s2}(a) \neq \mathbf{0}$, $a \in J$, $a \neq 1$; (iii) for every real number $a \in J$ we have $\mathbf{x}_{s2}(a)=T(a,\mathbf{x}_{s2}(a))$.

2. * Study the stability of the stationary states along the bifurcating branches of Example 5.5.2.

3. * Let $\mathbf{x}_{n+1}=F(a,\mathbf{x}_n)$, where $F(a,\mathbf{x})=(ax(1-x-y), .5y+axy)$, $a \in (0,4]$. Find the stationary states and analyze their stability.

4. * Let $F(a,\mathbf{x})=(y, ay(1-x/2-y/2))$. Consider the dynamical system governed by F. Find its stationary states and determine their stability. Does Hopf' bifurcation take place at a=3?

5. * Let $F(a,\mathbf{x})=(ax(1-x-y), axy)$. Verify the details mentioned in Example 5.5.5 about the eigenvalues of $F_\mathbf{x}$ along the branch $(a,\mathbf{x}_{s3}(a))$, where $\mathbf{x}_{s3}(a) =(a^{-1}, 1-2a^{-1})$.

6. Study the presence of Hopf's bifurcation along the branches of fixed points of the dynamical system defined by the map of Problem 3.

CHAPTER 6

CHAOTIC BEHAVIOR

SUMMARY

In the first section of this chapter we introduce the notion of attractor of a dynamical system. In Section 2 we propose a definition of discrete chaotic system and we compare it with some definitions given by other authors. In the third section we talk about the capacity and the correlation dimension of an invariant set. In Section 4 we present the Lyapunov exponents.

Section 1. **ATTRACTORS**

We have already encountered the word **attractor**. We used it as another name for sink in the case of a stationary state or a periodic orbit.

Example 6.1.1 Let $F(x)=2x(1-x)$. The stationary state $x_s=.5$ is an attractor since $F'(.5)=0$. Orbits starting at points close to .5 converge to .5. Actually, in this case it can be shown that every orbit $O(x_0)$, $x_0 \in (0,1)$ converges to .5.

Example 6.1.2 Consider the dynamical system $x_{n+1}=F(a,x_n)=ax_n(1-x_n)$. Recall that for a>3 there is the unique periodic orbit of period 2:

$$x_1(a)=\frac{a+1+\sqrt{a^2-2a-3}}{2a}, \qquad x_2(a)=\frac{a+1-\sqrt{a^2-2a-3}}{2a}.$$

Let a=3.2. Then

$$x_1(3.2)=\frac{2.1+\sqrt{.21}}{3.2}, \qquad x_2(3.2)=\frac{2.1-\sqrt{.21}}{3.2},$$

and $F_x(3.2,x_1(3.2))F_x(3.2,x_2(3.2))=.16$. The periodic orbit (see Fig. 6.1.1) is an attractor. Orbits with initial state close to either $x_1(3.2)$ or $x_2(3.2)$ converge to the periodic orbit (see Remark 2.2.3).

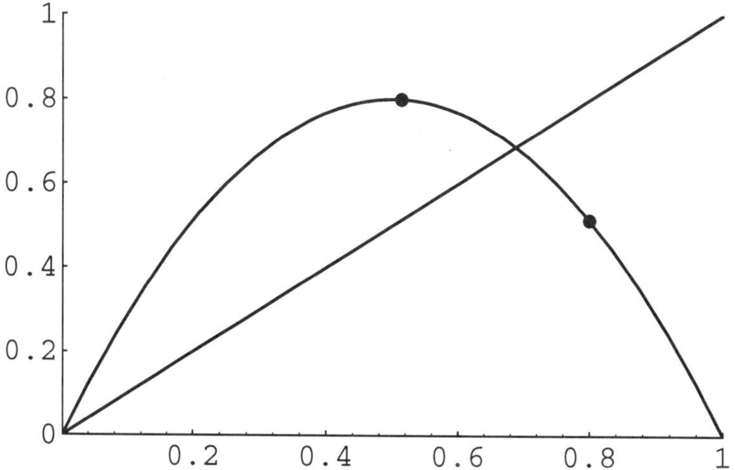

Fig. 6.1.1 Attracting periodic orbit of period 2 of the logistic map (a=3.2). Starting at $x=.3$, we plotted (x_n,x_{n+1}), n=180,...,200.

The analysis of the long-range behavior of a dynamical system requires a definition of attractor, which is applicable to sets more general than stationary states and periodic orbits. It does not require the stringent conditions we selected in the previous definitions of sinks in **R** and $\mathbf{R}^q$. Until now we have used **sink** and

attractor interchangeably. Example 6.1.3 (see [Martelli-Marshall, 1995]) suggests that a different definition of attractor is needed, since every orbit of the system converges to an unstable stationary state. Hence, it appears that attractivity and instability should not be mutually exclusive.

Example 6.1.3 Let F : $\mathbf{R}^2 \to \mathbf{R}^2$ be defined by $F(x,y)=(h_1(x,y), h_2(x,y))$, where

$$h_1(x,y) = g_1(x,y)\cos(k(x,y)) - g_2(x,y)\sin(k(x,y))$$
$$h_2(x,y) = g_1(x,y)\sin(k(x,y)) + g_2(x,y)\cos(k(x,y))$$

with

$$k(x,y) = \begin{cases} .1((x-1)^2+y^2)\arccos(\frac{y^2}{(x-1)^2+y^2}) & (x-1)^2+y^2>0 \\ 0 & (x,y)=(1,0) \end{cases}$$

$$g_1(x,y) = x + d(x,y)(x-1),$$
$$g_2(x,y) = y + d(x,y)\,y,$$

and

$$d(x,y) = .5 - (x^2+y^2)((x^2+y^2+1).$$

The function F is C^1 and every orbit of the system governed by F converges to the unique fixed point $\mathbf{x}_s=(1,0)$. More precisely, the orbits of points inside $D(\mathbf{0},1)$ approach $S(\mathbf{0},1)$ from the inside while rotating counterclockwise toward (1,0). The orbits of points outside $D(\mathbf{0},1)$ approach $S(\mathbf{0},1)$ from the outside while rotating counterclockwise. They may rotate more than 2π before converging to (1, 0).

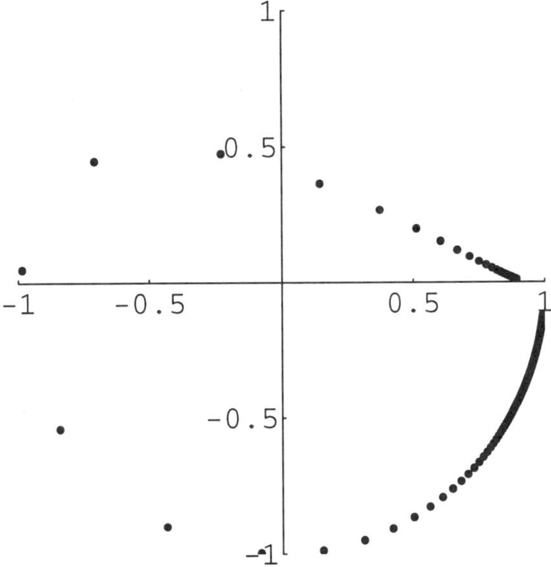

Fig. 6.1.2 First 1000 states of O((.9,.01)). The orbit is "moving away" from the fixed point in the short run, but it "comes back to it" in the long run.

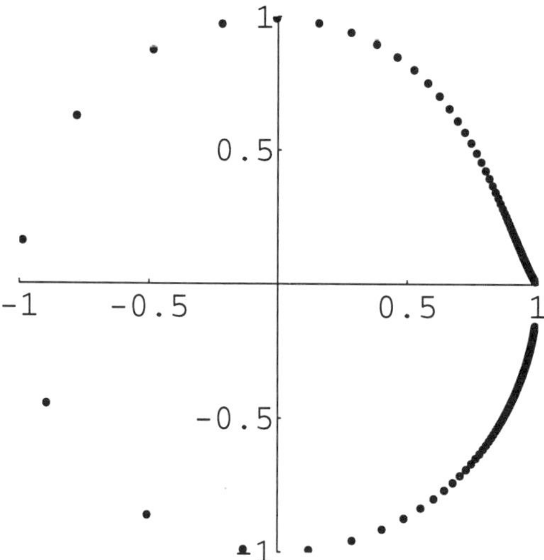

Fig. 6.1.3 First 1000 states of O((.99,.01)). We observe the same behavior as in the previous case.

In Fig. 6.1.2 we have plotted the first 1000 states of $O(\mathbf{x}_0)$, where $\mathbf{x}_0$=(.9,.01). In Fig. 6.1.3 we have plotted the first 1000 states of $O(\mathbf{y}_0)$, where $\mathbf{y}_0$=(.99,.01). Although every orbit of the system converges to the point $\mathbf{x}_s$=(1,0), we see that $\mathbf{x}_s$ is not a sink since it is unstable. In fact, given r=.5 and for every $\delta>0$ we can find an initial condition $\mathbf{y}_0 \in B(\mathbf{x}_s,\delta)$ and a positive integer $m \geq 1$ such that $\|\mathbf{y}_m - \mathbf{x}_s\| \geq .5$. Hence, $\mathbf{x}_s$ is unstable.

Example 6.1.3 shows that the properties of attractivity and instability may coexist. One deals with the long-term behavior of the orbits (attractivity), while the other deals with their short-term evolution. Consequently, we propose a more general definition of attractor. As a consequence of this definition a sink remains an attractor, but an attractor may not be a sink. We first introduce the idea of distance between a point and a set.

Given a subset Z of $\mathbf{R}^q$ and $\mathbf{x} \in \mathbf{R}^q$, we define the distance from $\mathbf{x}$ to Z by

$$d(\mathbf{x},Z) = \inf\{ \|\mathbf{x} - \mathbf{z}\| : \mathbf{z} \in Z\}. \tag{6.1.1}$$

Notice that $d(\mathbf{x},Z)$ is the **greatest lower bound** of the set of nonnegative real numbers

$$K=\{\|\mathbf{x} - \mathbf{z}\| : \mathbf{z} \in Z\}.$$

In other words, $d(\mathbf{x},Z)$ is not larger than $\|\mathbf{x}-\mathbf{z}\|$ for every $\mathbf{z} \in Z$ and it is the largest number having this property.

Example 6.1.4 Let $Z=\{\mathbf{x} \in \mathbf{R}^2: |x|+|y| \leq 1\}$ and let $\mathbf{x}_1=(1,2)$, $\mathbf{x}_2=(1,1)$. Then $d(\mathbf{x}_1, Z)=2^{.5}$ while $d(\mathbf{x}_2,Z)=1/2^{.5}$. The reader can easily verify the two results graphically.

Definition 6.1.1
Let $U \subset \mathbf{R}^q$ be open and let $F : U \to U$. A closed and bounded set $\mathbf{A} \subset U$ is an **attractor** if $F(\mathbf{A})=\mathbf{A}$ and there exists $r>0$ such that $d(\mathbf{x}_0,\mathbf{A}) \leq r$ implies that $d(\mathbf{x}_n,\mathbf{A}) \to 0$ as $n \to \infty$.

We require that $d(\mathbf{x}_n,\mathbf{A}) \to 0$ as $n \to \infty$, but we do not impose any restriction on the behavior of the orbit for n small. In other words, the orbit may go away from **A** in the short run, as long as it comes back to it in the long run. Consequently, a sink is an attractor, but an attractor need not be a sink.

Attractors are of primary importance in the study of the long-range behavior of a system. It is not unusual for their geometry to be complicated to the point that some of them have been called **strange attractors**. The name strange attractor first appeared in print in a paper of Ruelle-Takens (see [Ruelle-Takens, 1971]). It is an expression well-suited for describing these astonishing and poorly understood objects. The following example in **R** gives a taste of their "complicated" structure.

Example 6.1.5 Consider the dynamical system $x_{n+1}=F(x_n)=3.67x_n(1-x_n)$ in the invariant set $U=[0,1]$. We now investigate numerically the attractor of F by selecting two different initial conditions $x_0=.3$ and $y_0=.7$, running the orbits starting from them for a long time (2000 iterates), and plotting (x_n,x_{n+1}), (y_n,y_{n+1}) for n=1401, ...,2000 (see Fig. 6.1.4). The purpose of not plotting the first 1400 iterates is to avoid the "transient states" of the system and to look only at the long-term behavior of its orbits.

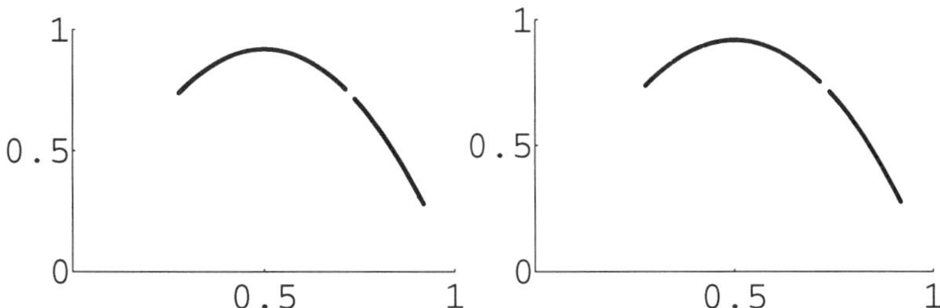

Fig. 6.1.4 On the left is the orbit of .3 and on the right the one of .7. To visualize the orbits we have plotted the points (x_n,x_{n+1}), (y_n,y_{n+1}) for n=1401,...,2000. They belong to the graph of F.

As we see, the two sets of points are very similar. An approximation to the attractor of F in [0,1] is obtained by considering only the first coordinates of the points. Although we cannot provide an exact configuration of it, we see that its geometry is not simple.

To have a better feeling for the attractor we "blow up" a portion of the second graph (see Fig. 6.1.5), by focusing our attention only on the two intervals [.3,.4] (left) and [.34,.38] (right).

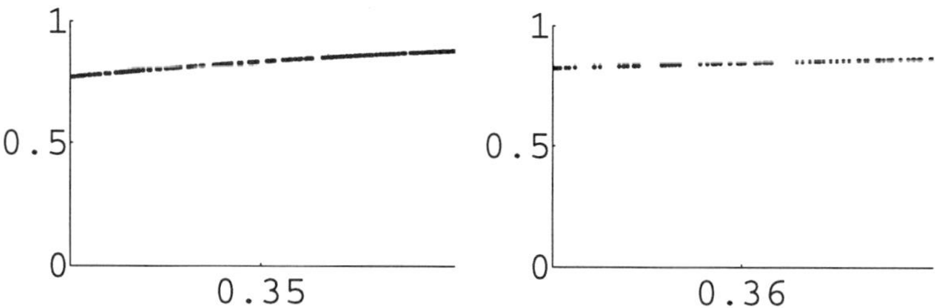

Fig. 6.1.5 A blowup of the second graph of Fig. 6.1.4 in the intervals [.3,.4] (left) and [.34,.38] (right) shows the fine structure of the attractor of $F(x)=3.67x(1-x)$.

Definition 6.1.2
Given an attractor **A**, we call the **basin of attraction** of **A** the set of all initial conditions $\mathbf{x}_0$, such that $d(\mathbf{x}_n, \mathbf{A})\to 0$ as $n\to\infty$.

Different basins of attraction are separated by **basin boundaries**. The geometry of these boundaries is frequently as complex as the geometry of the attractors themselves.

Example 6.1.6 Let

$$F(x) = \begin{cases} x^2 & -3\le x\le 1 \\ 4\sqrt{x}-3 & 1<x\le 9. \end{cases}$$

F maps $I=[-3,9]$ into itself. The fixed point $x_{s1}=0$ is an attractor and its basin of attraction is the interval $(-1,1)$. Another attractor is $x_{s2}=9$ and its basin of attraction is the union of the two intervals $[-3,-1)$ and $(1,9]$. The fixed point $x_{s3}=1$ is a source(see Fig. 6.1.6)

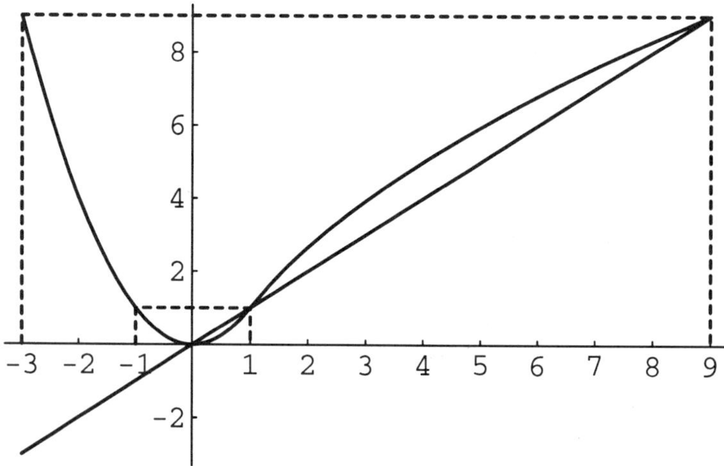

Fig. 6.1.6 The graph shows that the domain of attraction of $x_{s2}=9$ is $[-3,-1)\cup(1,9]$, while the domain of attraction of the origin is $(-1,1)$.

The study of dynamical systems frequently focuses on the attractor(s) of the system, whenever we are interested in its long-range behavior. Frequently, chaos is studied on attractors. However, as we shall see in the next section, when we say that a certain dynamical system is chaotic on an invariant set $X \subset \mathbf{R}^q$, we should not assume that X is an attractor.

Problems

1. Investigate the domain of attraction of the origin for the map

$$F(x) = \begin{cases} x/2 & 0 \le x \le .2 \\ 3x-1/2 & .2 < x \le 1/2 \\ 2-2x & 1/2 < x \le 1. \end{cases}$$

2. Show that the map of Problem 1 has a periodic orbit of period 3. What can be said as a consequence of Theorem 1.3.1 (Li-Yorke)?

3. * Let $F(x) = (1/4)x^2-1$. Find an interval with a nonempty interior that is mapped into itself by F and find the attractor(s) of F in the interval.

4. Prove that the set of all limit points of a system governed by a function F : $[0,1] \to [0,1]$ where F is continuous cannot be finite if the system has a periodic orbit of period 6. (Hint: Use Theorem 2.2.2.)

5. Assume that the limit set of an orbit of a continuous dynamical system is made of two points. Is it true that the set is a periodic orbit of period 2?

6. Assume that the set of limit points of an orbit has exactly three points. Is it possible that one of these points is a fixed point?

7. * a) Let $F(x)=2x(1-x)$. Verify that $x_s=1/2$ is a stationary state of F.
b) Choose $x_0=1/2+h$ with $|h|<1/2$. Verify that $x_n \to 1/2$ as $n \to \infty$. Find an attractor of F in [0,1].

8. * Investigate the attractor(s) of the system $x_{n+1}=F(x_n)$ with $F(x)=1.8x^2-1$.

Section 2. CHAOTIC DYNAMICAL SYSTEMS

Before presenting our definition of chaotic systems, we would like to discuss some recent developments that have motivated the attention reserved today to chaotic dynamics.

In 1963, **E. N. Lorenz** [Lorenz, 1963], a meteorologist at MIT, proposed the system of differential equations presented in Chapter 1 as a crude model of atmospheric behavior. In Chapter 5 we proved that the discrete dynamical system derived from it with the one-step Euler method has a bounded invariant region in $\mathbf{R}^3$ that attracts all orbits. The same is true for the orbits of the continuous model.

By selecting realistic values of the control parameters a, b, and r and solving the system numerically, Lorenz noticed that very minor changes in the initial conditions could produce significantly different outcomes. More precisely, the distance between corresponding states of orbits very close at t=0 could oscillate aperiodically in an open interval (0,b). Lorenz published the paper *Deterministic Nonperiodic Flow* to make the scientific community aware of this unexpected behavior. It was suggested that this discovery implied that no long-range prediction of atmospheric changes would ever be possible.

In 1975, the mathematicians **T. Li** and **J. A. Yorke** [Li-Yorke, 1975] published a paper in which they proved a surprising theorem. The presence of a periodic orbit of period 3 in a dynamical system defined by a continuous map F of an interval I into itself implies the presence of uncountably many aperiodic and unstable orbits. The title of their paper was *Period Three Implies Chaos*. The word chaos appeared in this paper for the first time, as suggestive of the dynamic complexity of the system studied by Li-Yorke. We can safely say that it has been a remarkably successful name.

In 1976, just one year after the paper of Li-Yorke, the biologist **R. M. May** [May, 1976] in the paper *Simple Mathematical Models with Very Complicated Dynamics* illustrated the surprisingly complex orbits of a one-dimensional dynamical system in the real line governed by a quadratic map.

Around this time the interest in dynamical systems having characteristics similar to the one proposed by Lorenz and to those analyzed by Li-Yorke and May started picking up considerably. The theory of chaotic systems emerged. As R. Devaney [Devaney, 1989] points out, "the field of dynamical systems and especially the study of chaotic systems has been hailed as one of the important breakthroughs in science in this century."

Three observations need to be made here. The **first** is that the use of computer-aided investigation was fundamental in the discovery of Lorenz and has been important in subsequent developments. On the one hand, all major ideas and theorems would certainly have been detected and established without the aid of numerical evidence. On the other, the use of computers has certainly played a significant role in providing the theory with supporting evidence.

The **second** is that **H. Poincaré** should be credited as the first to propose, about one hundred years ago, a global rather than a local analysis of dynamical systems, and to realize that the presence of homoclinic or heteroclinic trajectories was accompanied by very complex behavior. His footsteps were followed in the United States by **G. D. Birkhoff** [Birkhoff, 1927] with his classical book *Dynamical Systems*, and in the Soviet Union by **L. Pontrjagin** and others. While the attention of these researchers was focused primarily on continuous systems, the French mathematicians **G. Julia** [Julia, 1918] and **P. Fatou** [Fatou, 1926] were studying the extremely complicated dynamics provided by the iteration of rational functions in the complex plane.

Around 1960, several scientists gave renewed impulse to the study of dynamical systems. Among them we find **S. Smale**, **J. Moser**, **A. N. Kolmogorov**, **V. I. Arnold**, and **Y. G. Sinai**. Therefore, the field was ready for the major developments that have taken place in the last thirty years.

The **third** and final observation is that chaotic dynamical systems have always been an interdisciplinary field. Scientists from areas other than mathematics have played and continue to play a significant role in the development of the theory. Hardly any mathematical topic has sparked so much cross-cultural interest as this one and the trend continues unabated. Side by side with technical studies we find less formal presentations (see, for example, [Stewart, 1989]) and books written for a more general audience (see [Gleick, 1987]).

We now define chaotic discrete dynamical systems in the way we consider most appropriate for the intended audience of this book. Later, definitions proposed by other authors will be discussed and compared with the one we have selected. Two preliminary concepts are needed: instability of an orbit $O(\mathbf{x}_0)$ with respect to an invariant set X and density of $O(\mathbf{x}_0)$ in X.

Definition 6.2.1
Let $X \subset \mathbf{R}^q$ be invariant under the action of F, i.e., $F(X) \subset X$ and let $\mathbf{x}_0 \in X$. The orbit of $\mathbf{x}_0$ is said to be **unstable with respect to** X (or simply **unstable in** X) if in the definition of instability given in Chapter 1, the points $\mathbf{y}_0$ are required to be in X.

Definition 6.2.1 is more restrictive than the definition of instability introduced in Chapter 1. First, no orbit $O(\mathbf{x}_0)$ on a finite set X can satisfy it, since for every $\varepsilon>0$ we must find $\mathbf{y}_{0\varepsilon} \in X$, $\mathbf{y}_{0\varepsilon} \neq \mathbf{x}_0$ such that $\|\mathbf{y}_{0\varepsilon} - \mathbf{x}_0\| \leq \varepsilon$. Second, even in infinite invariant sets X we may find orbits that are unstable in the sense of Chapter 1, but not according to Definition 6.2.1. The following example illustrates this occurrence.

Example 6.2.1 Let $F(x)=x(1-x)$ and let $X=[0,1]$. The stationary state $x_s=0$ is unstable in the sense of Chapter 1. In fact if $x<0$, then $x-x^2<x$ and $O(x) \to -\infty$. However, $x_s=0$ is not unstable in X since $x_0 \in [0,1]$ implies that $x_n \to 0$ as $n \to \infty$. Hence, the orbit O(0) is actually a sink in X!

An additional concept is needed before the introduction of chaotic systems.

Definition 6.2.2
Let U be open in $\mathbf{R}^q$, $F : U \to \mathbf{R}^q$ be continuous, and $X \subset U$ be closed, bounded, and invariant. An orbit $O(\mathbf{x}_0)$, $\mathbf{x}_0 \in X$ is said to be **dense** in X if $L(\mathbf{x}_0)=X$.

Example 6.2.2 Let $F(x)=x(1-x)$. The interval [0,1] is invariant under the action of F. Since $x_n \to 0$ for every $x_0 \in [0,1]$, no orbit of F in [0,1] has the property $L(x_0)=[0,1]$.

Example 6.2.3 Let $F(x,y)=(x\cos 1 - y\sin 1,\ x\sin 1 + y\cos 1)$. We have seen that $X=\{(x,y) : x^2+y^2=1\}$ is invariant under the action of F and that no matter how small an arc of X we choose, we can find points of the orbit $O(\mathbf{x}_0)$, $\mathbf{x}_0=(1,0)$ in the arc. Hence, $L(\mathbf{x}_0)=X$.

We are now ready to propose our definition of discrete chaotic systems.

Definition 6.2.3
Let U be open in $\mathbf{R}^q$, $F : U \to \mathbf{R}^q$ and $X \subset U$ be closed, bounded, and invariant. Assume that F is continuous in X.
Then F is **chaotic** in X if there exists $\mathbf{x}_0 \in X$ such that

(i) $O(\mathbf{x}_0)$ is dense in X.
(ii) $O(\mathbf{x}_0)$ is unstable in X.

Before providing some examples of chaotic maps, we remark that no dynamical system can be chaotic on a finite set X in the sense of Definition 6.2.3. This follows from condition (ii) and the observation we made after Definition 6.2.1. We now quote a theorem (see [Martelli-Dang-Seph, 1998]) which will be of great help in establishing that a function F has property (i) on an invariant set X.

Theorem 6.2.1 *Assume that for every* $\mathbf{x}\in X$ *and every* $r>0$, *there exists* $n\geq 1$ *such that* $F^n(B(\mathbf{x},r)\cap X)=X$. *Then there exists* $\mathbf{x}_0\in X$ *such that* $L(\mathbf{x}_0)=X$.

Using Theorem 6.2.1 we provide our readers with a first example of a chaotic map.

Example 6.2.4 Let $F(x)=2|x|-1$ and $X=[-1,1]$. It is easily seen that $F(X)=X$. The derivative of F is either 2 or -2 for all $x\neq 0$. Hence, we can conclude that every orbit of F is unstable (see Fig. 6.2.1).

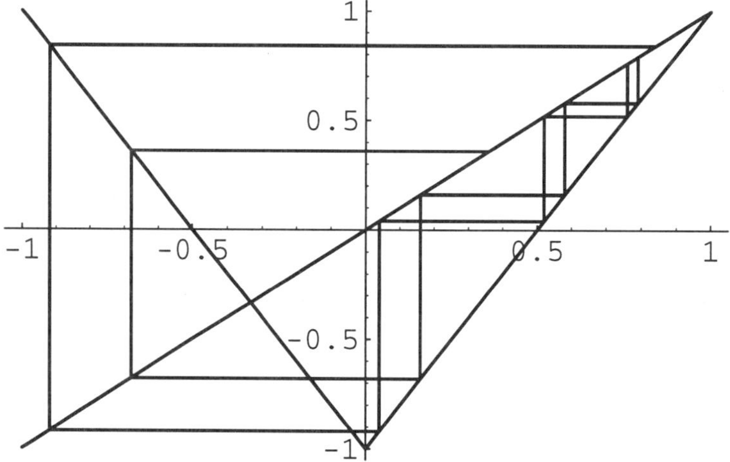

Fig. 6.2.1 Graph of the function F of Example 6.2.4. The instability is graphically illustrated with the cobweb method. Two orbits that start very close are quite far from each other just after a few iterations.

To obtain the existence of $x_0\in[-1,1]$ such that $L(x_0)=[-1,1]$ we can reason as follows. Let $x\in[-1,1]$ and $r>0$ be given. Define $I=(x-r,x+r)\cap[-1,1]$. As long as the interval I or one of its images $F^m(I)$, $m=1,2,...,p-1$ do not contain 0, the length is doubled at every iteration. Hence, for some $p\geq 1$ we must have $0\in F^p(I)$. Since $F(0)=-1$, $F(-1)=1$, and $F(1)=1$ we obtain that $1\in F^{p+j}(I)$ for every $j=2,3,...$. Let $J=F^{p+2}(I)$. As long as either J or one of its images does not contain 0, the length is doubled at every iteration. Hence, for some $q\geq 1$ we must have $0\in F^q(J)$. Then $F^{q+1}(J)=[-1,1]$, which implies that $F^{p+q+3}(I)=[-1,1]$. By Theorem 6.2.1 there is $x_0\in[-1,1]$ such that $L(x_0)=[-1,1]$. Putting together this conclusion and the instability established before, we obtain that F is chaotic in $[-1,1]$.

We know that F has a periodic orbit of period 3 (see Example 2.5.1). Since F is continuous we can apply the theorem of Sarkovskii. Hence, F has periodic orbits of every period. Moreover, F is chaotic in [–1,1] in the Li-Yorke sense.

With our mind on the statement made in the previous section, we notice that the interval I=[–1,1] is not an attractor of F if we identify U with the entire domain of F, namely **R**. In fact, for every $x_0 \notin [-1,1]$, the orbit $O(x_0)$ goes to infinity. Consequently, F is chaotic on a set X that is not an attractor. We could change F outside [–1,1] and define $F(x)=1$ for $x \notin [-1,1]$. This definition does not affect the behavior of F in [–1,1] and now [–1,1] is an attractor. However, this approach may not be possible for other maps. Hence, we leave F unchanged and consider it chaotic on a set X that is not an attractor.

Conjugacy can be used to obtain that other systems are chaotic. Here is why. Let X be closed, bounded, and invariant under the action of a continuous map F. Let Y be closed, bounded, and invariant under the action of a continuous map G. Assume that there exists a continuous map $\phi : X \to Y$ which is invertible with continuous inverse and such that $\phi(F(\mathbf{x}))=G(\phi(\mathbf{x}))$ for every $\mathbf{x} \in X$. Then G is chaotic in Y if and only if F is chaotic in X. This conclusion is intuitive, but we are going to prove it theoretically with Theorems 6.2.2 and 6.2.3.

Theorem 6.2.2 *The orbit* $O(\mathbf{y}_0)$ *is dense in* Y *if and only if the orbit* $O(\mathbf{x}_0)$, $\mathbf{y}_0=\phi(\mathbf{x}_0)$, *is dense in* X.

Proof. Assume that $O(\mathbf{x}_0)$ is dense in X and let $\mathbf{w}_0 \in Y$. Set $\mathbf{z}_0=\phi^{-1}(\mathbf{w}_0)$. Since $\mathbf{z}_0$ is a limit point of $O(\mathbf{x}_0)$ there is a subsequence of this orbit that converges to $\mathbf{z}_0$. We denote this subsequence by $\{\mathbf{z}_p\text{: } p=1,2,\ldots\}$. The continuity of ϕ implies that the sequence $\{\phi(\mathbf{z}_p)\} : p=1,2,\ldots\}$ converges to $\mathbf{w}_0$. Since this sequence is a subsequence of $O(\mathbf{y}_0)$ we obtain that $\mathbf{z}_0 \in L(\mathbf{y}_0)$. Hence, every element of Y is an element of $L(\mathbf{y}_0)$ and $O(\mathbf{y}_0)$ is dense in Y.

This argument is obviously symmetric in $O(\mathbf{x}_0)$ and $O(\mathbf{y}_0)$. Hence, $O(\mathbf{y}_0)$ is dense in Y if and only if $O(\mathbf{x}_0)$ is dense in X. QED

To prove Theorem 6.2.3 we need the following lemma, which is usually presented in advanced calculus.

Lemma 6.2.1 *Let* $X \subset \mathbf{R}^q$ *be closed and bounded and* $T:X \to \mathbf{R}^p$ *be continuous. Then for every* $r>0$ *there exists* $d>0$ *such that* $\|\mathbf{x}-\mathbf{y}\| \le d$ *implies* that $\|T(\mathbf{x})-T(\mathbf{y})\| \le r$.

Theorem 6.2.3 *The orbit* $O(\mathbf{y}_0) \subset Y$ *is stable if and only if the orbit* $O(\mathbf{x}_0)$, $\mathbf{y}_0=\phi(\mathbf{x}_0)$, *is stable.*

Proof. Also in this case the argument is symmetric in $O(\mathbf{x}_0)$ and $O(\mathbf{y}_0)$. Hence, we shall simply prove that the stability of $O(\mathbf{x}_0)$ implies the stability of $O(\mathbf{y}_0)$. Let $r>0$ be given and let d be the real positive number associated to r by Lemma 6.2.1 applied to ϕ. Since $O(\mathbf{x}_0)$ is stable, there is $\varepsilon>0$ such that

$$\|\mathbf{z}_0-\mathbf{x}_0\| \le \varepsilon \quad \text{implies that} \quad \|\mathbf{z}_n-\mathbf{x}_n\| \le d,\ n=1,2,\ldots .$$

Given this $\varepsilon>0$, and using the continuity of ϕ^{-1}, we can find δ such that $\|\mathbf{w}_0-\mathbf{y}_0\|\leq\delta$ implies that $\|\mathbf{z}_0-\mathbf{x}_0\|\leq\varepsilon$ with $\mathbf{w}_0=\phi(\mathbf{z}_0)$.

Now we show that $\|\mathbf{w}_0-\mathbf{y}_0\|\leq\delta$ implies $\|\mathbf{w}_n-\mathbf{y}_n\|\leq r$, which ensures the stability of $O(\mathbf{y}_0)$. From $\|\mathbf{w}_0-\mathbf{y}_0\|\leq\delta$ we derive $\|\mathbf{z}_0-\mathbf{x}_0\|\leq\varepsilon$. This implies that $\|\mathbf{z}_n-\mathbf{x}_n\|\leq d$. By Lemma 6.2.1, $\|\phi(\mathbf{z}_n)-\phi(\mathbf{x}_n)\|\leq r$. From $\phi(\mathbf{x}_n)=\mathbf{y}_n$ and $\phi(\mathbf{z}_n)=\mathbf{w}_n$ we derive $\|\mathbf{w}_n-\mathbf{y}_n\|\leq r$. QED

Putting together Theorems 6.2.2 and 6.2.3, we conclude that G is chaotic in Y if and only if F is chaotic in X.

Example 6.2.5 Consider the system governed by H : $[0,1]\to[0,1]$,

$$H(x)=\begin{cases}2x & 0\leq x<1/2\\ 2-2x & 1/2\leq x\leq 1.\end{cases}$$

The map H is conjugate to the map F of Example 6.2.4 by $\psi:[-1,1]\to[0,1]$, $\psi(x)=(1-x)/2$. Consequently, H is chaotic in [0,1], and for this map we can repeat the remark made above regarding the function F, namely the interval [0,1] is not an attractor for H.

In Chapter 2 (see Example 2.1.5) we verified that F and the map

$$G:[-1,1]\to[-1,1],\quad G(x)=2x^2-1$$

are conjugate by the function

$$\phi(x)=(2/\pi)\arcsin x.$$

Hence, G is chaotic in [–1, 1].

The function K : $[0,1]\to[0,1]$, $K(x)=4x(1-x)$ is conjugate to G with the map ψ. Thus K is chaotic on [0,1].

All maps of Example 6.2.5 have a periodic orbit of period 3 in their respective intervals. Hence, they are chaotic in the Li-Yorke sense. Moreover, according to the result of Sarkovskii, they have periodic orbits of every period. The following is an additional example of a chaotic system and has a significant theoretical value. The function governing the system is discontinuous at x = .5 and x=1.

Example 6.2.6 Let B : $[0,1]\to[0,1]$ be defined by $B(x)=2x-[2x]$, where $[x]$ denotes the largest integer less than or equal to x. Hence,

$$B(x)=\begin{cases}2x & 0\leq x<1/2\\ 2x-1 & 1/2\leq x<1\\ 0 & x=1.\end{cases}$$

We now prove that B is chaotic in [0,1] according to Definition 6.2.3, with the continuity of F removed from the required properties. First, we show that every orbit of B is unstable. This result is easily obtained from the derivative of B, which is always 2 for $x\neq1/2$ or 1. At x=1/2,1 the derivative does not exist but B(1/2)=B(1)

=0=B(0). Hence, every orbit that contains 1/2 or 1 ends up at 0, which is an unstable fixed point (a source).

Select the point $x_0 \in [0,1]$ such that its expansion in base 2 is

$$x_0 = .\ 0\ 1\ 00\ 01\ 10\ 11\ 000\ 001\ 010\ 100\ 011\ 101\ 110\ 111\ 0000\ 0001\ \ldots .$$

To see why $L(x_0)=[0,1]$ we look at the action of B on the elements of (0,1) written in base 2. Let $x=.a_1a_2a_3...$, where a_i is either 0 or 1. There are two cases to consider: $a_1=0$ and $a_1=1$. In the first case $x<1/2$ and $B(x)=.a_2a_3...$, since the action of B on x, written in base 2, has the same effect as the multiplication by 10 when x is written in base 10. For the same reason, when $a_1=1$ we obtain $B(x)=1.a_2a_3...-1=.a_2a_3...$. Now we can understand why $L(x_0)=[0,1]$. For example, to see why 0 is a limit point of $O(x_0)$, consider the subsequence

$$\begin{aligned} x_2 &= .00\ 01\ 10\ 11... \\ x_{10} &= .000\ 001\ 010... \\ x_{34} &= .0000\ 0001\ 0010... . \end{aligned}$$

We have $x_2<1/8$, $x_{10}<1/32$, $x_{34}<1/128$ etc. Hence, 0 is a limit point of $O(x_0)$.

Summarizing, B satisfies both requirements of Definition 6.2.3 and the dynamical system defined by B is chaotic in [0,1]. The map B is called the **baker's transformation** (see Fig. 6.2.2) Neither the definition of chaos of Li-Yorke nor the theorem of Sarkovskii can be applied to B since B has two points of discontinuity. However, it is not difficult to see that B has periodic orbits of every period. For example, to obtain a periodic point of period 4, we can select the initial condition $x=.111011101110...$.

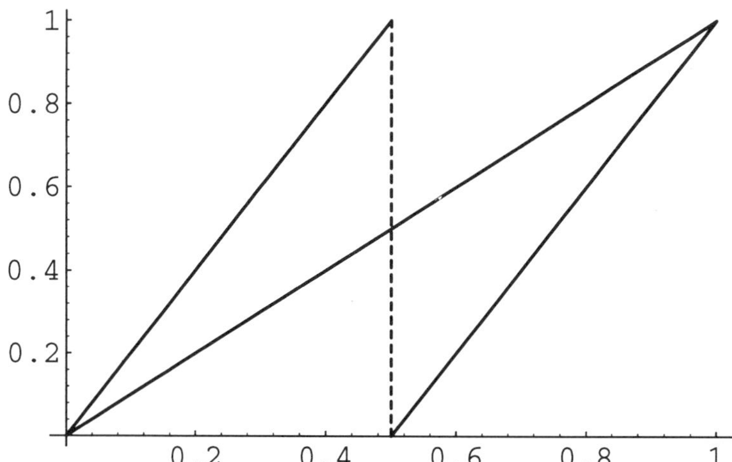

Fig. 6.2.2 The baker's transformation on the unit interval provides one of the most quoted examples of chaotic behavior.

Li-Yorke chaos in $\mathbf{R}^q$

We mentioned in Chapter 2 that the Li-Yorke definition of chaos is applicable only to continuous and one-dimensional systems (see Chapter 2, Section

5). An extension of their result to higher-dimensional systems was obtained in 1979 by F.R. Marotto [Marotto, 1979]. He proved that a dynamical system $F : \mathbf{R}^q \rightarrow \mathbf{R}^q$, which is of class C^1 in some open ball $B(\mathbf{x}_S, r)$ centered at a stationary state $\mathbf{x}_S$ has uncountably many aperiodic and unstable orbits if $\rho(F'(\mathbf{x}_S))>1$, and there exists $\mathbf{z}_1$ in the unstable manifold $\mathbf{M}_U$ of $\mathbf{x}_S$ such that

$$\det((F^m)'(\mathbf{z}_1)) \neq 0 \quad \text{and} \quad F^m(\mathbf{z}_1) = \mathbf{x}_S$$

for some positive integer m.

In addition, Marotto proved that there exists a positive integer N such that for every p≥N the dynamical system has a periodic orbit of period p. Marotto called $\mathbf{x}_S$ a **snapback repeller**. It can be shown that the existence of a periodic point of period 3 for continuous and one-dimensional maps is equivalent to the existence of a snapback repeller.

It is legitimate to ask if the presence of a snapback repeller in a dynamical system authorizes us to say that the systems is chaotic in the sense of Definition 6.2.3. The following example illustrates the difficulty of answering this question.

Example 6.2.7 Let (see Fig. 6.2.3)

$$F(x) = \begin{cases} 0 & 0 \leq x \leq .25 \\ 4x-1 & .25 < x \leq .5 \\ 3-4x & .5 < x \leq .75 \\ 0 & .75 < x \leq 1. \end{cases}$$

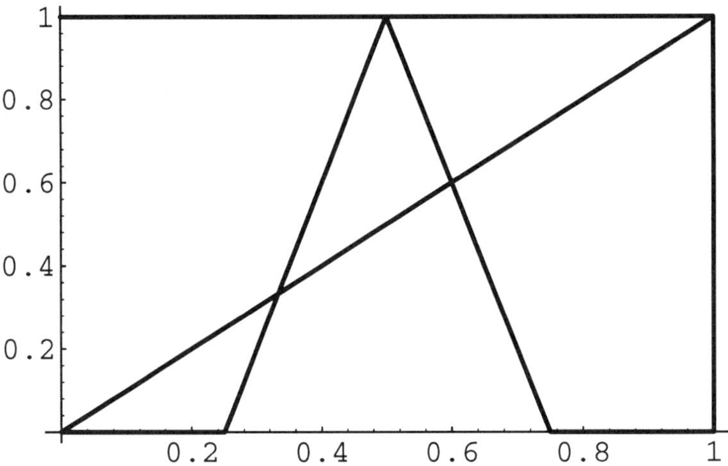

Fig. 6.2.3 Graph of the map of Example 6.2.7.

The origin is an attractive stationary state. The point x=23/65 is periodic of period 3. Hence, by the result of Marotto, F has a snapback repeller and the dynamical system defined by F in [0,1] is chaotic in the Li-Yorke sense. However, the system is not chaotic in the interval [0,1] according to Definition 6.2.3, since it does not have an unstable and dense orbit in [0,1]. However, it can be shown that there is a closed and invariant subset X of [0,1] such that F has in X a dense and unstable orbit. This set is obviously contained in the interval [1/3,2/3] since the orbit of every $x \in [0,1/3) \cup (2/3,1]$ converges to 0. However, this information does not provide a clear picture of X. In fact, consider the union U of all disjoint intervals of [0,1] such that $O(x)$ converges to 0 for every $x \in U$. It is not hard to prove (see [Martelli-Dang-Seph, 1998]) that the sum of the lengths of all these intervals is 1. For this reason X is called a **negligible** set and it does not contain any interval. It appears that X cannot easily be visualized.

We shall now present a definition of chaos (proposed by R.L. Devaney [Devaney, 1989]) which has received a lot of attention in recent years. It requires a property, defined below, which at the outset seems to be stronger than instability.

Definition 6.2.4
F has (in X) **sensitive dependence on initial conditions** if there exists a positive real number r such that no matter what initial condition $\mathbf{x}_0$ is selected in X and no matter how small is d>0, we can find a state $\mathbf{y}_0 \in B(\mathbf{x}_0, d)$ ($\mathbf{y}_0$ may not be necessarily in X) such that for some integer $m \geq 1$ we have $\|\mathbf{y}_m - \mathbf{x}_m\| > r$ (see Fig. 6.2.4).

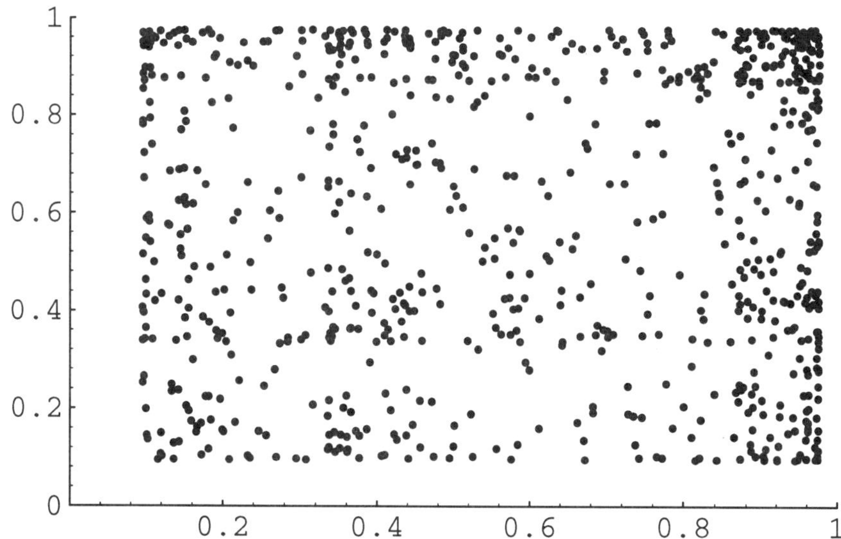

Fig. 6.2.4 The sensitivity with respect to initial conditions of the dynamical system governed by the map $F(x) = 3.9x(1 - x)$ is dramatically underlined by this graph, in which corresponding states of the trajectories starting from x_0=.3 and y_0=.31 have been plotted versus each other. We see a lot of states close to the diagonal $y=x$ (the corresponding states are close to each other) and a lot of corresponding states as far away as .9 from each other.

Whenever the additional restriction $\mathbf{y}_0 \in B(\mathbf{x}_0, r) \cap X$ is imposed, we say that F has sensitive dependence on initial conditions **with respect to** X. It can be shown (see Theorem 6.2.4 below) that for continuous functions F in a bounded, closed, and invariant set X, sensitive dependence on initial conditions with respect to

X is equivalent to the instability of $O(\mathbf{x}_0)$ in X, when $O(\mathbf{x}_0)$ is dense in X. Therefore, in this case the new requirement is not stronger than instability.

Definition 6.2.5
Let $F : V \subset \mathbf{R}^q \to V$ be continuous. F is **chaotic on** V in the sense of Devaney if

i. F is topologically transitive in V
ii. Periodic points are dense in V.
iii. F has sensitive dependence on initial conditions.

Topological transitivity is easily explained when V is a disk. In this case it means that for every pair of open balls B_1 and B_2, contained in V, there exists an integer n such that

$$F^n(B_1) \cap B_2 \neq \varnothing.$$

The density of periodic orbits in V means that for every $\mathbf{x} \in V$ and every $r>0$ we can find a periodic point of F in the ball $B(\mathbf{x},r)$.

In 1992, J. Banks, J. Brooks , G. Cairns, G. Davis, and P. Stacey [Banks et al., 1992] proved that property 3 can be removed from Devaney's list since it is a consequence of the first two. Later (see [Vellekoop-Berglund, 1994]), it was shown that in an interval J the only property which is needed is topological transitivity.

In 1997, P. Touhey proved that a system is chaotic in X in the sense of Devaney if and only if for every pair of open sets in X there exists a periodic orbit which visits both sets (see [Touhey, 1997]).

The relation between properties (i) and (ii) of Definitions 6.2.3 and properties (i) and (iii) of Definition 6.2.5 is contained in the following theorem (see [Martelli-Dang-Seph, 1998]), which we state without proof.

Theorem 6.2.4 *Let* $X \subset \mathbf{R}^q$ *be closed and bounded and* $F : X \to X$ *be continuous. Then* F *is topologically transitive in* X *if and only if there exists* $\mathbf{x}_0 \in X$ *such that* $O(\mathbf{x}_0)$ *is dense in* X. *In addition,* F *has sensitive dependence on initial conditions with respect to* X *if and only if* $O(\mathbf{x}_0)$ *is unstable in X.*

Remark 6.2.1 It can be shown that the Theorem 6.2.4 remains true for a class of functions, called quasi-continuous, which is larger than the class of continuous functions. We will not venture into this topic, but simply mention that the baker's transformation (see Example 6.2.6) belongs to this class.

Is the baker's transformation $B : [0,1] \to [0,1]$ chaotic in the sense of Definition 6.2.5 with continuity removed? By Theorem 6.2.4 and Remark 6.2.1 it remains to show that periodic orbits are dense in [0,1]. Since only all rational numbers of [0,1] whose binary expansion contains infinitely many 0's and 1's are periodic, we must show that every element of [0,1] is a limit point of a sequence of rationals of this type. This part is left as an exercise and it can be worked out with the same idea we used before (see Example 6.2.6) to find a dense orbit of B.

We close this section with an example of a two-dimensional chaotic system.

Example 6.2.8 The function

$$T(\rho,\theta) = \begin{cases} (2\rho,\theta+1) & 0\le\rho<.5 \\ (2-2\rho,\theta+1) & .5\le\rho\le 1 \end{cases}$$

maps the unit disk X in the (ρ,θ) plane into itself. The instability of T in X derives from its instability in the ρ coordinate.

To prove that T has a dense orbit in X we could reason as follows. Every ρ interval, no matter how small, is stretched to the full length 1 of the radius in finitely many iterations, since the ρ-coordinate part is conjugate to the map F of Example 6.2.4, and we have seen that F has this property (see Example 6.2.5). At this point the rotation of an angle of 1 radian spreads the images of this radius in a dense manner on X since the ratio between 1 and π is irrational (see Fig. 6.2.5).

Hence, for every open and nonempty disk D in X, there is a positive integer n such that $T^n(D)=X$. By Theorem 6.2.1 T has a dense orbit in X and it is chaotic in X in the sense of Definition 6.2.3. Notice that T does not have any periodic orbit of period p for $p\ge 2$, and T does not have any snapback repeller. Hence, T is not chaotic in X in the sense of Devaney or in the sense of Marotto.

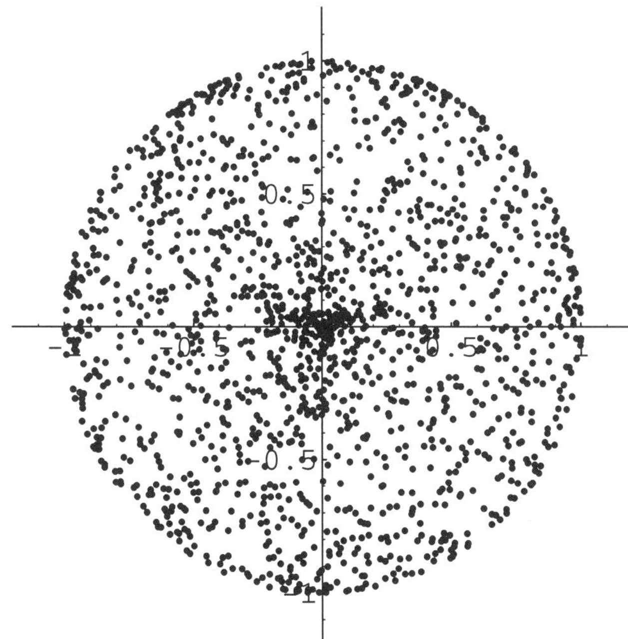

Fig. 6.2.5 After 1600 iterations the orbit of (.8,.4) is clearly on its way to satisfy the first requirement of Definition 6.2.3.

Problems

1. Let $F(\rho,\theta) = (\rho,\rho+\theta)$. Show that F has sensitive dependence on initial conditions in the set $X=\{(\rho,\theta): 1\le\rho\le 2\}$ and that there is no orbit of F that is dense in X. (This problem shows that chaos cannot be guaranteed by instability alone.)

2. Let F be the baker transformation of the unit interval [0,1]. Given $y_0 \in [0,1]$, $y_0 = .a_1a_2a_3a_4...$, prove that the subsequence $\{x_0, x_1, x_3, x_6, x_{10}, x_{15}, x_{21}, ...\}$ of the orbit starting at $x_0 = .a_1a_1a_2a_1a_2a_3a_1a_2a_3a_4a_1 ...a_5a_1...a_6...$ converges to y_0.

3. Show that the rational numbers of [0,1] whose decimal expansion is periodic with infinitely many digits equal to 1 are dense in [0,1].

4. * With B : [0,1]→[0,1] , F : [−1,1]→[−1,1], and ϕ : [0,1]→ [−1,1] defined by $B(x)=2x-[2x]$, $F(y)=-2y^2+1$, and $\phi(x)=-\cos 2\pi x$ prove that $F(\phi(x))=\phi(B(x))$. Verify that the equality above implies that $F^n(y_0)=\phi(B^n(x_0))$ with $\phi(x_0)=y_0$.

5. From the results of Problem 4 and knowing that B has a dense orbit in [0,1] derive that F has a dense orbit in [−1,1].

6. * Let $F(x)=xe^{x^3-x}$. Verify that F maps the interval [−2,1] into itself. Verify also that −1, 0, and 1 are stationary states of F.

7. * Let $F(x)=xe^{x^3-x}$. Verify that for every x_0,

$$0 < x_0 < 1, \text{ we have } x_n \to 0 \text{ as } n \to \infty.$$

Verify that 0 is not stable.

8. * Which are the points $x \in [0,1]$ such that $F^2(x)=0$ where F is the function of Example 6.2.7 and F^2 is the second iterate of F?

9. * Let F be the map of Example 6.2.7. Show that the fixed point $x_s=1/3$ is a snapback repeller.

10. Let

$$F(x)=\begin{cases} 3x+2 & -1\leq x<-1/3 \\ -3x & -1/3\leq x<1/3 \\ 3x-2 & 1/3\leq x\leq 1. \end{cases}$$

Show that F is chaotic in [−1,1].

11. *Using $\phi(x)=(2/\pi)\arcsin x$ and the result of Problem 10 show that $T(x)=4x^3-3x$ is chaotic in [−1,1].

Section 3. FRACTAL DIMENSION

The study of an invariant, closed, and bounded set $X \subset \mathbf{R}^q$ of a dynamical system F frequently includes the determination of a number representing the **dimension** of X according to a suitable definition of this property. We shall see later that this number may be quite important when we have reason to believe that F is chaotic in X.

Our intuitive idea of dimension assigns an integer to common geometric objects. For example, a point has dimension 0, a line segment has dimension 1, a full square (interior and boundary) has dimension 2, a full cube (interior and boundary) has dimension 3, etc. This intuitive idea is not sophisticated enough for complex geometric objects, including strange attractors. More elaborate definitions of dimension have been proposed, and two of them will be presented below. Every definition of dimension assigns the same number to common geometric objects such as points, segments, lines, planes, squares, spheres, etc. For example, no matter what definition we use, a line will always have dimension 1.

It may happen that different definitions of dimension assign different numbers to any one of the more complex subsets of $\mathbf{R}^q$, for which the intuitive idea of dimension fails. Moreover, these numbers may not be integers. We would like our readers not be discouraged by these surprising outcomes. They should see them as clear evidence that the objects are complex, with different properties and aspects quantified by the different definitions.

A geometric object whose dimension (in some suitable definition) is not an integer is called a **fractal**. Chaos and fractals are strongly related since frequently, although not always, an invariant set X where F is chaotic turns out to be a fractal. Conversely, when an invariant set X is a fractal, we have a clear although not definite indication that F is chaotic in X. At the end of this section we shall make some additional comments on the relevance of fractal dimension with respect to chaotic behavior.

Two different definitions of dimension will be discussed in this section: capacity and correlation dimension. **Capacity** belongs to a group of definitions which are simply concerned with the geometric properties of the set. **Correlation dimension** belongs to a group of definitions that try to capture the dynamical nature of an invariant set by looking at how the states of the orbits of the system are distributed in its regions, exploring the relations between successive states, and measuring the strength of these relations.

Capacity

The fundamental steps needed to arrive at the capacity of a set X are illustrated by the following examples.

Example 6.3.1 Consider the segment S=[0,L]. The minimum number of segments of length L/2 needed to cover S is obviously 2. They are [0,L/2] and [L/2,L]. The minimum number of segments of length L/n needed to cover S is n. In general, given $\varepsilon>0$, the minimum number of segments of length εL needed to cover S is $1/\varepsilon$. When $1/\varepsilon$ is not an integer, we should round it up to the nearest integer.

Denote by $N(\varepsilon)$ the minimum number of segments of length εL that are needed to cover the interval [0,L]. We find

$$\lim_{\varepsilon\to 0} \varepsilon N(\varepsilon) = \text{constant} > 0. \tag{6.3.1}$$

From the formula above we derive, when ε is very small,

$$\ln N(\varepsilon) \approx \text{constant} - \ln \varepsilon \; .$$

Keeping in mind that $-\ln\varepsilon\to\infty$ as $\varepsilon\to 0$, we obtain

$$\lim_{\varepsilon\to 0} \frac{\ln N(\varepsilon)}{-\ln \varepsilon} = 1. \qquad (6.3.2)$$

The limit of (6.3.2) is called **capacity** of the interval [0,L]. As expected, the limit is 1, namely the same number representing the "intuitive" dimension of [0,L].

Let us now analyze a more complex subset of the real line, called the **Cantor set**, introduced in 1893 by **G. Cantor**, 1845-1918.

Example 6.3.2 The Cantor set is generated from the interval [0,1] in the following way. First we remove from [0,1] the interval (1/3,2/3). Then we remove from what is left the intervals (1/9,2/9) and (7/9,8/9). This procedure is iterated. For every integer n we remove the center part from each of the 2^{n-1} intervals left. The length of each part removed is 3^{-n}. What is left at the end of the procedure is the **Cantor set K**. It can be shown that an element x of [0,1] belongs to **K** if and only if we can write x in base 3 using only the digits 0 and 2. The total length of all segments we have removed is

$$\frac{1}{3}+\frac{2}{9}+\frac{4}{27}+\ldots+\frac{2^n}{3^{n+1}}+\ldots=\frac{1}{3}\left[1+\frac{2}{3}+\frac{4}{9}+\ldots+(\frac{2}{3})^n+\ldots\right]=1.$$

Hence, on intuitive grounds we could say that the dimension of **K** is zero. This statement is not, however, our best description of the size of **K**.

Using the idea illustrated in Example 6.3.1, we first cover **K** with the minimum number of segments of length 1/3. Two such segments are needed: [0,1/3] and [2/3,1]. Then we cover **K** with the minimum number of segments of length 1/9. Four such segments are needed, namely [0,1/9], [2/9,1/3], [2/3,7/9], and [8/9,1]. In general, covering **K** with the minimum number of segments of length 3^{-n} requires 2^n segments. Hence, assuming that

$$\lim_{\varepsilon\to 0} \ln N(\varepsilon)/(-\ln\varepsilon)$$

exists, we can use the particular sequence $\varepsilon_n=2^{-n}$ to compute it and we obtain

$$\lim_{\varepsilon\to 0} \ln N(\varepsilon)/(-\ln\varepsilon) = \lim_{n\to\infty} n\ln 2/(n\ln 3) = \ln 2/\ln 3. \qquad (6.3.3)$$

The capacity of **K** is ln2/ln3.

Example 6.3.3 We construct a set called the Sierpinski triangle (see Fig. 6.3.1), introduced in 1916 by **W. Sierpinski**, 1892-1969. The first step of our construction is to take an equilateral triangle T, join the middle points of the three sides with line segments, and remove from T the interior of the equilateral triangle so obtained.

The second step is the repetition of the first step, performed on the three equilateral triangles left in T after the removal of the central part. The iteration of this process for every integer n produces, at the end, a geometric figure in the plane which is called the Sierpinski triangle. Its capacity is found as follows. We need only one square of size 1 to cover the Sierpinski triangle. Hence, N(1)=1. Three squares of size 1/2 are needed to obtain the same result. Hence, N(1/2)=3. With squares of size 1/4, the minimum number needed to cover the Sierpinski triangle climbs to 9. Therefore, N(1/4)=9. In general, we obtain $N(1/2^n)=3^n$. Assuming once again that

$$\lim_{\varepsilon \to 0} \ln N(\varepsilon)/(-\ln\varepsilon)$$

exists, we can use the particular sequence $\varepsilon_n=2^{-n}$ to compute it and we find that the capacity of the Sierpinski triangle is ln3/ln2.

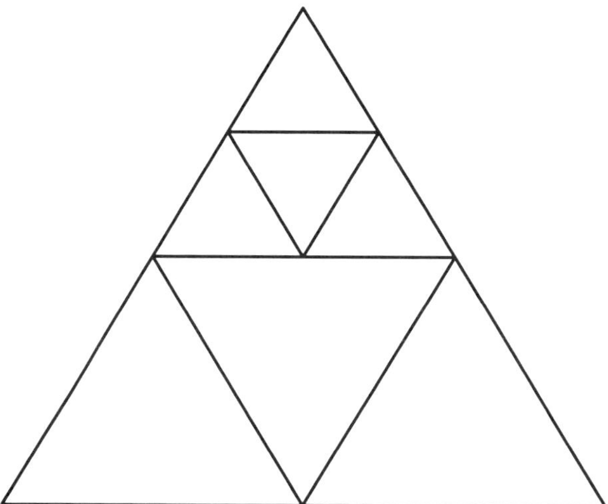

Fig. 6.3.1 Sierpinski triangle.

The idea of capacity is therefore the following. Suppose that A is a bounded subset of $\mathbf{R}^q$. Choose $\varepsilon>0$ and select hyper cubes (segments in **R**, squares in $\mathbf{R}^2$, cubes in $\mathbf{R}^3$, etc.) of side ε. Denote by $N(\varepsilon)$ the **smallest number** needed to cover A. Let

$$c(A) = \lim_{\varepsilon \to 0} \ln N(\varepsilon)/(-\ln\varepsilon) \qquad (6.3.4)$$

Definition 6.3.1
c(A) is called the **capacity** of the set A.

Assuming that the limit in (6.3.4) exists, we can use a suitable sequence $\{\varepsilon_n: n=1,2,...\}$ to compute it, just as we did in the three preceding examples.

The capacity of a set A is strongly dependent on the geometry of the set. The possible dynamical properties of a map F having A as an invariant set do not play any role in calculating its capacity. Moreover, we need quite a good knowledge of A to estimate $N(\varepsilon)$ accurately. This proves to be nontrivially difficult when dealing with invariant sets of dynamical systems, since frequently we cannot describe them with equations, inequalities, or geometrical procedures. Our next definition of dimension avoids this pitfall.

Correlation dimension

This type of dimension makes good sense for a set that is generated by a dynamical system $F : U \to U$, where U is some open and bounded set in $\mathbf{R}^q$. We start with an orbit $O(\mathbf{x}_1)=\{\mathbf{x}_1,\mathbf{x}_2,\dots,\mathbf{x}_n,\dots\}$, frequently called a **time series**. Notice that the initial state is indicated with $\mathbf{x}_1$ rather than $\mathbf{x}_0$. The change is made to simplify the notation in this and the next section. The correlation dimension of the set $O(\mathbf{x}_1)$ is computed in the following way. Given a positive real number r, we form the "correlation integral "

$$C(r) = \lim_{n\to\infty} \frac{1}{n^2-n} \sum_{i\neq j}^{n} H(r - \|\mathbf{x}_i - \mathbf{x}_j\|) \tag{6.3.5}$$

where

$$H(x) = \begin{cases} 0 & x < 0 \\ 1 & x \geq 0 \end{cases} \tag{6.3.6}$$

is the unit-step function.

The summation counts how many pairs of vectors are closer than r when $1 \leq i,\ j \leq n$, and $i \neq j$, while n^2-n is the total number of pairs with $i \neq j$. Hence, the ratio between the two represents the **fraction** of pairs that are closer than r, and C(r) measures the **density** of pairs of distinct vectors $\mathbf{x}_i$ and $\mathbf{x}_j$ that are closer than r. The standard Euclidean norm is used in (6.3.5). This, however, can be replaced by any other norm.

Definition 6.3.2

The **correlation dimension** D_C of $O(\mathbf{x}_1)$ is defined as

$$D_C = \lim_{r\to 0} \ln C(r)/\ln r. \tag{6.3.7}$$

In practice, when a numerical estimate of D_C is desired, we select a large number n of states of the orbit $O(\mathbf{x}_1)$: $\mathbf{x}_1, \mathbf{x}_2,\dots,\mathbf{x}_n$ and we approximate C(r) for several different values of r.

$$C(r) \approx \frac{1}{n^2-n} \sum_{i\neq j}^{n} H(r - \|\mathbf{x}_i - \mathbf{x}_j\|) \tag{6.3.8}$$

Equality (6.3.7) tells us that (6.3.8) is not going to provide any useful information on D_C for large values of r. At the same time, since we are using only n states, we should avoid values of r which are so small that the estimate of C(r) is meaningless. Thus the values of r should be selected so as to avoid these two extremes. The ratio $\ln C(r)/\ln r$ is plotted versus r and the y-intercept of the least squares line is regarded as a reasonable estimate of D_C when the slope is sufficiently small. By using different values of n, we can check how stable the estimate is when n is increased.

The correlation dimension of an orbit is 0 when the orbit is periodic or asymptotically periodic. In fact, assume that $\{\mathbf{x}_1,\ldots,\mathbf{x}_p,\ldots\}$ is a periodic orbit of period p. Let $d=\min\{\|\mathbf{x}_i-\mathbf{x}_j\|: i\neq j,\ i,j=1,2,\ldots,p\}$. We see that d is the smallest distance between pairs of different states of the periodic orbit. Given $r<d$ and a positive integer $n>p$, write $n=kp+i$, $i=1,2,\ldots,p$. We can easily verify the following inequalities (see Problem 8):

$$\frac{k-1}{kp+i-1} \leq \frac{1}{n^2-n}\sum_{i\neq j}^{n} H(r-\|\mathbf{x}_i-\mathbf{x}_j\|) \leq \frac{k}{kp+i-1}. \tag{6.3.9}$$

As $k\to\infty$ the first and last terms of (6.3.9) converge to $1/p$. Thus $C(r)=1/p$ for every $r<d$. Using this value in computing D_C, we obtain

$$D_C = \lim_{r\to 0}\frac{\ln C(r)}{\ln r} = \lim_{r\to 0}\frac{-\ln p}{\ln r} = 0.$$

Similarly, it can be shown that the correlation dimension of two orbits $O(\mathbf{x}_1)$ and $O(\mathbf{y}_1)$ such that

$$\lim_{n\to\infty}\|\mathbf{x}_n-\mathbf{y}_n\| = 0$$

is the same. Therefore, every asymptotically periodic orbit has 0 correlation dimension and a strictly positive correlation dimension indicates that the orbit is **aperiodic**.

Example 6.3.4 Consider the dynamical system governed by the function

$$F(x) = \begin{cases} x^2 & \text{if } -3 \leq x \leq 1 \\ 4\sqrt{x} - 3 & \text{if } 1 < x \leq 9 \end{cases}$$

in the interval [–3, 9]. Suppose that we start from the point $x_0 = -2$. We have

$$x_1 = 4,\ x_2 = 5,\ x_3 = 4\sqrt{5} - 3,\ \ldots,$$

and the sequence $\{x_n\}$ is increasing and converges to 9. For every $r>0$ there is an index $n(r)$ such that all elements of the sequence with index larger than $n(r)$ are closer than r. This implies that $C(r)=1$ and $D_C=0$, a result we expected since the set of limit points of $O(-2)$ is simply $L(-2)=\{9\}$.

Example 6.3.5 Consider the dynamical system $x_{n+1}=2x_n^2-1$. We know that the system is chaotic in the interval [–1,1]. Hence, there is an orbit $O(x_0)$ such that $L(x_0)=[-1,1]$. It turns out, although we do not prove it, that if we choose an initial state y_0 at random in the interval [−1,1], then, with probability 1, L(y_0)=[−1,1]. Suppose that our initial statc is y_0=.3 . Can we say that L(0.3)=[−1,1]?

The graph of Fig. 6.3.2 seems to suggest an affirmative answer. We are plotting the states (y_n, y_{n+1}) for n=1000,...,2000. It appears that the orbit is "reconstructing" the entire graph of $F(x)=2x^2-1$, a feature that requires the equality L(.3)=[−1,1].

Therefore, if we compute the correlation dimension of the orbit of .3 we should get 1. The evaluation was performed by considering the states 501,...,600 of the orbit of .3. Figure 6.3.3 gives the plot of the 10 points representing the ratios ln(C(r)/lnr for r=.01, .02,...,.1. The least squares line has equation y=.995−.055x and it suggests that the correlation dimension is indeed 1.

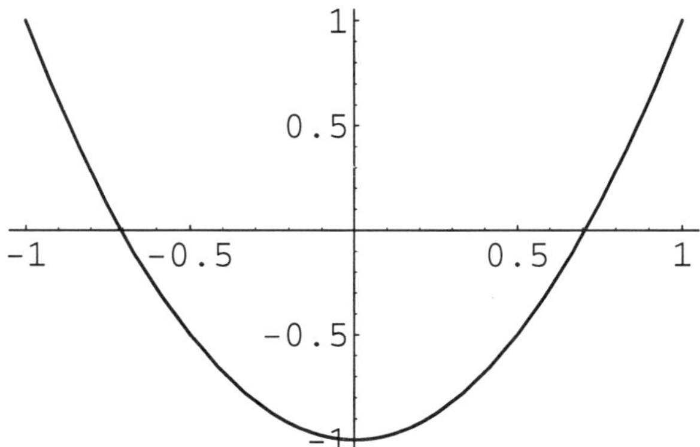

Fig. 6.3.2 The graph is obtained by plotting (y_n, y_{n+1}) for n=1000,...,2000. The initial condition is y_0=.3 and the dynamical system is governed by $F(x)=2x^2$-1.

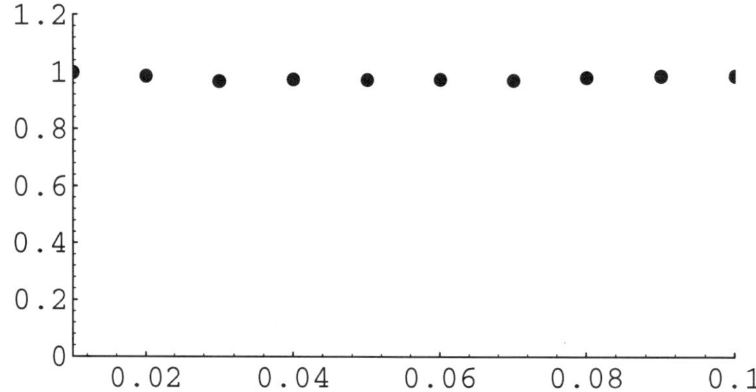

Fig. 6.3.3 The correlation dimension of the orbit O(.3) of the dynamical system governed by the map $F(x)=2x^2-1$ appears to be 1. The result is expected since the set of limit points of the orbit is the entire interval [−1,1].

With suitable changes, the method we have just described can be applied to a dynamical system of which only observable quantities are available and no governing equations are known with sufficient reliability. In general, the state of the system is a q-dimensional vector **x**, although we may not know how big q is. In this case, one of the observable quantities (one of the components of **x**) is measured at fixed time intervals and a long **time series** $\{x_i : i=0,1,2,...,n\}$ is produced. Notice that the elements of the time series are not the first n states of an orbit of the system. The measured quantity may represent, among the variables of the system, the one that is easier to estimate or that is more relevant to the dynamical process. It is advisable to discard some terms at the beginning of the sequence to avoid interferences with the transient of the system.

For convenience we rename the elements of $\{x_i : i=0,1,2,...,n\}$ as follows:

$$x(i) = x_i. \tag{6.3.10}$$

We now select three positive integers: m (embedding dimension), L (window gap), and J (shift), and we form the vectors

$$\begin{aligned}
\mathbf{v}_1 &= (x(1), x(1+L),..., x(1+(m-1)L),\\
\mathbf{v}_2 &= (x(1+J), x(1+J+L),..., x(1+J+(m-1)L),\\
&\cdots\cdots\cdots\cdots\cdots\cdots\\
\mathbf{v}_p &= (x(1+(p-1)J),x(1+(p-1)J+L),...,x(1+(p-1)J+(m-1)L),\\
&\cdots\cdots\cdots\cdots .
\end{aligned} \tag{6.3.11}$$

For example, with m=3, L=4, and J =2, we have

$$\begin{aligned}
\mathbf{v}_1 &= (x(1),x(5),x(9)),\\
\mathbf{v}_2 &= (x(3),x(7),x(11)),\\
&\cdots\cdots\cdots\cdots\cdots\cdots\\
\mathbf{v}_p &= (x(-1+2p),x(3+2p),x(7+2p)),\\
&\cdots\cdots\cdots\cdots .
\end{aligned}$$

Given a positive real number r and assuming that the sequence $\{x_n$: n=0,1,2,...,$\}$ is infinite, the "m-correlation integral" of the time series is

$$C(m,r) = \lim_{n\to\infty} \frac{1}{n^2-n} \sum_{i\neq j}^{n} H(r - \|\mathbf{v}_i - \mathbf{v}_j\|). \tag{6.3.12}$$

The **m-correlation dimension** $D_C(m)$ is defined as

$$D_C(m) = \lim_{r\to 0} \frac{\ln C(m,r)}{\ln r}, \tag{6.3.13}$$

and the **correlation dimension** D_C is

$$D_C = \lim_{m\to\infty} D_C(m). \tag{6.3.14}$$

In practice, (6.3.12), (6.3.13), and (6.3.14) must be estimated since we cannot choose n as large as we like, r very small, or m very big. For a given time series of length n and keeping m fixed, we estimate C(m,r) for different values of r as in (6.3.8). The values of ln C(m,r)/lnr are plotted versus r and the y-intercept of the least squares line is recorded. This number provides a first approximation to $D_C(m)$. As m increases, the y-intercept should converge to an acceptable value for D_C and the slope should become sufficiently small. The parameters L and J are used to improve the reliability of the numerical scheme.

Sometimes the constraint $i \neq j$, which is equivalent to $|i-j| \geq 1$, is replaced by the stronger inequality $|i-j| \geq K$, where K is an integer larger than 1, to avoid artificial correlates. In fact, dynamical systems arise from modeling real processes, and the time step used for the observations may be too small with respect to the time interval the system needs to undergo measurable changes.

Important merits of the procedure just described, proposed by P. Grassberger and I. Procaccia [Grassberger-Procaccia, 1983], are the following:

1. The computational effort is not prohibitive, although it could be very delicate and does not require a full knowledge of the system.
2. If no equations of the dynamical process are known and only observable quantities are available, a time series can be constructed using the observable that is easiest to measure and considered most relevant to the system.
3. When there is uncertainty about the deterministic nature of the dynamical process and doubts that random components are predominant in it, the correlation dimension can help in deciding the issue. In fact, in a truly random time series $D_C(m) \to \infty$ as $m \to \infty$. This feature of the real $D_C(m)$ will be detected by the numerical scheme.

In the two Examples 6.3.3 and 6.3.4 we used (6.3.8) to compute the correlation dimension since the two dynamical systems were perfectly determined. In the following example we use both (6.3.8) and (6.3.13). The two results seem to agree.

Example 6.3.6 Using the Mathematica program listed in Appendix 1, Section 3, we estimate the correlation dimension of $O(\mathbf{x}_0)$, $\mathbf{x}_0=(.4,.3)$ for the system $\mathbf{x}_{n+1}= F(\mathbf{x}_n)=(3.7x_n(1-x_n-y_n),\ 3.8x_ny_n)$. From the time series $\{(x_n,y_n)\colon n= 1,2,...\}$ we select the points (x_i,y_i) for i=401,...,600, we start with r=.02 and use 8 increments of .01. The least square line is y=1.22–.28x. Thus, it appears that 1.22 can be accepted as a reasonable value for the correlation dimension of $O(\mathbf{x}_0)$ (see Fig. 6.3.4).

Next, we select the time series $\{x_n\colon n=1,...\}$ with $x_0=.4$. We estimate the correlation dimension by considering the 200 entries $x_{401},...,x_{600}$, and use m=2, J=1, and L=2. We start with r=.02 with 8 increments of .01 each. We find that the correlation dimension of the time series made up only with the x-entries is again 1.2. More precisely, the least squares line is now y=1.23–.23x. Since we have used only 200 entries in both cases, it appears that the values we have obtained are quite reasonable (see Fig. 6.3.4). We can conclude that the correlation dimension of the system is about 1.2, and it can be safely estimated using one of the variables rather than both.

Some final remarks are necessary. Frequently, we have reason to believe that a map F is chaotic, but we do not know exactly the invariant, closed, and bounded set X where F has this property. Then, we select an initial condition $\mathbf{x}_1$ which we estimate to belong to X. If the orbit $O(\mathbf{x}_1)$ is asymptotically periodic, we obtain that $L(\mathbf{x}_1)$ is finite and its correlation dimension is 0. Therefore, we search for initial conditions $\mathbf{x}_1$ such that the estimated correlation dimension of L($\mathbf{x}_1$) is strictly positive. Now we can safely conclude that O($\mathbf{x}_1$) is aperiodic and L($\mathbf{x}_1$) is infinite. It may happen that O($\mathbf{x}_1$)$\subset$L($\mathbf{x}_1$) and we may have enough evidence to support this assumption. Moreover, assuming that F is continuous and O($\mathbf{x}_1$) is bounded we obtain that the first condition for F to be chaotic in X=L($\mathbf{x}_1$) is satisfied, since O($\mathbf{x}_1$) is obviously dense in X. Hence, X appears to be the set we are looking for. To be more comfortable with this conclusion, we should apply the procedure just described to a large number of different orbits and make sure that the results we obtain point toward the same conclusion, namely that the set of limit points is infinite, and it coincides with X=L($\mathbf{x}_1$). In the next section we present a strategy for estimating that the second condition (instability) is also satisfied in X.

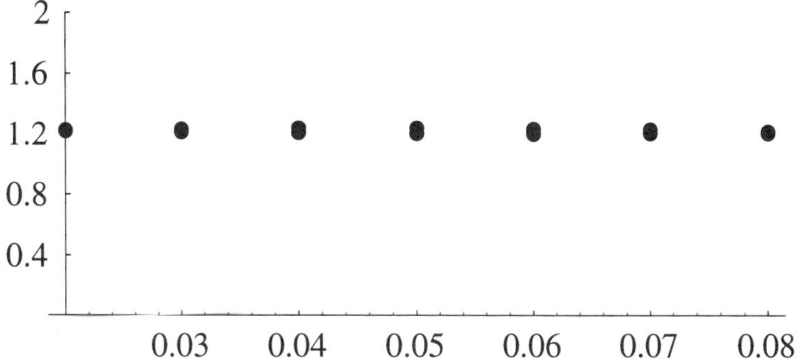

Fig. 6.3.4 In this plot we have combined the results found by estimating the correlation dimension with the states $\{(x_i,y_i)$: i=401,...,600}, and with $\{x_i$: i=401,...,600} as described in Example 6.3.6. The horizontal coordinate has the values of r from .02 to .08 in increments of .01. We see that the two sets of points almost overlap, thus showing that the correlation dimension can be computed with the full orbit or with only one of its coordinates.

We close this section by repeating an important observation we made before. We could be tempted to consider a map F chaotic in an invariant, closed, and bounded set X only in the case when the dimension of X is not an integer in some suitable definition of dimension, such as the capacity or the correlation dimension. This issue, however, must be approached with precaution. For example, the baker's transformation is one of the most classical examples of chaotic systems in [0,1], whose dimension is known to be 1 in all accepted definitions of dimension. Similarly, the map $2x^2-1$ is universally accepted as chaotic in [−1,1] and the correlation dimension of its orbits is 1. The fractional dimension (capacity, correlation, or others) of an invariant, closed, and bounded set X can certainly be regarded as an indication that F is chaotic in X, but the converse is not true.

Problems

1. * Estimate numerically the correlation dimension of O(.3) for the dynamical system $F(x)=3.2x(1-x)$.

2. * Estimate numerically the correlation dimension of the attractor of the dynamical system governed by the map $F(x)=3.8\,x(1-x)$.

3. * Estimate numerically the correlation dimension of the attractor of the dynamical system governed in $[-1,1]$ by $T(x)=4x^3-3x$.

4. * Consider the two dynamical systems $F(x)=2x^2-1$, $x\in[-1,1]$ and $G(x)=4x(1-x)$, $x\in[0,1]$. The two systems are conjugate by $\phi(x)=(1-x)/2$. Explore the correlation dimension of several corresponding orbits and derive some conclusion about the role of conjugacy in estimating the correlation dimension of corresponding orbits.

5. * Show that the cubic map $Q(x)=16x^3-24x^2+9x$, $x\in[0,1]$ is conjugate to the map $T(x)$ of Problem 3. Repeat for these two maps the investigation of Problem 4 regarding the correlation dimension of corresponding orbits.

6. Prove that the total area of the parts removed to obtain the Sierpinski triangle equals the entire area of the original triangle.

7. Again start with an equilateral triangle T. Remove the center third of each side and substitute it as in the included figure. Keep repeating this procedure. The limiting perimeter is called a Koch snowflake (see Fig. 6.3.5). Prove that not only is its total length infinite, but the distance along the perimeter between any pair of points on the snowflake is infinite. Prove that the capacity of the snowflake is ln4/ln3. Find the area inside the snowflake.

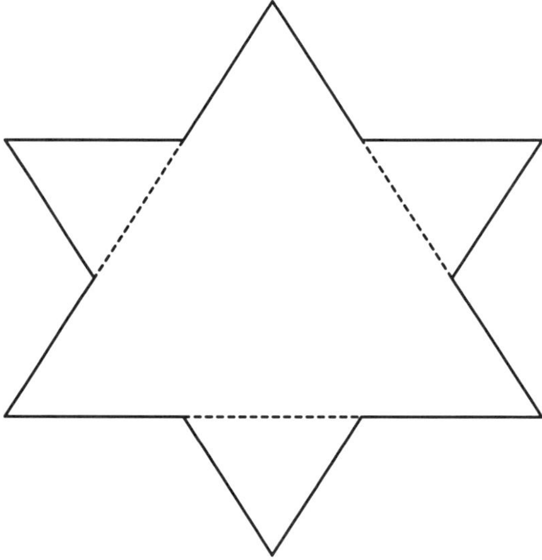

Fig. 6.3.5 Koch snowflake (introduced by **H. von Koch** in 1904).

8. Justify inequality (6.3.9).

Section 4. **LYAPUNOV EXPONENTS**

The name of the topic of this section recognizes the seminal contributions of **Alexander Michailowitsch Lyapunov** (1857-1918) to the study of stability of solutions for systems of differential equations (see [Lyapunov, 1907]). Our definition of sink for discrete dynamical systems is the *asymptotic stability* in the sense of Lyapunov for equilibria of continuous systems (see [Hartman, 1973]).

As a way of introducing Lyapunov exponents, let us first talk about the Lyapunov number $N(\mathbf{x}_s)$ of a fixed point $\mathbf{x}_s$. This is simply the spectral radius of the derivative of F at $\mathbf{x}_s$, namely

$$N(\mathbf{x}_s) = \rho(F'(\mathbf{x}_s)). \tag{6.4.1}$$

Notice that when the system is one-dimensional, the spectral radius is the absolute value of the derivative, $\rho(F'(x_s))=|F'(x_s)|$.

To go one step further, the Lyapunov number $N(\mathbf{x}_1)$ of a periodic point of period p (following the convention used in Section 3 we denote by $\mathbf{x}_1$ the initial state of the orbit) is

$$N(\mathbf{x}_1) = \left[\rho\left(\frac{d}{d\mathbf{x}} F^p(\mathbf{x}_1)\right)\right]^{1/p}. \tag{6.4.2}$$

F^p denotes the pth iterate of F. We compute its derivative at $\mathbf{x}_1$, and take the pth root of its spectral radius. The reader can easily verify the following two facts.

First, for every point $\mathbf{x}_2,...,\mathbf{x}_p$, $\mathbf{x}_{p+1}=\mathbf{x}_1,...$ of the periodic orbit $O(\mathbf{x}_1)$, we have

$$N(\mathbf{x}_i) = N(\mathbf{x}_1), \qquad i=2,3,...,p. \tag{6.4.3}$$

Consequently, we should call $N(\mathbf{x}_1)$ the Lyapunov number of the orbit $O(\mathbf{x}_1)$ rather than the Lyapunov number of the point $\mathbf{x}_1$.

Second, when the system is one-dimensional we have

$$\rho\left(\frac{d}{dx}F^p(x_1)\right) = |F'(x_p)...F'(x_1)|. \tag{6.4.4}$$

Therefore, the Lyapunov number of the orbit $O(x_1)$ is the geometric mean of the absolute values of the derivatives of F at every point of the periodic orbit.

From the definitions above we see that when the Lyapunov number is smaller than 1, the fixed point and the periodic orbit are sinks; and when it is larger than 1, the fixed point and the periodic orbit are unstable (sources in dimension 1). Hence, the Lyapunov number "measures" the stability or instability of fixed points and periodic orbits. The Lyapunov exponent is the natural logarithm of the Lyapunov number, and as we shall see, is frequently more suitable for computational purposes.

Let us now elaborate briefly on the introduction of this new concept. As mentioned before, an invariant set X of a dynamical system governed by a function F may have integer dimension, even though we suspect that the system is chaotic in X. Thus, a chaotic dynamical system in X cannot be characterized by the fractional dimension of the invariant set X. Something more is needed. Hence, let us assume that from our numerical investigations we can safely say that $L(\mathbf{x}_1)$ is infinite, and $O(\mathbf{x}_1) \subset L(\mathbf{x}_1)$. Then $X = L(\mathbf{x}_1)$ satisfies property 1 of Definition 6.2.3. Lyapunov numbers and exponents can now be used to test the instability of $O(\mathbf{x}_1)$, which is the other property required by our definition of chaos.

In (6.4.9) we give the definition of Lyapunov number for an arbitrary orbit $O(\mathbf{x}_1)$. When (6.4.9) is applied to a stationary or periodic orbit, it requires an approach slightly different from the one outlined in (6.4.1) and (6.4.2). It is easy to verify that the two approaches are equivalent when $O(\mathbf{x}_1)$ is a stationary orbit. A little work is needed to show the equivalence when $O(\mathbf{x}_1)$ is a periodic orbit. We shall not prove this result here, but simply illustrate the equivalence in the examples provided after Definition 6.4.1.

Recall that $O(\mathbf{x}_1)$ is the sequence of states

$$\mathbf{x}_1 \, , \mathbf{x}_2 = F(\mathbf{x}_1), \, \mathbf{x}_3 = F(\mathbf{x}_2), \ldots \, . \tag{6.4.5}$$

Consider the matrices

$$A_1 = \frac{d}{d\mathbf{x}} F(\mathbf{x}_1), \; A_2 = \frac{d}{d\mathbf{x}} F^2(\mathbf{x}_1), \ldots, \; A_n = \frac{d}{d\mathbf{x}} F^n(\mathbf{x}_1), \ldots \tag{6.4.6}$$

and let

$$\rho_1(\mathbf{x}_1), \, \rho_2(\mathbf{x}_1), \ldots, \, \rho_n(\mathbf{x}_1), \ldots \tag{6.4.7}$$

be their spectral radii. By the chain rule, we have

$$A_m = \frac{d}{d\mathbf{x}} F^m(\mathbf{x}_1) = F'(\mathbf{x}_m) F'(\mathbf{x}_{m-1}) \ldots F'(\mathbf{x}_1), \tag{6.4.8}$$

i.e., A_m is the product (in the given order) of the matrices representing the derivative of F at all points of the orbit from $\mathbf{x}_1$ to $\mathbf{x}_m$. The reader should keep in mind that such a product is noncommutative.

Definition 6.4.1

The Lyapunov number $N(\mathbf{x}_1)$ of $O(\mathbf{x}_1)$ is

$$N(\mathbf{x}_1) = \lim_{n \to \infty} [\rho_n(\mathbf{x}_1)]^{1/n}. \tag{6.4.9}$$

The Lyapunov exponent is

$$\Lambda(\mathbf{x}_1) = \ln N(\mathbf{x}_1), \tag{6.4.10}$$

i.e., the natural logarithm of the Lyapunov number $N(\mathbf{x}_1)$.

From (6.4.9) we understand why the Lyapunov exponent is computationally preferable to the Lyapunov number, at least when dealing with one-dimensional systems. In fact, since $\rho_n(x_1) = |F'(x_n) \ldots F'(x_1)|$, we see that as n grows, $\rho_n(x_1)$ may become very large. Combining this possibility with taking its nth root may result in nonnegligible errors. The Lyapunov exponent avoids this pitfall, since we can exchange the limit with the logarithm and consider

$$\Lambda(x_1) = \lim_{n\to\infty} \frac{1}{n} (\ln|F'(x_n)|+\ln|F'(x_{n-1})|+...+\ln|F'(x_1)|). \qquad (6.4.11)$$

Clearly, (6.4.11) is preferable to (6.4.9) from a computational point of view.

Like the correlation dimension presented earlier, Lyapunov numbers and exponents are based on the full length of the orbit $O(\mathbf{x}_1)$. However, in practice we **estimate** Lyapunov numbers and exponents using only a finite section of the orbit since we cannot use its full length.

Example 6.4.1 Let $F(x)=5x(1-x)/2$ and let $x_1=0$. Since $F(0)=0$ and $F_x(0)=5/2$, we obtain, according to (6.4.1), $N(0)=5/2$. Let us now compute $N(0)$ using (6.4.9). Since $(d/dx)F^n(0)=(5/2)^n$, we have $\rho_n(0)=(5/2)^n$. Thus, the Lyapunov number of $O(0)$ is

$$N(0) = \lim_{n\to\infty} [(5/2)^n]^{1/n} = 5/2$$

and $\Lambda(0)=\ln 5-\ln 2>0$. We see that the two definitions agree. Notice that $x=0$ is an unstable fixed point. Now let $x_1=3/5$. Since $F(3/5)=3/5$ and $F'(3/5)=-1/2$, we obtain $N(3/5)=1/2$ according to (6.4.1). Using (6.4.9) and taking into account that $(d/dx)F^n(3/5)=(-1)^n2^{-n}$ we obtain $\rho_n(3/5)= 2^{-n}$ and

$$N(3/5) = \lim_{n\to\infty} (2^{-n})^{1/n} = 1/2.$$

Thus $\Lambda(3/5)=-\ln 2<0$. Notice that 3/5 is a sink. Again, (6.4.1) and (6.4.9) provide the same answer.

Example 6.4.2 Let $F(x)=3.2x(1-x)$. The point $x_s=11/16$ is a fixed point. From $F'(11/16)=-1.2$ we derive $N(11/16)=1.2$ and $\Lambda(11/16)=\ln 1.2>0$. The fixed point is a source. For the periodic orbit of period 2,

$$x_1 = \frac{21+\sqrt{21}}{32}, \quad x_2 = \frac{21-\sqrt{21}}{32},$$

we find

$$F'(x_1) = -1 - .2\sqrt{21} \qquad F'(x_2) = -1 + .2\sqrt{21},$$

and we obtain $F'(x_1)F'(x_2)=.16$. Hence, according to (6.4.2), we have $N(x_1)=\sqrt{.16}=.4$. The periodic orbit is a sink. Let us now use (6.4.9). From

$$\frac{d}{dx}F^n(x_1) = \begin{cases} (.16)^m & n = 2m \\ (.16)^m(-1-.2\sqrt{21}) & n = 2m+1 \end{cases}$$

we obtain

$$\rho_n(x_1) = \begin{cases} (.16)^m & n = 2m \\ (.16)^m(1 + .2\sqrt{21}) & n = 2m + 1. \end{cases}$$

Using the result

$$\lim_{n\to\infty} a^{1/n} = 1, \ a>0$$

we arrive at

$$N(x_1) = \lim_{n\to\infty} (\rho_n(x_1))^{1/n} = .4 \ , \ \Lambda(x_1) = \ln 2 - \ln 5<0.$$

We see that (6.4.2) and (6.4.9) agree.

In Figure 6.4.1 we numerically estimate the Lyapunov exponent of the orbit O(.3) for the dynamical system governed by the map F(x)=3.8x(1–x). The orbit is aperiodic, and the estimate is obtained using (6.4.11) with n=100.

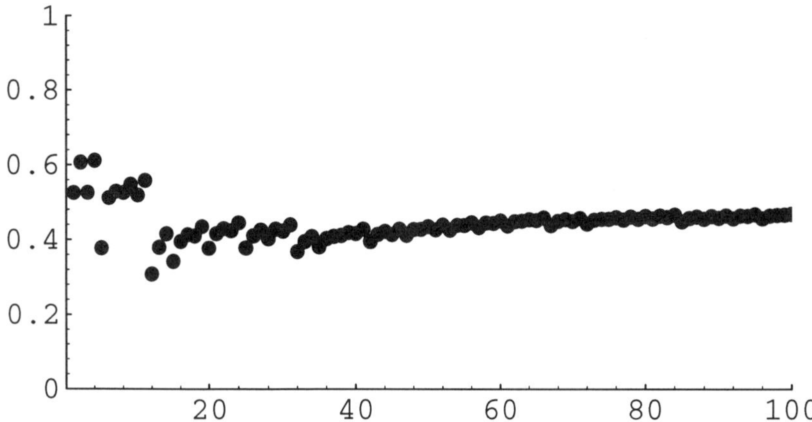

Fig. 6.4.1 The Lyapunov exponent of the orbit O(.3) of the map $F(x)=3.8x(1-x)$ is numerically estimated with 100 iterations. It seems that the sequence converges to .45. The orbit appears to be unstable since the exponent is approaching a positive value.

As mentioned previously, the Lyapunov number defined in (6.4.9) is the same as the one defined in (6.4.1) and (6.4.2) when $O(\mathbf{x}_1)$ is a stationary or a periodic orbit. Moreover, we see that the Lyapunov exponent is negative when the orbit is stable and positive when the orbit is unstable. However, to make this statement fully accurate we need a small adjustment to (6.4.9) [not to (6.4.1) or (6.4.2)]. To see why this adjustment is needed, consider the following example.

Example 6.4.3 Let $F(x)=4x(1-x)$ and select the initial condition $x_1=.25\ (2-\sqrt{2}\ )$. It can be easily verified that

$$x_2 = .5,\ x_3 = 0,\ x_4 = 0,\ \ldots .$$

Since $F'(.5)=0$, the Lyapunov number of the orbit $O(x_1)$ is 0. The result suggests that the orbit is stable, whereas we have established previously that the dynamical system governed by F has no stable orbits. Moreover, the stationary state $x_s=0$ is a source since $F'(0)=4$. The presence of the state $x_2=.5$ is responsible for the wrong information provided by the Lyapunov number of the orbit $O(x_1)$. To avoid this inconvenience we propose a slight variation of (6.4.9).

Given the orbit $\mathbf{x}_1, \mathbf{x}_2=F(\mathbf{x}_1),\ldots$ and a positive integer k, define the k-Lyapunov number $N(\mathbf{x}_1,k)$ of the orbit $O(\mathbf{x}_1)$ as

$$N(\mathbf{x}_1,k) = N(\mathbf{x}_k). \tag{6.4.12}$$

In other words, the k-Lyapunov number of $O(\mathbf{x}_1)$ is the Lyapunov number of the orbit starting at $\mathbf{x}_k$. The Lyapunov number $N(\mathbf{x}_1)$ and the Lyapunov exponent $\Lambda(\mathbf{x}_1)$ are now defined by

$$N(\mathbf{x}_1) = \lim_{k\to\infty} N(\mathbf{x}_1,k), \quad \Lambda(\mathbf{x}_1) = \ln N(\mathbf{x}_1). \tag{6.4.13}$$

With this new definition we obtain that both quantities are not affected by any particular state of the orbit $O(\mathbf{x}_1)$.

For one-dimensional systems the Lyapunov number $N(x_1)$ given by (6.4.12) is the geometric mean of the absolute values of the derivative of F **along the tails** of the orbit $O(x_1)$. Again we remark that $N(\mathbf{x}_1)>1$ $[\Lambda(\mathbf{x}_1)>0]$ as defined in (6.4.13) indicates that the orbit is unstable, whereas $N(\mathbf{x}_1)<1$ $[\Lambda(\mathbf{x}_1)<0]$ indicates that the orbit is stable.

Example 6.4.4 With F and x_1 as in Example 6.4.3 we obtain for $k>2$,

$$\rho_n(x_k) = [F'(0)]^n = 4^n \quad \text{and} \quad N(x_k) = 4.$$

Thus

$$N(x_1) = \lim_{k\to\infty} N(x_k) = 4, \quad \Lambda(x_1) = 2\ln 2.$$

The effect of $x_2=1/2$ in the computation of $N(x_1)$ has been eliminated.

Example 6.4.5 Let

$$F(x) = \begin{cases} 2x & 0 \le x < 1/2 \\ 2 - 2x & 1/2 \le x \le 1. \end{cases}$$

Since $|F'(x)|=2$ for every $x\neq 1/2$, we see that the Lyapunov number of every orbit which does not contain the state 1/2 is 2. For an orbit $O(x_1)$ that contains 1/2 we can choose k so large that the state 1/2 is eliminated from the computation of $N(x_1, k)$, since $F(1/2)=1$, $F(1)=0$, and $F(0)=0$. Hence, $N(x_1)=2$ also in this case.

When the initial state $\mathbf{x}_1$ is a periodic point of period $p \geq 2$ (for $p=1$ this is evident), the k-Lyapunov number is independent of k, and one can use the simpler approach (6.4.2).

Example 6.4.6 Let $F : [0,1] \to [0,1]$ be defined by

$$F(x) = \begin{cases} 4x & 0 \leq x \leq .25 \\ -5x + 2.25 & .25 < x \leq .375 \\ -.6x + .6 & .375 < x \leq 1. \end{cases}$$

Then $\{3/17, 12/17\}$ is a periodic orbit of period 2. Moreover,

$$F'(3/17) = 4, \quad \text{and} \quad |F'(12/17)| = .6.$$

Hence,

$$N(3/17) = N(12/17) = \sqrt{4 \times .6} = 2\sqrt{.6}$$

and

$$\Lambda(3/17) = \Lambda(12/17) = 0.5(\ln 12 - \ln 5) > 0.$$

We now discuss briefly the relationship between the Lyapunov numbers of two maps that are conjugate. We start with the particular case of $G, F : [0,1] \to [0,1]$, $G(x)=4x(1-x)$ and

$$F(x) = \begin{cases} 2x & 0 \leq x < 1/2 \\ 2-2x & 1/2 \leq x \leq 1. \end{cases}$$

The map $h : [0,1] \to [0,1]$, $h(x)=\sin^2 \pi x/2$ is a conjugacy between F and G since $h(F(x))=G(h(x))$ for every $x \in [0,1]$. It is natural to ask if h can be used to evaluate the Lyapunov exponent of an orbit $O(h(x_1))$ of G using the equality

$$F^n(x_1) = h^{-1}G^n h(x_1), \tag{6.4.14}$$

which is valid for every n, and the result $\Lambda(x_1)=\ln 2$ established in Example 6.4.5 for F. The answer to this question is affirmative and we can say that for the dynamical system governed by G, we have $\Lambda(h(x_1))=\ln 2$ as well. There are exceptions. For example, the Lyapunov exponent of the stationary orbit O(0) is 2ln2 when the dynamical system is governed by G, and it is ln2 when it is governed by F. Notice that $h'(0)=0$.

In general, the Lyapunov number and the Lyapunov exponent of corresponding orbits of two one-dimensional dynamical systems that are conjugate is the same, provided that the derivative of the conjugacy is different from 0 and

$$\lim_{n \to \infty} \frac{\ln|h'(x_n)|}{n} = 0.$$

In fact, from (6.4.14) and the chain rule, we derive

$$h'(x_{n+1})F'(x_n)...F'(x_1) = G'(y_n)...G'(y_1)h'(x_1). \tag{6.4.15}$$

Our assumptions on h imply that

$$\Lambda(x_1) = \lim_{n\to\infty} \frac{1}{n} (\ln|F'(x_n)| + ... + \ln|F'(x_1)|)$$

$$= \lim_{n\to\infty} \frac{1}{n} [\ln|F'(x_n)| + ... + \ln|F'(x_1)| + \ln|h'(x_{n+1})|]$$

$$= \lim_{n\to\infty} \frac{1}{n} [\ln|G'(y_n)| + ... + \ln|G'(y_1)| + \ln|h'(x_1)|]$$

$$= \lim_{n\to\infty} \frac{1}{n} [\ln|G'(y_n)| + ... + \ln|G'(y_1)|] = \Lambda(y_1).$$

In Fig. 6.4.2 we have plotted the difference between ln2, which is the Lyapunov exponent of O(.3) according to F, and the elements of the sequence $(1/n)(\ln|G'(y_n)|+ ...+\ln|G'(y_1)|)$ for $y_1=\sin^2(.3\pi/2)$, n=1,...,200. We see that the difference goes to 0. The reader is encouraged to try some additional orbits using the program on Lyapunov exponents contained in Section 3 of Appendix 1.

In Fig. 6.4.3 we have plotted the Lyapunov exponents of the one-parameter family of maps $F(a,x)=ax(1-x)$ (logistic) when the parameter a ranges in the interval [3,4]. Remarkable in the plot is the window of stability around the value a=3.83. In this window the system has a stable periodic orbit of period 3.

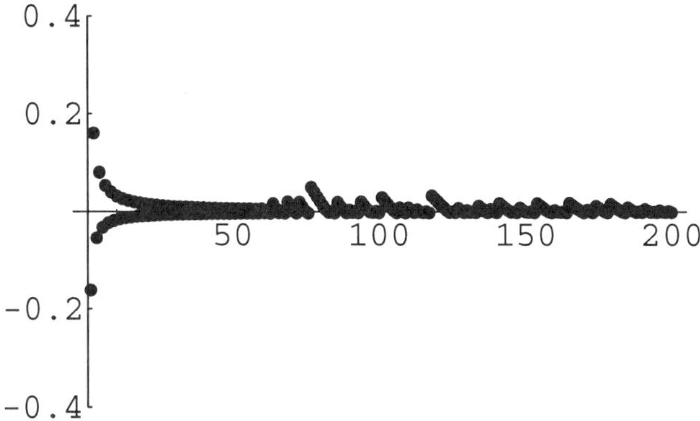

Fig. 6.4.2 Plot of the difference between successive approximations of $\Lambda(\sin^2(.3\pi/2))$ and $\Lambda(.3)=\ln 2$ according, respectively, to the maps $G(x)=4x(1-x)$ and $F(x)=1-|2x-1|$.

There are two remarks to be made before concluding this section and the chapter. First, we just talked about the Lyapunov exponent of the map $F(a,x)=ax(1-x)$ and not about the Lyapunov exponent of an orbit of the map. The reader may have questions about the validity of such "extension." We have good reasons for doing so. It can be shown that for every fixed value of the parameter a, the Lyapunov exponent is independent of the orbit we consider, in the sense that if we choose x_1 at random in [0,1] the Lyapunov exponent of $O(x_1)$ is independent of x_1. There are exceptions, namely there are points for which this is not true, but the probability that the random choice will pick one of those points is 0. These considerations are valid for every dynamical system that is chaotic on a set X according to Definition 6.2.3. The first condition for chaotic behavior, namely the presence of a dense orbit, provides the theoretical foundation for the independence of the Lyapunov exponent from the particular orbit.

The second remark regards the technical difficulty of computing the Lyapunov exponent of m-dimensional dynamical systems with m≥5. In this case the search for $\rho_n(\mathbf{x}_1)$ is a nontrivial task. Suitable numerical methods can help. Since we simply need the spectral radius of the matrix $A_n=F'(\mathbf{x}_n)...F'(\mathbf{x}_1)$, the quantity can be estimated without computing the eigenvalues of A_n. The **power method** illustrated below is a possible way to do it. Assume that a q×q matrix M is such that:

1. All eigenvalues of M are semisimple.
2. The ratio of the modulus (absolute value) of the second largest eigenvalue of M to ρ(M) is smaller than 1.

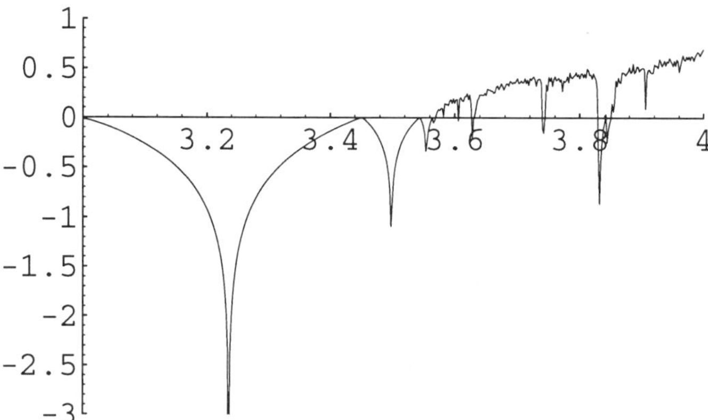

Fig. 6.4.3 Lyapunov exponent of the logistic map $F(a,x)=ax(1-x)$ for a ∈ [3,4].

Then, for almost all vectors $\mathbf{v} \in \mathbf{R}^q$, the sequence

$$\frac{|M\mathbf{u}_p \cdot \mathbf{u}_p|}{\|\mathbf{u}_p\|^2} \tag{6.4.16}$$

converges to ρ(M). The elements of the sequence are defined as follows:

$$\mathbf{u}_1 = \mathbf{v}, \quad \mathbf{u}_2 = \frac{1}{r_1} M\mathbf{u}_1, \quad \mathbf{u}_3 = \frac{1}{r_2} M\mathbf{u}_2 \ldots, \tag{6.4.17}$$

with $r_p = \|M\mathbf{u}_p\|_\infty$. Recall that $\|\mathbf{x}\|_\infty = \max\{|x_i|: i=1,2,\ldots,q\}$.

Example 6.4.7 The eigenvalues of the matrix

$$M = \begin{pmatrix} 2 & 2 \\ 1 & 3 \end{pmatrix}$$

are $r_1=1$ and $r_2=4$. Hence, $\rho(M)=4$. We select $\mathbf{v}=(1,0)$ and apply the power method to obtain

$$\mathbf{u}_2 = (1,0.5),\ \mathbf{u}_3 = (1,\ 5/6),\ \mathbf{u}_4 = (1,\ 21/22),\ M\mathbf{u}_4 = (43/11,\ 85/22)\ ).$$

The approximate value of $\rho(M)$ after these four iterations is 3.98.

Computing $N(k,\mathbf{x}_1)$ and $N(\mathbf{x}_1)$ is time consuming, and margins of uncertainty cannot be eliminated due to the presence of the limits (6.4.9) and (6.4.13). For obvious reasons, they can only be estimated in any numerical scheme. Moreover, when q is large, these problems are compounded by the task of finding the spectral radius. Therefore, only after extensive and careful investigation we feel authorized to derive conclusions on the instability of the dynamical process. The conclusion that F is chaotic in an invariant set X, when based only on numerical evidence, should always be supported by the analysis of several orbits and by the essential agreement among the results obtained from the study of each orbit.

Problems

1. * Let F be as in Example 6.4.6. Verify that {3/17,12/17} is a periodic orbit. Verify that the definitions of Lyapunov number of (6.4.2), (6.4.9), and (6.4.13) all agree and we have $N(3/17)=N\ (12/17)=2\sqrt{.6}$.

2. Let $B(x)=2x-[2x]$. Verify that $x_1=(1/7)$ is a periodic point of period 3. Find the Lyapunov number of the orbit.

3. Let B be as in Problem 2. Consider the orbit with initial state 1/28. Find the Lyapunov k-number of this orbit. Find the Lyapunov number of the orbit.

4. Let F be a dynamical system in the real line and assume that $\{x_1, x_2,\ldots,x_p,\ldots\}$ is a periodic orbit of period p of F . Prove that $N(x_1,k)=N(x_1)$ for all k.

5. Let F be a dynamical system in the real line and assume that $\{x_1, x_1,\ldots,x_p,\ldots\}$ is a periodic orbit of period p. Prove that $N(x_1)=N(x_2) =\ldots=N(x_p)$.

6. * Let $F(x)=x^2-1$. Find $N(0)$.

7. * Let $F(x)=(1/4)x^2-1$. Find $N(0,1)$ and $N(0)$.

8. * Let $F(x)=(1/4)x^2-1$. Find $N(2+2\sqrt{2},2)$ and $N(2+2\sqrt{2})$.

9. * Let $G(x)=2x^2-1$ and $x_1=1/3$. Verify numerically that the orbit starting from x_1 is not periodic. Find $N(x_1)$ using the conjugacy between G and the map $F(x)=2|x|-1$.

10. * Let

$$M = \begin{pmatrix} -3 & -2 & 1 \\ -2 & -6 & -2 \\ 1 & 2 & -3 \end{pmatrix}.$$

Estimate the spectral radius of M using the power method. Then find the spectral radius of M and compare the two results.

CHAPTER 7

ANALYSIS OF FOUR DYNAMICAL SYSTEMS

SUMMARY

In this last chapter we analyze the dynamical systems that were introduced in the fourth section of Chapter 1. We are not presenting a comprehensive and detailed study of these systems, since such a task would require a book of its own. Rather, we are focusing on certain important features of each system. Sections 1, 2, 3, and 4 are devoted to the one-, two-, three-, and multidimensional model, respectively.

Section 1. BLOOD-CELL POPULATION MODEL

As mentioned in Chapter 1, the population of red blood cells at time n+1 can be written as

$$x_{n+1} = x_n - d_n + p_n, \tag{7.1.1}$$

where d_n and p_n account, respectively, for the number of cells destroyed and released into the bloodstream during the time interval from n to n+1. The two functions d_n and p_n are assumed to depend only on x_n and on some control parameters.

The form of this dependence can obviously be a matter of debate. There are several possibilities, and none of them can claim to be the best. The form we propose is due to A. Lasota [Lasota, 1977] (see Example 1.1.2 for a different approach). He assumed that

$$d_n = ax_n, \quad p_n = b(x_n)^r e^{-sx_n}, \qquad)(7.1.2)$$

where a∈ (0,1] is the destruction rate, and b,r,s>0 are to be determined. Our model takes the form

$$x_{n+1} = (1 - a)x_n + b(x_n)^r e^{-sx_n}, \tag{7.1.3}$$

and the function governing the dynamical process is

$$F(\mathbf{a},x) = (1 - a)x + bx^r e^{-sx}. \tag{7.1.4}$$

The state variable, x, is one-dimensional and the control parameter $\mathbf{a}$= (a,b,r,s) is an element of $\mathbf{R}^4$. Notice that $F(\mathbf{a},0)=0$ for every $\mathbf{a}$. Thus $x_{s1}(\mathbf{a})=0$ is always a fixed point. We expect for this point to be a sink since, from a clinical point of view, a person cannot recover when the population of blood cells falls below a certain critical level. Since $F_x(\mathbf{a},x)=1-a+bx^{r-1}e^{-sx}(r-sx)$, we see that the origin is a sink if r>1. In this case $F_x(\mathbf{a},0)=1-a<1$. Hence, we shall assume that r>1.

Experimental evidence (see [Gearhart-Martelli, 1990]) suggests the presence of a maximum point x_M such that the corresponding maximum value $F(\mathbf{a},x_M)$ is larger than x_M. To account for this feature we first set the derivative of F equal 0. We obtain

$$bx^r e^{-sx} = x\,\frac{1 - a}{sx - r}\,. \tag{7.1.5}$$

Hence, at the maximum point x_M, we must have

$$r < sx_M \tag{7.1.6}$$

since 1–a>0. From the required inequality

$$F(\mathbf{a},x_M) > x_M, \tag{7.1.7}$$

and from (7.1.3), (7.1.5), and (7.1.6), we derive

$$r < sx_M < r + (1-a)/a. \tag{7.1.8}$$

At the fixed points where the state variable is strictly positive, we have $a=bx^{r-1}e^{-sx}$. Substituting into the derivative we obtain $F_x(\mathbf{a},x)=1-a+a(r-sx)=1+a(r-sx-1)$. For obvious clinical reasons the smallest positive fixed point $x_{s2}(\mathbf{a})$ must be a repeller. In fact, it acts as a threshold level for the blood-cell population. Recovery is possible only above this threshold. Below it, the person will not survive without external help. Consequently, we must have

$$r > sx_{s2}(\mathbf{a})+1. \tag{7.1.9}$$

There is also a third fixed point $x_{s3}(\mathbf{a})$. In fact, from (7.1.7) we derive that the graph of $F(\mathbf{a},x)$ is above the line $y=x$ at x_M. Moreover, since $bx^re^{-sx}\to 0$ as $x\to+\infty$ and since $a<1$, the graph of $F(\mathbf{a},x)$ is below the line $y=x$ for x sufficiently large. By the intermediate value theorem the graph of $F(\mathbf{a},x)$ intersects the line $y=x$ at $x_{s3}(\mathbf{a})>x_M$. This third fixed point represents the state at which the blood-cell population of a healthy person should be. For small values of the destruction rate the state should be a sink, while for values close to 1 it should be a source. Consequently, for a close to 0 we need

$$r - sx_{s3}(\mathbf{a}) - 1 < 0, \tag{7.1.10}$$

which implies that $r<sx_{s3}(\mathbf{a})+1$. The instability is achieved when $1+a(r-sx_{s3}(\mathbf{a})-1)=-1$, which gives $x_{s3}(\mathbf{a})=(2+a(r-1))/as$. Hence, r and s should be selected so that this value of $x_{s3}(\mathbf{a})$ is within expected ranges.

At this point the best thing to do is to use reliable data and estimates to determine acceptable values of the parameters b, r, and s. The reader can consult the paper of W. Gearhart and M. Martelli [Gearhart-Martelli, 1990] for this approach. Suggested values for some cases are $r=8$, $s=16$, and $b=1.1\times10^6$, which we shall use in the sequel. Hence, we study the dynamical system governed by

$$F(a,x) = (1-a)x + 1.1\times10^6x^8e^{-16x} \tag{7.1.11}$$

and denote its three fixed points by

$$x_{s1}(a) = 0 < x_{s2}(a) < x_{s3}(a).$$

Notice that **a** has been replaced by a since the three remaining parameters have been given specific numerical values. We know that 0 is a sink since r=8. More precisely, every orbit $O(x)$ such that $x<x_{s2}(a)$ converges to 0. The second fixed point $x_{s2}(a)$ is always a source and there exists $a_0\in(0,1)$ such that the third $x_{s3}(a)$ is a sink for $a<a_0$ and a source for $a>a_0$. We now determine a_0.

In Fig. 7.1.1 we present the graph of the function F for two different values of a, namely a=.2 and a=.78 (dashed). The derivative of F at $x_{s3}(a)$ is negative in both cases. For a=.2 the absolute value of the derivative is smaller than 1, while for a=.78 it is larger than 1. Thus $a_0\in(.2,.78)$. The stability is lost for those values of a for which the derivative of F at $x_{s3}(a)$ is smaller than -1. Since $x_{s3}(a)$ satisfies the equation

$$1.1\times10^6\,(x_{s3}(a))^7 e^{-16x_{s3}(a)} = a, \tag{7.1.12}$$

and we must have

$$-1 = F_x(a,x_{s3}(a)) = 1 + 7a - 16a x_{s3}(a), \tag{7.1.13}$$

we obtain $x_{s3}(a)=(2+7a)/16a$, in agreement with $x_{s3}(\mathbf{a})=(2+a(r-1))/as$. Substituting into (7.1.12) gives

$$1.1\times10^6\left(\frac{2+7a}{16a}\right)^7 e^{(-2+7a)/a} = a, \tag{7.1.14}$$

or

$$1.1\times10^6(2+7a)^7\, e^{(-2+7a)/a} = 2^{28}a^8. \tag{7.1.15}$$

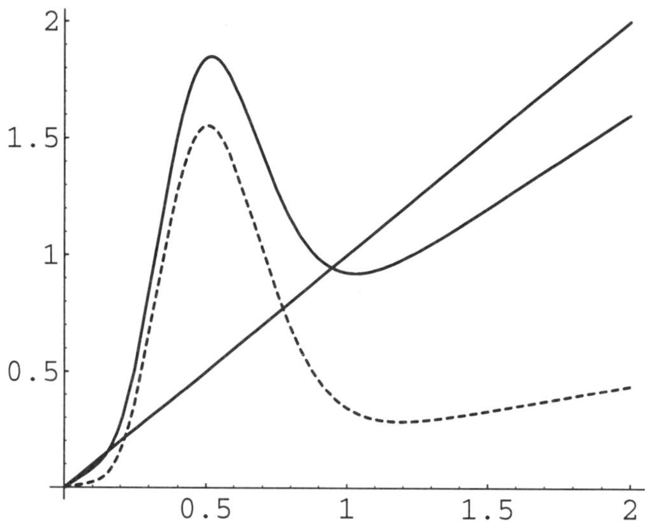

Fig. 7.1.1 Graph of $F(a,x)=(1-a)x+1.1\times10^6x^8e^{-16x}$ for a=.2 and a=.78 .The straight line is $y=x$.

Taking the logarithms on both sides, we arrive at the equation

$$a \ln 11 + 5a \ln 10 + 7a \ln(2+7a) - 2 - 7a - 28a \ln 2 - 8a \ln a = 0.$$

The solution we are looking for is a=.262727... . Hence, $x_{s3}(.262727...)=(2+7a)/16a$ =.91327904... .

In Fig. 7.1.2 we use the value a=.3. We see that a periodic orbit of period 2 is arising, thus showing that the parameter has crossed a period-doubling bifurcation point. As mentioned in the paper of Gearhart-Martelli, the appearance of this periodic orbit of period 2 is in agreement with experimental findings.

For the value a=.78 we present the graph of F and the graph of its third iterate (see Fig. 7.1.3). We find two periodic orbits of period 3. Hence, for this value of the parameter, the system is chaotic in the sense of Li and Yorke.

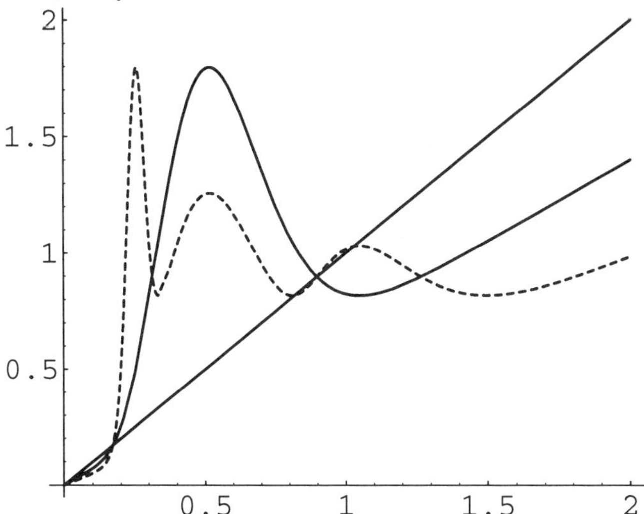

Fig. 7.1.2 Graph of F and F^2 for a=.3.

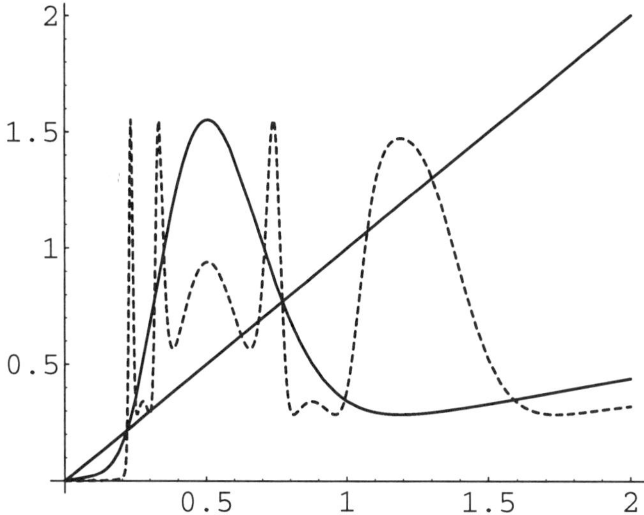

Fig. 7.1.3 Graph of F and F^3 for a=.78.

For a=.81 the system has the invariant interval I≈[.25,1.54]. Numerical investigations suggest that there exists $x_0 \in$ I such that L(x_0)=I. A test on the Lyapunov exponents of the system reveals that the orbits are unstable at a=.81. For example, we obtain the following successive approximations to the Lyapunov exponent of O(.3):

4996	4997	4998	4999	5000
.208	.208	.208	.208	.208.

On the top row we see the iterate's number, and on the bottom row, the corresponding value of the numerical estimate of the Lyapunov exponent along the orbit. Starting from x_0=.7, we get the following table:

4996	4997	4998	4999	5000
.208	.209	.209	.208	.209.

The agreement is pretty remarkable and it suggests that the system does not have any stable orbit. Hence, at least on numerical grounds, we can say that it satisfies the two conditions of Definition 6.2.3.

For values of a close to 1 we have $F(a,F(a,x_M))<x_{s2}(a)$. This might be expected. When $a\approx 1$ the body tries to compensate for the high destruction rate by increasing the production to its maximum. Consequently, a very large number of cells enter the bloodstream, causing an excessive accumulation. This "crowding" is sensed by the automatic controls of cells level, and during the next time interval only a small number of cells is released. Since the destruction rate is unchanged, the population falls below the critical level $x_{s2}(a)$ and the orbit goes to 0.

Problems

1. Consider the production function

$$P(x) = bp^m \frac{x}{p^m+x^m}, \ m > 1.$$

Show that there is a single maximum on $(0,\infty)$ and a single inflection point. Find their value and sketch the graph of the function.

2. Show that with r=8, s=16, and $b=1.1\times 10^6$, and for every $a\in(0,1)$, the first nonzero fixed point $x_{s2}(a)$ of the dynamical system defined by (7.1.3) is always unstable.

3. * What are the values of a for which the function F of (7.1.11) has a minimum point x_m after the largest fixed point $x_{s3}(a)$? Estimate x_m numerically and check if there are any values of a such that $F(a,x_m)<x_{s2}(a)$ (see Problem 2). What interpretation can be given to this occurrence?

4. * Let F be as in (7.1.11). Compare x_m (see Problem 3) with the maximum value of F, i.e., with $F(a,x_M)$. Is there any value of a for which $F(a,x_M)>x_m$ and $F(a,F(a,x_M))<x_{s2}(a)$?

5. * Do the cobweb diagram of O(.5) and O(.7) for a=.81. Make some comments about the behavior of the system based on the two diagrams.

6. * Do a histogram of O(.5) and O(.7) for a=.81.

7. * Find the fixed points of the system $x_{n+1}=b(x_n)^r e^{-sx_n}$ and study their stability.

Section 2. PREDATOR-PREY MODELS

Recall that V. Volterra [Volterra, 1931] proposed a model of predator-prey interaction to explain the apparent anomalies found by his son-in-law, the biologist U. D'Ancona, in the fish population of the North Adriatic Sea during World War I. D'Ancona observed that the average number of sharks captured by the few fishermen operating in the area during the years 1915-1920 was up to three times as large as the average number captured before 1915 and after 1920. Volterra suggested the following system of differential equations in which x and y represent prey and predators, respectively, and a, b, c, and d are positive parameters:

$$\begin{cases} x' = ax - bxy \\ y' = -cy + dxy. \end{cases} \tag{7.2.1}$$

Volterra assumed, although not very realistically, that in absence of y, the prey population grows exponentially. In fact, from

$$x' = ax \quad \text{we derive} \quad x(t) = x_0 e^{at}.$$

Thus $x(t)\to\infty$ as $t\to\infty$.

In absence of prey, the predators will not survive, since

$$y' = -cy \quad \text{implies that} \quad y(t) = y_0 e^{-ct}.$$

Thus $y(t)\to 0$ as $t\to\infty$. This result is in line with expectations. Prey-predator encounters are obviously favorable to predators ($+dxy$) and fatal to prey ($-bxy$).

The presence of outside harvesting activity can be added to the model by writing

$$\begin{cases} x' = ax - bxy - \varepsilon x = (a - \varepsilon)x - bxy \\ y' = -cy + dxy - \varepsilon y = -(c + \varepsilon)y + dxy. \end{cases} \tag{7.2.2}$$

The steady states of systems (7.2.1) and (7.2.2) are obtained by setting

$$x' = y' = 0.$$

The origin satisfies this condition for both systems. A second steady state is given by

$$(c/d,a/b) \text{ for } (7.2.1) \qquad ((c+\varepsilon)/d,(a-\varepsilon)/b) \text{ for } (7.2.2). \tag{7.2.3}$$

According to (7.2.3), the presence of a harvesting activity favors the prey population. Their steady-state value increases while the predators' decreases. Volterra argued that the absence of sustained fishing activity during the war was the source of the dramatic increase in the number of sharks captured during the war period.

System (7.2.1) cannot be solved by explicitly finding x(t) and y(t). One can solve it numerically. However, the scheme adopted is extremely important. In Chapter 1 we used the one-step Euler method. This approach is acceptable if a small timestep is used and we are not interested in the long-range evolution of the system. We arrive at

$$\begin{cases} x_{n+1} = (1 + ah)x_n - bhx_ny_n \\ y_{n+1} = (1 - ch)y_n + dhx_ny_n. \end{cases} \tag{7.2.4}$$

The stationary states of our system are again $\mathbf{x}_{s1}(\mathbf{a})$=(0,0) and $\mathbf{x}_{s2}(\mathbf{a})$= (c/d,a/b). We want to analyze their stability. Since F(**a**,**x**)=((1+ah)x–bhxy, (1–ch)y +dhxy) we obtain

$$F_{\mathbf{x}}(\mathbf{a},\mathbf{x}) = \begin{pmatrix} 1+ah-bhy & -bhx \\ dhy & 1-ch+dhx \end{pmatrix}.$$

Hence,

$$F_{\mathbf{x}}(\mathbf{a},\mathbf{0}) = \begin{pmatrix} 1+ah & 0 \\ 0 & 1-ch \end{pmatrix},$$

whose eigenvalues are 1+ah and 1–ch. The first is larger than 1 and the second is smaller than 1. Therefore, the origin is a saddle.

Similarly, we obtain

$$F_{\mathbf{x}}(\mathbf{a},\mathbf{x}_{s2}(\mathbf{a})) = \begin{pmatrix} 1 & -bch/d \\ adh/b & 1 \end{pmatrix},$$

whose eigenvalues are $1+ih(ac)^{.5}$ and $1-ih(ac)^{.5}$. This fixed point is a repeller. To visualize the behavior of the system let us choose the numerical values ah=bh=ch=dh =.2. The fixed point is (1,1).

Starting at (1.3,1), the two populations (see Fig. 7.2.1) display an out-of-phase oscillatory behavior and the amplitudes of the oscillations are increasing. This behavior is expected, since the eigenvalues of $F_{\mathbf{x}}$ are complex with modulus larger than 1.

It can be shown that the orbit starting from (1.3,1) in the continuous system is periodic. This fact underlines the importance of a careful qualitative analysis of a system before using numerical methods for investigating its evolution patterns.

Improvements to Volterra's model have been proposed by requiring, for example, that the prey's growth be governed by a logistic map of the form ax(1–x). Consequently, our discrete one-step Euler version of the new model is

$$\begin{cases} x_{n+1} = (1 + ah)x_n - ah(x_n)^2 - bhx_ny_n \\ y_{n+1} = (1 - ch)y_n + dhx_ny_n. \end{cases} \tag{7.2.5}$$

The time step h should be considered a very small parameter. The origin is again a stationary state. Since

$$F(\mathbf{a}, \mathbf{x}) = ((1 + ah)x - ahx^2 - bhxy, (1 - ch)y + dhxy),$$

we have

$$F_{\mathbf{x}}(\mathbf{a}, \mathbf{0}) = \begin{pmatrix} 1+ha & 0 \\ 0 & 1-hc \end{pmatrix},$$

whose eigenvalues are 1+ha and 1–hc. Therefore, the origin is again a saddle.

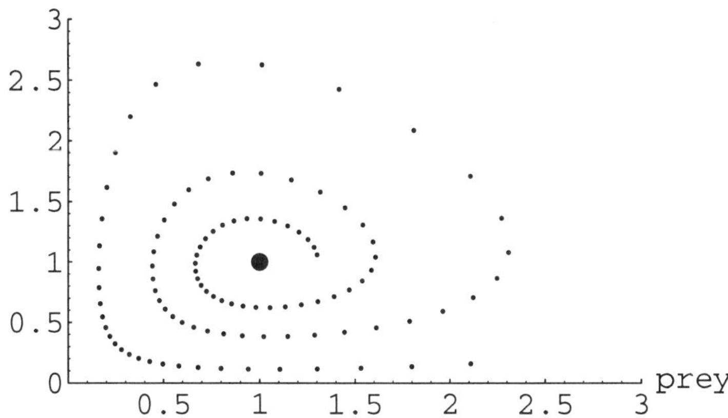

Fig. 7.2.1 Fixed point of system (7.2.4) and orbit starting at (1.3,1).

There are two more fixed points. One has coordinates (1,0). Assuming that c<d we have a third fixed point,

$$\mathbf{x}_{S3}(\mathbf{a}) = (c/d, a(d-c)/bd).$$

The fixed point (1,0) is a saddle since

$$F_{\mathbf{x}}(\mathbf{a},(1,0)) = \begin{pmatrix} 1-ha & -bh \\ 0 & 1-hc+dh \end{pmatrix},$$

whose eigenvalues are 1–ha and 1+h(d–c)>1 since d>c.

The analysis of the stability of the third fixed point is a bit more complicated. Notice that F=I+hG, where

$$G(\mathbf{a},\mathbf{x}) = (ax(1-x) - bxy, -cy + dxy).$$

Therefore, the eigenvalues of the derivative of F are found by adding 1 to the eigenvalues of G multiplied by h. We have

$$G_{\mathbf{x}}(\mathbf{a},\mathbf{x}_{s3}(\mathbf{a})) = \begin{pmatrix} -ac/d & -bc/d \\ a(d-c)/b & 0 \end{pmatrix}.$$

Since the trace of this matrix is negative and the determinant is positive, its eigenvalues have a negative real part. Therefore, for h sufficiently small, the modulus of the eigenvalues of $F_{\mathbf{x}}(\mathbf{a},\mathbf{x}_3)$ is smaller than 1 and the fixed point is a sink.

One may find reasons for criticism also in the logistic model of the predator-prey competition, arguing that the origin should be a sink, since both populations should become extinct if they reach a certain minimum level. The following predator-prey model presents this feature:

$$\begin{cases} x_{n+1} = ax_n(1 - x_n - y_n) \\ y_{n+1} = ax_n y_n \end{cases} \tag{7.2.6}$$

or $\mathbf{x}_{n+1} = F(a,\mathbf{x}_n)$, with $F(a,\mathbf{x})=F(a,(x,y))=(ax(1-x-y),axy)$ and $a\in(0,4]$. The stationary states of this system are

$$\mathbf{x}_{s1}(a) = (0,0), \quad \mathbf{x}_{s2}(a) = (1-a^{-1},0), \quad \mathbf{x}_{s3}(a) = (a^{-1},1-2a^{-1}).$$

At a=1 we have $\mathbf{x}_{s2}(1)=\mathbf{x}_{s1}(1)=(0,0)$, and at a=2 we have $\mathbf{x}_{s3}(2)=\mathbf{x}_{s2}(2)=(1/2,0)$. Hence, a=1 and a=2 are (transcritical) bifurcation points. Notice that $\mathbf{x}_{s2}(a)$ and $\mathbf{x}_{s3}(a)$ make sense for the model only when a>1 and a>2, respectively.

The eigenvalues of $F_{\mathbf{x}}(a,\mathbf{x}_{s1}(a))$ are a and 0. Hence, $\mathbf{x}_{s1}(a)$ is a sink for $a\in(0,1)$ and a saddle for $a\in(1,4]$. The eigenvalues of $F_{\mathbf{x}}(a,\mathbf{x}_{s2}(a))$ are 2–a and a–1. Thus, $\mathbf{x}_{s2}(a)$ is a sink for $a\in(1,2)$, a saddle for $a\in(2,3)$, and a source for $a\in(3,4]$. We see that at a=1 there is an exchange of stability between $\mathbf{x}_{s1}(a)$ and $\mathbf{x}_{s2}(a)$. The eigenvalues of $F_{\mathbf{x}}(a,\mathbf{x}_{s3}(a))$ are

$$\lambda_1 = \frac{1+\sqrt{9-4a}}{2}, \qquad \lambda_2 = \frac{1-\sqrt{9-4a}}{2}.$$

Both eigenvalues are real, positive, and smaller than 1 for $a\in(2,2.25)$. Hence, at a=2 the attracting character of the fixed points of the second branch is passed on to those of the third branch.

The eigenvalues become complex for a > 2.25 and their modulus is $(a-2)^{.5}$. They are inside the unit circle in the complex plane for $a\in(2.25,3)$. At a=3 the eigenvalues are

$$\lambda_1 = \frac{1+i\sqrt{3}}{2}, \qquad \lambda_2 = \frac{1-i\sqrt{3}}{2}.$$

The eigenvalues cross the unit circle as a crosses 3 since $(d/da)(a-2)^{.5}>0$ (at a=3). Moreover, the argument of the complex number λ_1 is $\theta(3)=\arctan 3^{.5}$, which implies that $\theta/(2\pi)=1/6$ and m=1, n=6 (see Chapter 5, Section 4). Thus, at a=3 we have Hopf bifurcation (see Fig. 7.2.2). The orbits close to $(a^{-1}, 1-2a^{-1})$ for a>3 are attracted by a periodic orbit of period 6, which is itself a function of a. For larger values of the control parameter, the system seems to be chaotic (see Fig. 7.2.3).

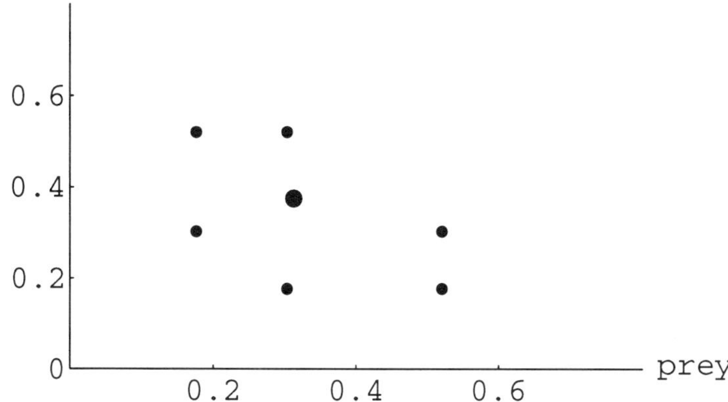

Fig. 7.2.2 Orbit starting from (.32,.34) for a=3.2. The first 50 iterations have been discarded and the next 50 have been used for the plot. The larger dot is the fixed point $x_3(3.2)$.

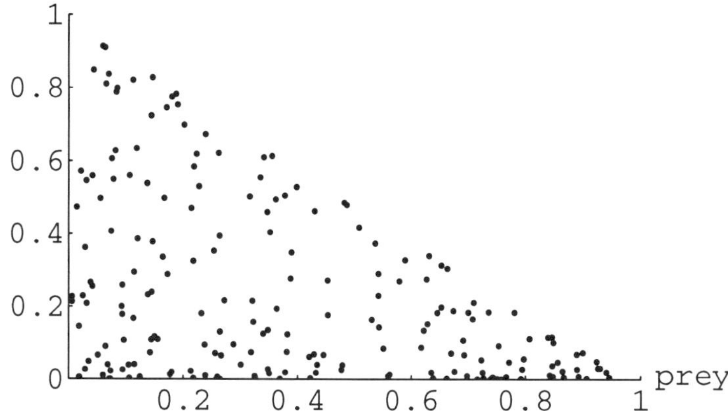

Fig. 7.2.3 Orbit starting at (.4,.2) for a=3.9. 250 iterations have been discarded and the next 200 have been plotted.

Let us now take a brief look at the system (see Chapter 1, Section 4)

$$\begin{cases} x_{n+1} &= 1.09x_n - .04x_n y_n \\ y_{n+1} &= .94y_n + .01x_n y_n(1.09 - .04y_n). \end{cases} \tag{7.2.7}$$

A similar model was proposed in Example 1.1.3 to study the interaction between whiteflies and black wasps. The system has exactly two fixed points: $\mathbf{x}_{s1}$=(0,0) and $\mathbf{x}_{s2}$=(6,2.25). The first is a saddle and its stable manifold is the y-axis. In fact, when the initial condition is $(0,y_0)$ we see that $x_n=0$ for every n and $y_n=(.94)^n y_0 \to 0$ as $n\to\infty$. It can be shown that the unstable manifold is the x-axis (see Problem 6).

The derivative of

$$F(\mathbf{x}) = (1.09x - .04xy,\ .94y + .01xy(1.09 - .04y))$$

at (6,2.25) is

$$F'(6,2.25) = \begin{pmatrix} 1 & -.24 \\ .0225 & .9946 \end{pmatrix},$$

whose characteristic polynomial is $r^2 - r(1.9946)+1$. The eigenvalues have modulus 1 and the theory we presented in Chapter 5 is not sophisticated enough to handle this situation.

Numerical investigation suggests that the orbits starting at points close to (6,2.25) are periodic. In Fig. 7.2.4 we represent the orbit starting at (5.9,2.4). In Fig. 7.2.5 we plot the first 200 states of the orbit with the iteration number n in the horizontal axis and the values of x_n and y_n in the vertical axis (y_n is the lower curve). We can clearly see the periodic behavior. The upper curve oscillates around the value x_{s2} =6 and the lower curve oscillates around the value y_{s2}=2.25. The period is about 100. With initial condition $\mathbf{x}_0$=(8,5), which is farther away from the fixed point, the qualitative behavior does not change (see Figs. 7.2.6 and 7.2.7).

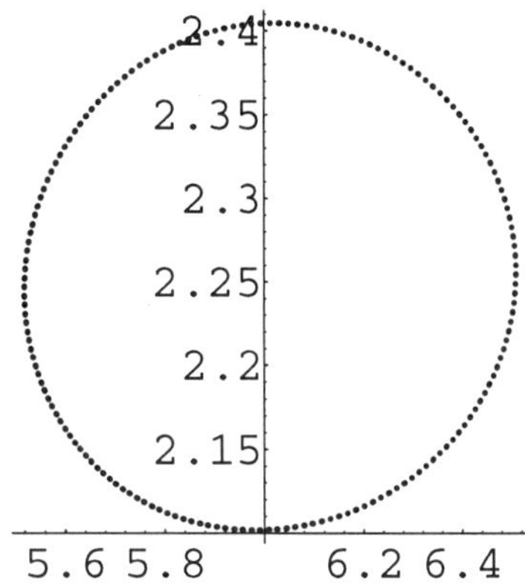

Fig. 7.2.4 Plot of the first 300 states of the orbit O(5.9,2.4) of system (7.2.7).

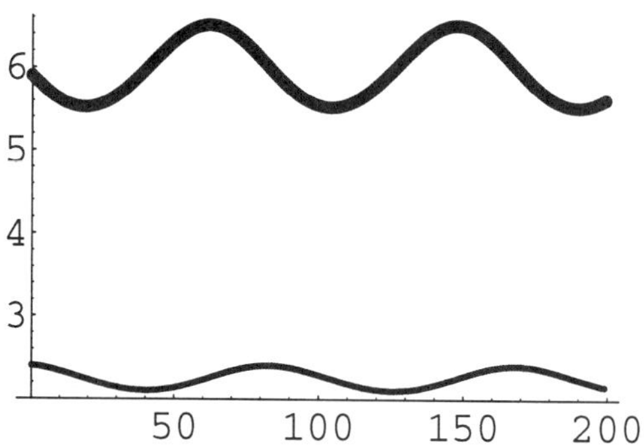

Fig. 7.2.5 The periodicity of the orbit O(5.9,2.4) is clear from this graph, which combines (n,x_n) and (n,y_n) for n=1,...,200 [system (7.2.7)].

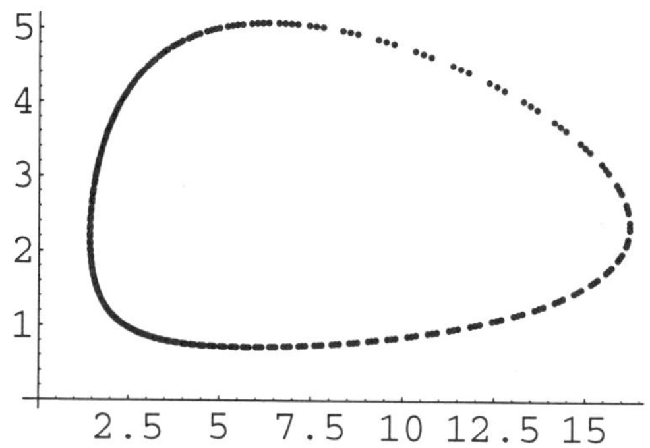

Fig. 7.2.6 Plot of the first 300 states of the orbit O(8,5) of system (7.2.7).

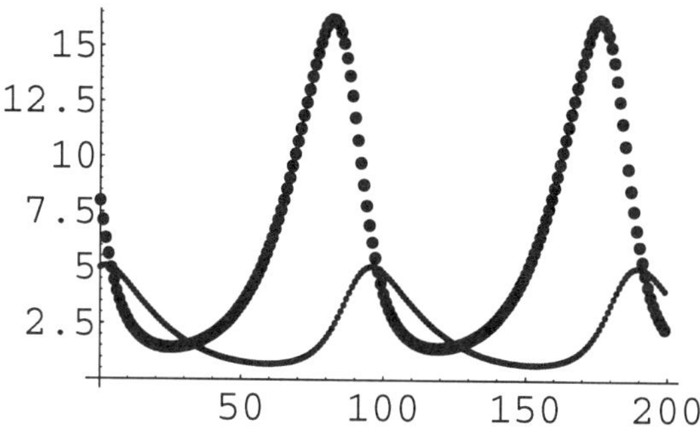

Fig. 7.2.7 The periodicity of the orbit O(8,5) is clear from this graph, which combines (n,x_n) and (n,y_n) for n=1,...,200 [system (7.2.7)].

Problems

1. A moderate harvesting activity modifies (7.2.5). Find the new fixed points. Is the activity in favor of the prey?

2. What is happening to the fixed points of (7.2.5) when $c > d$?

3. Let M be a square matrix of size 2. Justify why the eigenvalues of M have negative real part if the trace of M is negative and the determinant of M is positive.

4. Let $F(\mathbf{a},\mathbf{x})=(ax(1-x-y),bxy)$. Find the stationary states of the dynamical system $\mathbf{x}_{n+1}=F(\mathbf{a},\mathbf{x}_n)$.

5. * Let $F(a,\mathbf{x})=(ax(1-x-y),axy)$ and consider the dynamical system

$$\mathbf{x}_{n+1}=F(a,\mathbf{x}_n).$$

Investigate numerically the behavior of the system for a=1 and a=2 close to the fixed points (0,0) and (1/2,0), respectively.

6. Verify that the map which governs (7.2.7) is invertible and find its inverse. Use the inverse to determine the unstable manifold of the system at $\mathbf{0}$.

7. Let F be as in Problem 5. Verify that the derivative of the modulus of the eigenvalues of $F_{\mathbf{x}}(a,\mathbf{x}_{s3}(a))$ at the point a=3 is strictly positive.

8. * Let $F(a,\mathbf{x})=(ax-.1xy,.9y+.1xy(a-.1y))$. Find the stationary states of the system $\mathbf{x}_{n+1}=F(a,\mathbf{x}_n)$ and study their stability.

Section 3. LORENZ MODEL OF ATMOSPHERIC BEHAVIOR

As mentioned in Chapter 1, E.N. Lorenz [Lorenz, 1963] proposed a first-order system of differential equations as a crude model of weather patterns. The upper atmosphere was imagined as an infinite uniform layer of depth H, with the temperature difference between the lower and the upper surface maintained at a constant value ΔT. Denoting by x, y, and z, respectively, quantities proportional to the intensity of the convective motion between the two surfaces, to the temperature difference between the ascending and descending current, and to the deviation of the temperature from its linear profile, Lorenz arrived at the system

$$\begin{cases} x' = -ax + ay \\ y' = -xz + rx - y \\ z' = xy - bz. \end{cases} \tag{7.3.1}$$

We change the variable z to $u=z-a-r$. The qualitative behavior is not affected by this change and $u'=z'$. We obtain

$$\begin{cases} x' = -ax + ay \\ y' = -xu - ax - y \\ u' = xy - bu - ab - br. \end{cases} \tag{7.3.2}$$

The one-step Euler method applied to this system with $\Delta t=h$ gives (see Chapter 1, Section 4)

$$\begin{cases} x_{n+1} &= x_n + ah(y_n - x_n) \\ y_{n+1} &= y_n + h(-x_n u_n - a x_n - y_n) \\ u_{n+1} &= u_n + h(x_n y_n - b u_n - b(a + r)). \end{cases} \tag{7.3.3}$$

In the analysis of (7.3.3) we can use the ideas developed so far by considering it as a discrete dynamical system in $\mathbf{R}^3$. The system is governed by the function

$$F(\mathbf{a},\mathbf{x}) = (x + ah(y - x), y + h(-xu - ax - y), u + h(xy - bu - b(a + r)))$$
$$= \mathbf{x} + hG(\mathbf{a},\mathbf{x}),$$

with

$$G(\mathbf{a},\mathbf{x}) = (-ax + ay, -xu - ax - y, xy - bu - ba - br).$$

We see that $\mathbf{x}_{s1}(\mathbf{a})=(0,0,-a-r)$ is a fixed point of (7.3.3). The other fixed points are found by setting

$$x_1 = x_0, \quad y_1 = y_0, \quad \text{and} \quad u_1 = u_0 \tag{7.3.4}$$

in (7.3.3). From the first equation we obtain $x_0=y_0$. Substituting into the second and third equations and dropping the subscripts, we arrive at the system

$$\begin{cases} xu &= -ax - x \\ x^2 &= bu + b(a + r). \end{cases} \tag{7.3.5}$$

The first equality implies that $u=-(a+1)$. Thus $x^2=b(r-1)$ which requires that $r>1$ ($r=1$ gives again $x=y=u+a+r=0$) and we find two more stationary states:

$$\mathbf{x}_{s2}(\mathbf{a}) = (\sqrt{b(r-1)}\,, \sqrt{b(r-1)},-a-1)\,, \quad \mathbf{x}_{s3}(\mathbf{a}) = (-\sqrt{b(r-1)}\,, -\sqrt{b(r-1)}\,,-a-1)\,.$$

The derivative of F is the matrix

$$F_{\mathbf{x}}(\mathbf{a},\mathbf{x}) = \begin{pmatrix} 1-ha & ha & 0 \\ -hu-ha & 1-h & -hx \\ hy & hx & 1-bh \end{pmatrix}.$$

Hence,

$$F_{\mathbf{x}}(\mathbf{a},\mathbf{x}_{s1}(\mathbf{a})) = \begin{pmatrix} 1-ha & ha & 0 \\ hr & 1-h & 0 \\ 0 & 0 & 1-bh \end{pmatrix},$$

and we have $F_{\mathbf{x}}(\mathbf{a},\mathbf{x}_{S1}(\mathbf{a}))=I+hG_{\mathbf{x}}(\mathbf{a},\mathbf{x}_{S1}(\mathbf{a}))$, where

$$G_{\mathbf{x}}(\mathbf{a},\mathbf{x}_{S1}(\mathbf{a})) = \begin{pmatrix} -a & a & 0 \\ r & -1 & 0 \\ 0 & 0 & -b \end{pmatrix}.$$

The eigenvalues of $I+hG_{\mathbf{x}}(\mathbf{a},\mathbf{x}_{S1}(\mathbf{a}))$ can be found by adding 1 to the eigenvalues of $G_{\mathbf{x}}(\mathbf{a},\mathbf{x}_{S1}(\mathbf{a}))$ multiplied by h. Setting $\det(tI-G_{\mathbf{x}}(\mathbf{a},\mathbf{x}_{S1}(\mathbf{a})))=0$, we obtain the third-degree equation in t:

$$(t + b)[(t + a)(t + 1) - ar] = 0$$

whose solutions are

$$t_1 = -b,\ \ t_2 = (1/2)[-(a + 1) + \sqrt{(a - 1)^2 + 4ar}\,],$$

$$\text{and}\ \ t_3 = (1/2)[-(a + 1) - \sqrt{(a - 1)^2 + 4ar}\,].$$

Consequently,

$$\begin{aligned} \lambda_1 &= 1 - bh \\ \lambda_2 &= 1 - h(a + 1 + \sqrt{(a - 1)^2 + 4ar}\,)/2 \\ \lambda_3 &= 1 - h(a + 1 - \sqrt{(a - 1)^2 + 4ar}\,)/2. \end{aligned} \tag{7.3.6}$$

Since the time step h is very small, λ_1 and λ_2 can be considered positive and smaller than 1. The last eigenvalue λ_3 is also positive and it is smaller than 1 if

$$(a + 1)^2 > (a - 1)^2 + 4ar,$$

which is true for $r<1$.

Hence, $\mathbf{x}_{S1}(\mathbf{a})$ is a sink if $r<1$, and it becomes a saddle when $r>1$. The stable and unstable manifold of $\mathbf{x}_{S1}(\mathbf{a})$ (when $r>1$) are two- and one-dimensional, respectively. At $r=1$ we obtain $t_1=-b$, $t_2=-(a+1)$, and $t_3=0$. Thus

$$\begin{aligned} \lambda_1 &= 1 - bh \\ \lambda_2 &= 1 - h(a+1) \\ \lambda_3 &= 1. \end{aligned} \tag{7.3.7}$$

Let us now examine the stability of the two additional fixed points, which arise when r crosses the value 1. The two states are

$$\mathbf{x}_{S2}(\mathbf{a}) = (\alpha, \alpha, -a-1) \quad \text{and} \quad \mathbf{x}_{S3}(\mathbf{a}) = (-\alpha, -\alpha, -a-1) \quad \text{with} \quad \alpha = (b(r - 1))^{.5}$$

Notice that $F_{\mathbf{x}}(\mathbf{a},\mathbf{x}_{S2}(\mathbf{a}))=I+hG_{\mathbf{x}}(\mathbf{a},\mathbf{x}_{S2}(\mathbf{a}))$, where

$$G_{\mathbf{x}}(\mathbf{a},\mathbf{x}_{s2}(\mathbf{a})) = \begin{pmatrix} -a & a & 0 \\ 1 & -1 & -\alpha \\ \alpha & \alpha & -b \end{pmatrix}.$$

Setting $\det(tI - G_{\mathbf{x}}(\mathbf{a},\mathbf{x}_{s2}(\mathbf{a}))) = 0$, we obtain the third-degree equation

$$t^3 + t^2(a + b + 1) + t(ab + br) + 2ab(r - 1) = 0. \qquad (7.3.8)$$

As r approaches 1, the fixed points $\mathbf{x}_{s2}(\mathbf{a})$ and $\mathbf{x}_{s3}(\mathbf{a})$ approach $\mathbf{x}_{s1}(\mathbf{a})$ and the solutions of this cubic equation for r=1 are again $t_1=-b$, $t_2=-(a+1)$, and $t_3=0$.

We now describe what is happening to the three roots as r moves past 1. Remember that the roots are continuous functions of r and that a cubic equation has always one real root (negative in this case, since the coefficients are positive), while the other two could be real or complex conjugate. To visualize the behavior of the three roots, let us choose

$$a = 10 \quad \text{and} \quad b = \frac{8}{3},$$

and plot the function

$$P(t, r) = t^3 + \frac{41}{3}t^2 + \frac{8}{3}t(10 + r) + \frac{160}{3}(r - 1)$$

for r=1 and r=1.2. From Fig. 7.3.1 we see that as r increases past 1, the root $t_2(r)$, which is farthest from the origin, moves to the left, while $t_1(r)$ and $t_3(r)$ move towards each other.

Therefore, let us imagine the three roots as points in the complex plane, with the real part plotted along the horizontal axis and the imaginary part plotted along the vertical axis. The root $t_2(r)$ is always real and negative. The other two are real for r close to 1 and they get closer to each other as r increases.

Past the value of r for which $t_1(r)$ and $t_3(r)$ coincide, the two roots become complex conjugate with negative real part, and they start moving toward the right-hand side of the imaginary axis in the complex plane.

The fixed point $\mathbf{x}_{s2}(\mathbf{a})$ (and $\mathbf{x}_{s3}(\mathbf{a})$) becomes unstable after the two roots cross the imaginary axis since after that we obtain $t_1=c+id$ and $t_3=c-id$ with $c>0$. Hence

$$\lambda_1 = 1 + hc + ihd \quad \text{and} \quad \lambda_2 = 1 + hc - ihd,$$

whose modulus is larger than 1.

At the value of r when the two roots are on the imaginary axis, equation (7.3.8) must have two complex conjugate solutions of the form $t=\pm i\beta$. Substituting into (7.3.8), we get

$$-i\beta^3 - \beta^2(a + b + 1) + i\beta(ab + br) + 2ab(r - 1) = 0. \qquad (7.3.9)$$

Setting equal to 0 the real and imaginary part of this expression we obtain

$$\beta^2 = ab + br \qquad \text{and} \qquad \beta^2 = \frac{2ab(r-1)}{a+b+1}.$$

Hence,

$$a + r = \frac{2a(r-1)}{a+b+1}$$

which gives

$$r_2 = a\frac{a+b+3}{a-b-1}.$$

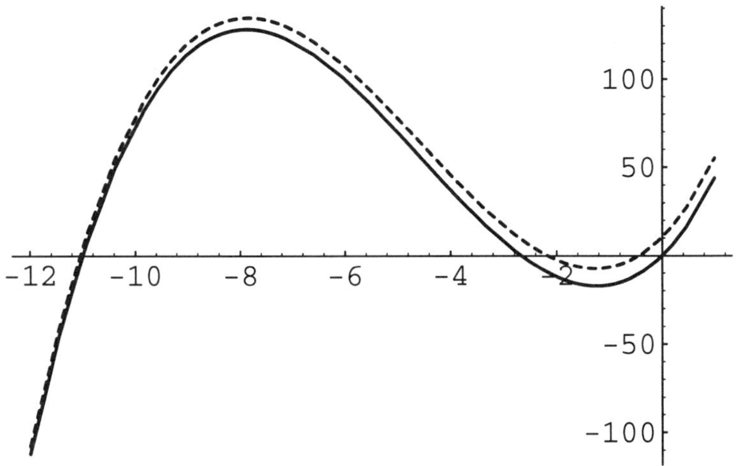

Fig. 7.3.1 Location of the roots of the cubic equation (7.3.8) for a=10, b=8/3, r=1, and r=1.2 (dashed).

Since r>1, we must have a>b+1. Thus the stationary states $\mathbf{x}_{s2}(\mathbf{a})$ and $\mathbf{x}_{s3}(\mathbf{a})$ are unstable for $r>r_2$. With the values a=10 and b=8/3 proposed by Lorenz, we obtain that the two fixed points $\mathbf{x}_{s2}(\mathbf{a})$ and $\mathbf{x}_{s3}(\mathbf{a})$ are unstable for r>24.74. Analysis of the system, after selecting the slightly supercritical value r=28, suggests that the dynamical process is chaotic. The point $\mathbf{x}_0$ is a saddle with a two-dimensional stable manifold and a one-dimensional unstable manifold, while the two fixed points $\mathbf{x}_{s2}(\mathbf{a})$ and $\mathbf{x}_{s3}(\mathbf{a})$ are saddles with one-dimensional stable and two-dimensional unstable manifolds. With a=10, b=8/3, and r=28, the ellipsoid

$$10x^2 + y^2 + \frac{8}{3}(u+19)^2 = \frac{2888}{3}$$

plays a determinant role in the dynamics of the system.

The three fixed points are sitting on the surface of the ellipsoid. Write

$$F(\mathbf{a},\mathbf{x}) = (x,y,u) + h(ay - ax, -xu - ax - y, xy - bu - ba - br) = \mathbf{x} + hG((\mathbf{a},\mathbf{x})$$

with a,b, and r given above. Then G(**x**)•**x**=0 on the ellipsoid, while it is positive inside and negative outside.

Let us follow the ideas of Theorem 5.1.2 but using ρ in place of r. The number ρ can be any value slightly larger than the norm of the point of the ellipsoid which is farthest from the origin. Thus, we can select ρ=40. We know that for every real number R>80 we can determine h so that every orbit starting at a point **x**, ||**x**||≤R, will remain inside the ball centered at the origin and with radius R. The orbit switches infinitely often between the region of $\mathbf{R}^3$ bounded by the ellipsoid and its complement.

In the original x, y, and z variables the fixed point $\mathbf{x}_{S1}(\mathbf{a})$ becomes the origin and the ellipsoid takes the form

$$10x^2 + y^2 + \frac{8}{3}(z - 19)^2 = \frac{2888}{3} \,. \tag{7.3.10}$$

Figure 7.3.2 shows the trajectory of an orbit starting from a point close to the origin, along the unstable manifold of the origin, and located on the surface of the ellipsoid. The time step is .01.

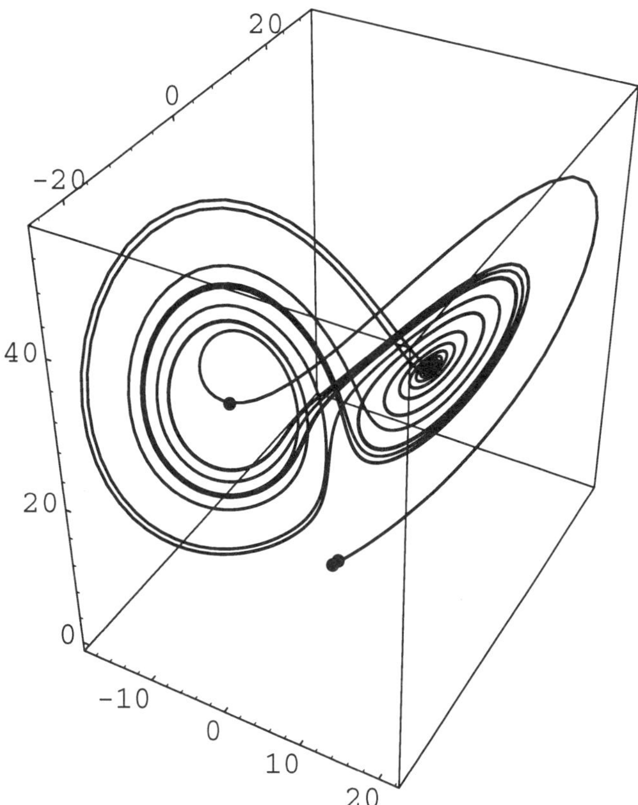

Fig. 7.3.2 Trajectory of (7.3.3) with h=.01, a=10, b=8/3, and r=28.

The chaotic behavior of the system seems to be emerging. Extensive numerical investigations have shown that there are orbits leaving the origin from points along the unstable manifold and arriving at one of the two other fixed points along their stable manifold. The presence of these heteroclinic orbits plays an important role in the chaotic behavior of the system. The conclusion suggested by this analysis is the instability and the chaotic behavior of weather patterns for certain values of the control parameters.

Problems

1. Verify that the analysis of the stability of $\mathbf{x}_{s3}(\mathbf{a})$ brings the same conclusions as the analysis of the stability of $\mathbf{x}_{s2}(\mathbf{a})$.

2. * Find the eigenvector of $F_{\mathbf{x}}(\mathbf{a},\mathbf{x}_{s1}(\mathbf{a}))$, $\mathbf{x}_{s1}(\mathbf{a})=(0,0,-a-r)$ corresponding to the eigenvalue larger than 1 using the values a=10, b=8/3, and r=28.

3. * Verify that for the same values of the parameters (see Problem 2), the surface of the ellipsoid (7.3.10) contains all three fixed points. Verify that the tangent plane to the ellipsoid at the origin is the xy plane.

4. * With a,b,r as in Problem 2, and h=0.01, verify that the eigenvector corresponding to the eigenvalue 1–bh (see (7.3.6)) is perpendicular to the tangent plane to the ellipsoid at (0,0,0), while the other two eigenvectors belong to the tangent plane. Are they pairwise orthogonal?

5. * Let a=10, b=8/3, r=28, and $\mathbf{a}$=(a,b,r), h=.01. Find the tangent plane to the ellipsoid (7.3.10) at $\mathbf{x}_{s2}(\mathbf{a})=(6(2^{.5}),6(2^{.5}),27)$. Find a vector perpendicular to this plane. Is this an eigenvector of the matrix $F_{\mathbf{x}}(\mathbf{a},\mathbf{x}_{s2}(\mathbf{a}))$?

6. * Study numerically the stability of the orbits of (7.3.3) for a=10, b=8/3, and r=28. Select the time step h=.01. Compute the Lyapunov exponent of some orbits and check their instability by plotting $(n, \|\mathbf{x}_n-\mathbf{y}_n\|)$ with $\mathbf{x}_0$ and $\mathbf{y}_0$ very close to each other.

Section 4. **NEURAL NETWORKS**

The neural network model presented in Chapter 1 is due to J.J. Hopfield [Hopfield, 1982] and in vector notation takes the form

$$\mathbf{x}_{n+1} = (I - C)\mathbf{x}_n + T\Phi(\mathbf{x}_n) + \mathbf{x}_I . \tag{7.4.1}$$

The entry x_i of $\mathbf{x}=(x_1,...,x_N)$ represents the potential of neuron i, i=1,2,...,N. The matrix C is diagonal:

$$C = \begin{pmatrix} c_1 & 0 & 0 & \dots & 0 & 0 \\ 0 & c_2 & 0 & \dots & 0 & 0 \\ \dots & \dots & \dots & \dots & \dots & \dots \\ 0 & 0 & 0 & \dots & 0 & c_N \end{pmatrix}$$

and $c_i \in [0,1]$ accounts for the leakage current of neuron i, i=1,2,...,N. The matrix $T=\{t_{ij}, i,j=1,...,N\}$ is the connectivity matrix, with t_{ij} denoting the action of neuron j on neuron i. This action can be excitatory ($t_{ij} > 0$) or inhibitory ($t_{ij}<0$). The vector function Φ is the neuron response function $\Phi(\mathbf{x})=(\phi_1(x_1),\phi_2(x_2),...,\phi_N(x_N))$. We can assume that all ϕ_i's are the same and have the form

$$\phi(x) = \begin{cases} 0 & x \le 0 \\ 1 & x \ge 1 \\ x & 0<x<1. \end{cases}$$

The vector $\mathbf{x}_I$ represents the signal received by the net (input).

Typically, but not always, the neurons of the net are divided into three groups: the **input**, **hidden**, and **output layers**. The neurons of the input layer receive the signal, and the neurons of the output layer provide the answer to the problem posed by the signal. This can be a recognition problem, such as identifying among a list of voices which one is represented by the signal just received. The incoming signal does not affect the hidden and output layers directly. Therefore, if we divide the components of the vector $\mathbf{x}$ into three groups, corresponding to the three different layers, we obtain that the components of $\mathbf{x}_I$ corresponding to the last two groups are zero.

Once the signal is received, the dynamical process starts from the state $\mathbf{x}_0$ in which the net is at that instant. We imagine that the signal remains active during the entire process. The initial state $\mathbf{x}_0$ of the net is usually unknown. From this state the net produces

$$\mathbf{x}_1 = (I - C)\mathbf{x}_0 + T\Phi(\mathbf{x}_0) + \mathbf{x}_I\,, \quad \mathbf{x}_2 = (I - C)\mathbf{x}_1 + T\Phi(\mathbf{x}_1) + \mathbf{x}_I \; \dots \,.$$

The process is considered completed when the net reaches a fixed point, and consequently, the potential of each neuron is fixed. This fixed point represents the answer of the net to the problem posed by the incoming signal.

In a typical situation the net is designed to deal with finitely many inputs $\mathbf{x}_I$, I=1,2, ...,m. Unfortunately, they may be received imperfectly for several reasons such as noise, distortion, or partial occlusion. We would like each perfectly received signal to be identified by the same fixed point regardless of the initial state $\mathbf{x}_0$ of the net when the signal is received. We would also like minor distortions of the perfect signal to generate fixed points close enough to the one produced by the perfect signal that the net will give the same answer. Finally, we need fixed points corresponding to different perfect signals to be different and as far away from each other as possible.

Notice that we don't need the components of the input and hidden layers to reach a stationary state as long as those of the output layer have done so, and their state provides the desired answer. This fact, however, is a mixed blessing since it makes it easier for the output layer to be the same even though the input signals are coming from different objects.

We now analyze some fundamental properties of the system defined by (7.4.1) on the basis of the theory developed in previous chapters. We first show that when $c_i>0$ for all i=1,2,...,N, every disk large enough is invariant under the dynamical system defined by (7.4.1). In fact, each component of $\Phi(\mathbf{x})$ is at most 1.

Hence, we obtain $\|\Phi(\mathbf{x})\|\leq N^{.5}$ and $\|T\Phi(\mathbf{x})\|\leq\|T\|N^{.5}$ for every $\mathbf{x}\in\mathbf{R}^N$. Moreover, with $\|\mathbf{x}\|=1$ and $c_0=\min\{c_i;\ i=1,2,\ldots,N\}$, we have

$$\|(I-C)\mathbf{x}\| \leq \max\{(1-c_i)\ ;\ i=1,2,\ldots,N\} = (1-c_0) < 1. \tag{7.4.2}$$

Thus,

$$\frac{\|(I-C)\mathbf{x} + T\Phi(\mathbf{x}) + \mathbf{x}_I\|}{\|\mathbf{x}\|} \leq 1 - c_0 + \frac{\|T\|\sqrt{N}}{\|\mathbf{x}\|} + \frac{\|\mathbf{x}_I\|}{\|\mathbf{x}\|}. \tag{7.4.3}$$

Choose r_0 so large that

$$\frac{\|T\|\sqrt{N}}{\|\mathbf{x}\|} + \frac{\|\mathbf{x}_I\|}{\|\mathbf{x}\|} \leq \frac{c_0}{2}$$

for every $\mathbf{x}$, $\|\mathbf{x}\|\geq r_0$. Under these circumstances, we have

$$\frac{\|(I-C)\mathbf{x} + T\Phi(\mathbf{x}) + \mathbf{x}_I\|}{\|\mathbf{x}\|} \leq 1 - \frac{c_0}{2} < 1 \tag{7.4.4}$$

for all $\mathbf{x}$, $\|\mathbf{x}\|\geq r_0$. Hence (see Theorem 5.1.3), $N(r_0)<1$, and the map that defines our dynamical system leaves invariant every sufficiently large disk.

We now show that different signals cannot produce the same fixed point, at least in the case when we require a stationary state in all three layers. Suppose, in fact, that two signals $\mathbf{x}_I$ and $\mathbf{y}_I$ have produced two stationary states, $\mathbf{x}_s$ and $\mathbf{y}_s$. From

$$\mathbf{x}_s = (I-C)\mathbf{x}_s + T\Phi(\mathbf{x}_s) + \mathbf{x}_I, \quad \text{and} \quad \mathbf{y}_s = (I-C)\mathbf{y}_s + T\Phi(\mathbf{y}_s) + \mathbf{y}_I,$$

we derive

$$C\mathbf{x}_s = T\Phi(\mathbf{x}_s) + \mathbf{x}_I\,, \qquad C\mathbf{y}_s = T\Phi(\mathbf{y}_s) + \mathbf{y}_I.$$

Therefore,

$$\|C\|\ \|\mathbf{x}_s - \mathbf{y}_s\| + \|T\|\ \|\Phi(\mathbf{x}_s) - \Phi(\mathbf{y}_s)\| \geq \|\mathbf{x}_I - \mathbf{y}_I\|.$$

Since $\|\Phi(\mathbf{x})-\Phi(\mathbf{y})\|\leq\|\mathbf{x}-\mathbf{y}\|$, we obtain

$$\|\mathbf{x}_s - \mathbf{y}_s\| \geq \frac{\|\mathbf{x}_I - \mathbf{y}_I\|}{\|C\|+\|T\|}. \tag{7.4.5}$$

This means that different inputs cannot produce the same output if the output is a fixed point **involving all layers**.

The existence of a unique fixed point $\mathbf{x}_s$ corresponding to an input $\mathbf{x}_I$ regardless of the initial state $\mathbf{x}_0$ of the net can be guaranteed when the map that defines the dynamical system is a contraction. In fact, according to Theorem 5.2.1, they have a unique fixed point and every sequence of iterates converges to it. Therefore, it is important to give a sufficient condition for the map defining our dynamical system to be a contraction. This condition is $c_0>\|T\|$. In fact, let

$$K(\mathbf{x}) = (I - C)\mathbf{x} + T\Phi(\mathbf{x}) + \mathbf{x}_I.$$

Then

$$K(\mathbf{x}) - K(\mathbf{y}) = (I - C)(\mathbf{x} - \mathbf{y}) + T(\Phi(\mathbf{x}) - \Phi(\mathbf{y}))$$

and (recall that $\|\Phi(\mathbf{x})-\Phi(\mathbf{y})\|\leq\|\mathbf{x}-\mathbf{y}\|$)

$$\|K(\mathbf{x}) - K(\mathbf{y})\| \leq \|I - C\| \, \|\mathbf{x} - \mathbf{y}\| + \|T\| \, \|\mathbf{x} - \mathbf{y}\|. \tag{7.4.6}$$

Since $\|I-C\|\leq 1-c_0$, we see that

$$\|K(\mathbf{x}) - K(\mathbf{y})\| \leq (1 - c_0 + \|T\|)\|\mathbf{x} - \mathbf{y}\|,$$

which shows that K is a contraction whenever $c_0>\|T\|$. Thus, in the presence of "large leakage and small interactions," the net, after receiving the input $\mathbf{x}_I$, will always reach the same fixed point regardless of its initial state $\mathbf{x}_0$.

The following example illustrates some of the properties and problems we have mentioned above.

Example 7.4.1 We are given a neural net with two neurons in the input, one in the hidden layer and two in the output layer. Then $\mathbf{x}=(x_1,x_2,x_3,x_4,x_5)$. Assume that the matrix C and the connectivity matrix T are

$$C = \begin{pmatrix} .1 & 0 & 0 & 0 & 0 \\ 0 & .2 & 0 & 0 & 0 \\ 0 & 0 & .3 & 0 & 0 \\ 0 & 0 & 0 & .2 & 0 \\ 0 & 0 & 0 & 0 & .4 \end{pmatrix},$$

$$T = \begin{pmatrix} 0 & .01 & 0 & 0 & 0 \\ -.01 & 0 & 0 & 0 & 0 \\ .05 & .3 & 0 & -.1 & .2 \\ -.1 & .2 & -.02 & 0 & .2 \\ 0 & 0 & -.1 & .5 & 0 \end{pmatrix},$$

The neuron response function is $\Phi(\mathbf{x})=(\phi(x_1),\phi(x_2),\ldots,\phi(x_5))$, with $\phi(x)$ as given at the beginning of the section. Let $\mathbf{x}_{I1}=(1,0,0,0,0)$ be the input and $\mathbf{x}_0=(1,0,.5,.4,0)$ be the initial condition when the input is received.

From $\mathbf{x}_{n+1}=(I-C)\mathbf{x}_n+T\Phi(\mathbf{x}_n)+(1,0,0,0,0)$ and after about 140 iterations, the system reaches the stationary state $\mathbf{x}_{s1}=(10,-.05,1/6,-31/60,-1/24)$. Changing the initial state $\mathbf{x}_0$ in many different ways does not affect the outcome. In every case the stationary state reached by the net is $\mathbf{x}_{s1}$. From this point of view the net operates as required, since the output does not depend on the initial state of the net, but only on the input vector.

Let us change the input vector to $\mathbf{x}_{I2}=(0,1,0,0,0)$. Starting again with $\mathbf{x}_0=(1,0,.5,.4,0)$ and after about 140 iterations we reach the stationary state $\mathbf{x}_{s2}=$

(.1,5,1.35,1.85,1), whose forth and fifth entries are quite different from those obtained with $\mathbf{x}_{I1}$. We pay attention only to these two entries since they are the only ones an observer can see. In fact, the state of any neuron of the input and hidden layers is usually inaccessible. With the input $\mathbf{x}_{I3}$=(1,1,0,0,0) and with the initial condition $\mathbf{x}_0$=(1,0,.5,.4,0) we arrive at the stationary state $\mathbf{x}_{s3}$=(10.1,4.95,1.5,1.4, 1). This is certainly different from $\mathbf{x}_{s1}$ and $\mathbf{x}_{s2}$. However, the output entries are too close to the corresponding entries of $\mathbf{x}_{s2}$. Therefore, it appears that the net is not able to provide a solid distinction between $\mathbf{x}_{I2}$ and $\mathbf{x}_{I3}$. This, in fact, may be too much to ask from such a small network.

Whenever possible a net is provided with orthogonal inputs, since this property may make their recognition easier. However, the net of Example 7.4.1 has a problem with the two orthogonal inputs $\mathbf{x}_{I4}$=(1,1,0,0,0) and $\mathbf{x}_{I5}$=(−1,1,0,0,0). The stationary states are $\mathbf{x}_{s4}$=(10.1,4.95,1.5,1.4,1) and $\mathbf{x}_{s5}$=(−9.9,5,4/3,1.9,1). The output entries are too close to each other. A different connectivity matrix is necessary to obtain better resolution between the two outputs.

Problems

1. Let $\phi(x)=\alpha x+\beta$ be the neuron response function. Show that the dynamical system (7.4.1) takes the form $\mathbf{x}_{n+1}=B\mathbf{x}_n+\mathbf{x}_J$, where B is the matrix $B=I-C+\alpha T$, and $\mathbf{x}_J$ is the vector $\mathbf{x}_J=\mathbf{x}_I+T\mathbf{b}$ with $\mathbf{b}=(\beta,...,\beta)$.

2. Assume that the neuron response function is as in Problem 1, but the parameters α and β are (possibly) different for each neuron, namely $\phi_i(x)=\alpha_i x+\beta_i$. Write the dynamical system (7.4.1) in this case.

3. When the neurons of the net are divided into the three groups: input layer (I = 1), hidden layer (H = 2), and output layer (O = 3), the connectivity matrix T can be considered as composed of blocks in the following way:

$$T=\begin{pmatrix} T_{11} & T_{12} & T_{13} \\ T_{21} & T_{22} & T_{23} \\ T_{31} & T_{32} & T_{33} \end{pmatrix}.$$

The T_{ij} block represents the action of the neurons of group j on the neurons of group i. The two blocks T_{13} and T_{31} are usually assumed to be zero. Can you justify this assumption? Also, what property of the connectivity matrix characterizes the assumption that no neuron has any influence on itself?

4. Assume that a neural net has three neurons in the output layer, four in the hidden layer, and two in the input layer. Find the size of each block of T.

5. * Consider a neural net with one neuron in the output, two in the hidden, and one in the input layer. Assume that $c_i=1$ for i=1,2,3,4. Also assume that the entries of the connectivity matrix T are as follows: $t_{13}=t_{23}=t_{31}$

$=t_{42}=t_{43}=1$, and all other entries are 0. Finally, let the neuron response function be the same for all neurons and be defined by

$$\phi(x) = \begin{cases} 0 & x \leq 3/4 \\ 4x-3 & 3/4 < x < 1 \\ 1 & x \geq 1. \end{cases}$$

Let

$$\mathbf{x}_I = (1/2,0,0,0) \quad \text{and} \quad \mathbf{x}_0 = (1/2,0,1,1).$$

Verify that the state of the output neuron is constant while the states of the three remaining neurons are periodic of period 2.

APPENDIX 1

MATHEMATICA PROGRAMS

SUMMARY

In this appendix we present some Mathematica programs aimed at the numerical and graphical investigation of discrete dynamical systems. One-dimensional systems are emphasized, but many techniques can be extended to higher dimension. Examples of such extensions are provided. The appendix is divided in several sections. It starts with graphing techniques, including the cobweb method. Study of the stability of orbits, density plots, and bifurcation diagrams follows. Lyapunov exponents, correlation dimension and odds and ends are presented in the last two sections.

Section 1. GRAPHING

1. Graphing Functions

1. *Graph of a function*

To graph $f(x) = x^2$ in the interval [0,2], we proceed as follows. We first define the function, and then we plot it.

f[x_]:=x^2 (hit the **enter** key)
Plot[f[x], {x,0,2}] (hit the **enter** key)

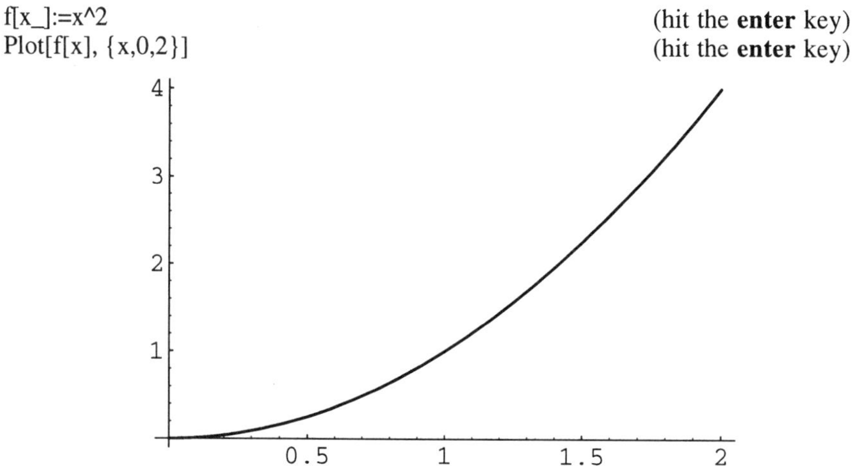

Always hit the **enter** not the **return key**, unless instructed otherwise.

2. *Different ticks*

Suppose that we want the tick marks in a position different from the one selected automatically by the program.

Plot[f[x],{x,0,2},Ticks->{Range[0,2,1],Automatic}] (hit the **enter** key)

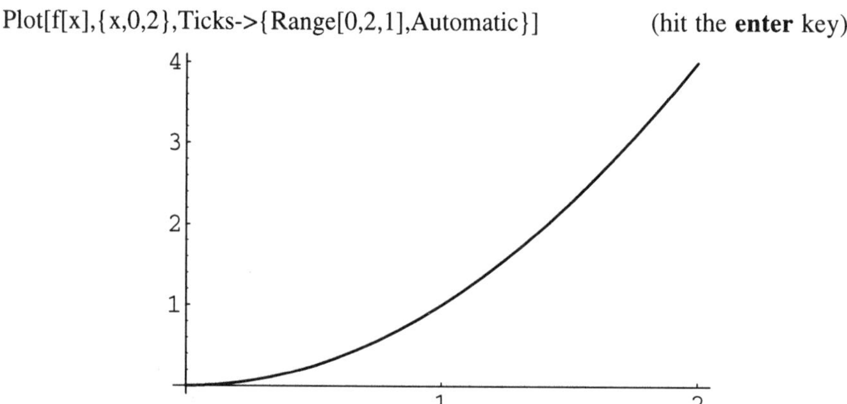

The string Range[0,2,1] starts the tick marks at 0 and ends them at 2, with increments of 1.

3. ***Different axes***

Suppose that we want the origin at the point (-1,0). This feature can be used to have a better view of the part of the graph we are interested in. The following string does the job. Since Mathematica does not recognize "option -" as a minus sign, we will never use this feature in Appendix 1.

Plot[f[x],{x,-1,2},AxesOrigin->{-1,0}] (hit the **enter** key)

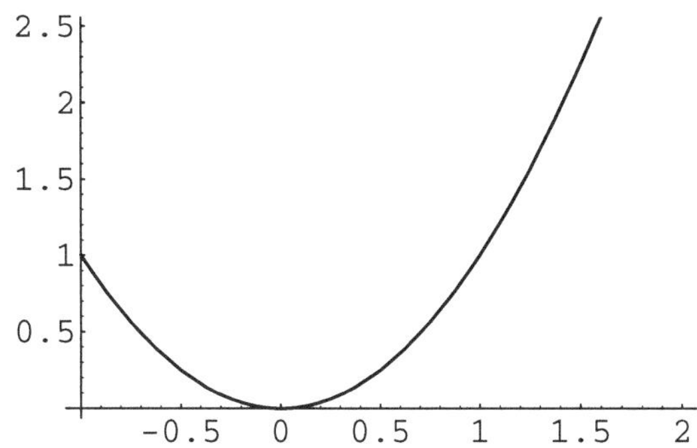

We combine the new origin with the different tick marks with both commands.

Plot[f[x],{x,-1,2}, AxesOrigin->{-1,0}, Ticks->{Range[0,2,1],Automatic}]

We do not show the graph.

4. ***Different texture: GrayLevel***

One can change the texture of a graph by using the GrayLevel.

Plot[f[x],{x,-1,2},AxesOrigin->{-1,0},PlotStyle->{GrayLevel[0.7]}] (**enter**)

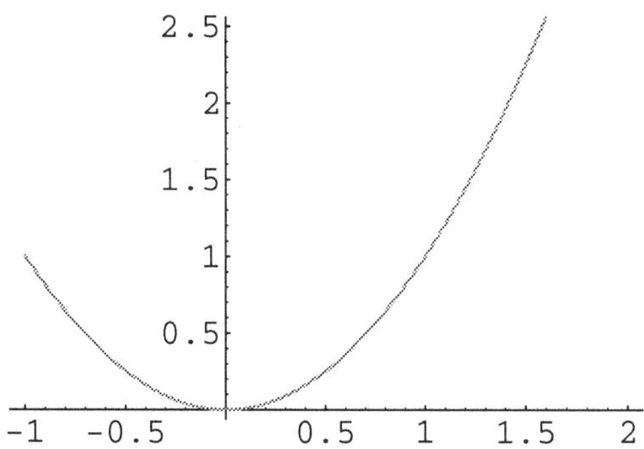

The GrayLevel must be between 0 and 1. The number 0 represents black, and it is the default used by the program.

5. ***Different texture: Thickness***

The Thickness command provides another way to change a graph.

Plot[f[x], {x,-1,2},AxesOrigin->{-1,0}, PlotStyle->Thickness[0.01]]

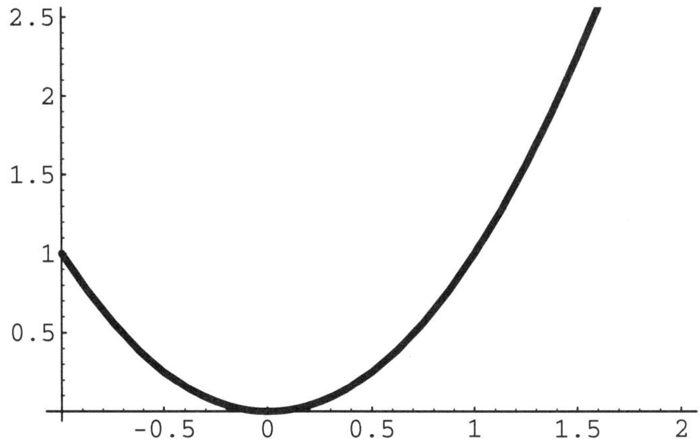

6. *Combining graphs*

To combine the graphs of two functions, say $f(x)=x^2$ and $g(x)=x^3-x$, we first define the two functions. The symbol ";" should separate the two definitions.

```
f[x_]:= x^2; g[x_]:=x^3 - x                                    (enter)
Plot[{f[x],g[x]},{x,-1,2},PlotStyle->{GrayLevel[0.4];          (return)
Dashing[{0.02,0.02}]}, Ticks->{Automatic, Range[-.5,1.5,.5]}]  (enter)
```

In the lines above we have summarized several features. We have moved the origin, used two different patterns for the two functions, and used different tick marks. The Dashing command can be changed to longer segments by increasing the number, or to shorter ones by decreasing it.

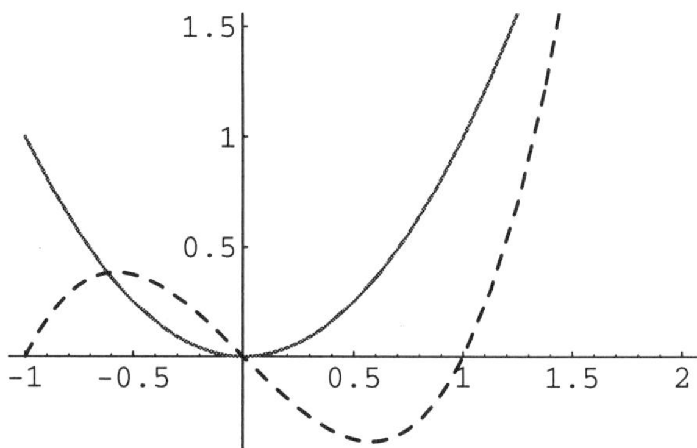

7. *Symbols for most common functions*

The most frequent functions you may want to graph are:

1. x^2 = x^2, $\sqrt{x}$ = Sqrt[x], $\dfrac{x}{x^2 + 1}$ = x/(x^2 + 1).
2. Sin[x], Cos[x], Tan[x].
3. Log[x] (natural logarithm), Log[2,x] (logarithm in base 2), Exp[x].
4. ArcSin[x], ArcCos[x], ArcTan[x].

5. Floor[x] (greatest integer contained in x), Ceiling[x] (greatest integer contained in x+1), Abs[x] (absolute value of x), Max[f[x],g[x]].

8. *Using the Max function*

Let us see how the Max function works. We plot the Max between two functions, namely $2x$ and x^2.

```
Plot[Max[2x,x^2],{x,-1,3}]                                  (enter)
```

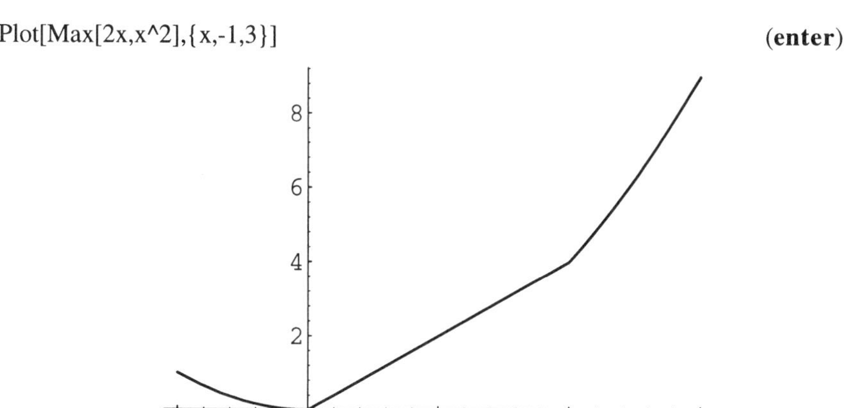

9. *Graphing functions not defined by a single formula*

Sometimes we have to plot functions defined differently in disjoint intervals. For example,

$$f(x) = \begin{cases} 2x & 0 \le x < .5 \\ 2 - 2x & .5 \le x \le 1. \end{cases}$$

We define the function f as follows, taking care to hit the **return** key between the two lines. After the second line we hit the **enter** key.

```
f[x_]:=2x/;0<=x<.5                                          (return)
f[x_]:=2-2x/;.5<=x<=1                                       (enter)
Plot[f[x],{x,0,1}]                                          (enter)
```

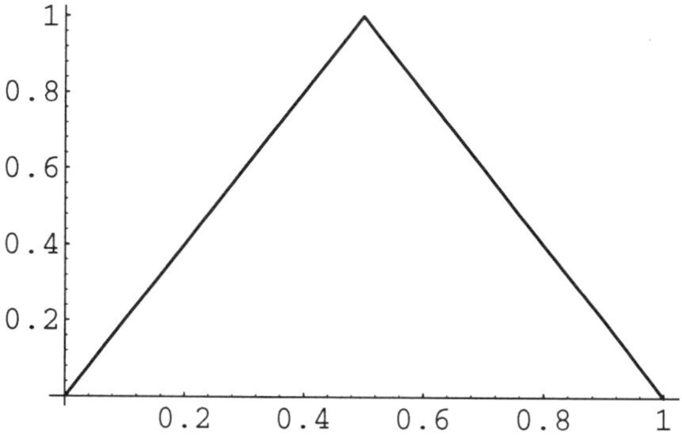

10. ***Graphing two or more functions defined in different intervals***

Suppose we want to plot two functions that are not defined in the same interval. For example:

$$f(x) = \sqrt{x - 1} \text{ and } g(x) = \ln(x),$$

in the interval [.5,2]. Notice that f(x) is defined for $x \geq 1$, while lnx is defined for $x > 0$. We first plot one of the functions, say f.

f[x_]:=Sqrt[x-1] **(enter)**
q1=Plot[f[x],{x,1,2}] **(enter)**

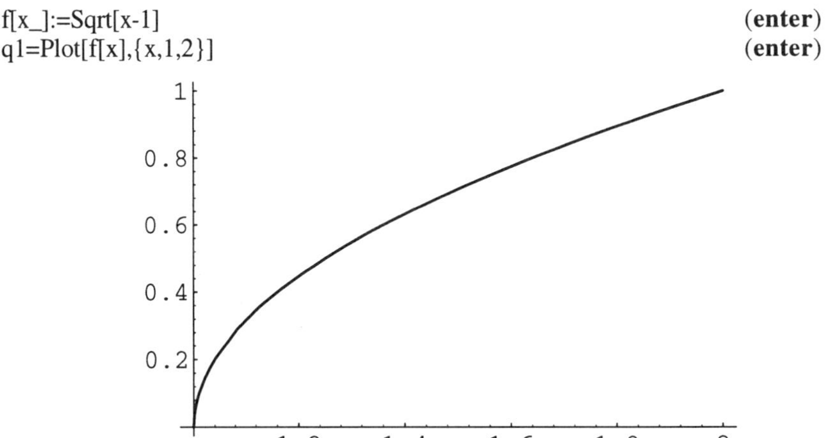

Then we plot the second function:

g[x_]:=Log[x] **(enter)**
q2=Plot[g[x],{x,.5,2}] **(enter)**

We put the two graphs together with the Show command.

Show[q1,q2, AxesOrigin->{0.5,0}, Ticks->{Range[.5,2,.5], Range[-.5,1,.5]}]
(enter)

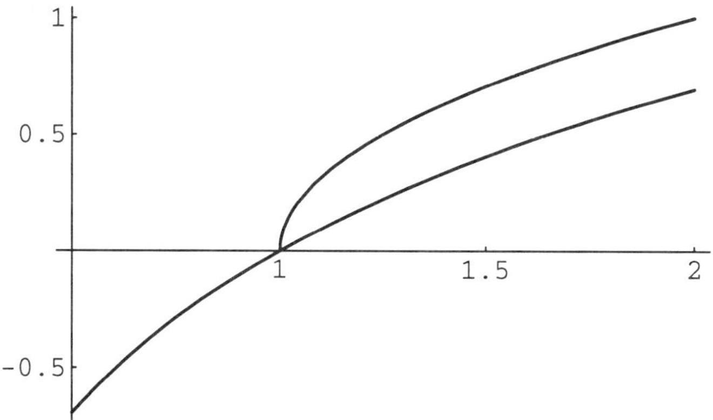

We can use a different technique that may sometimes be more useful.

Plot[f[x],{x,1,2}] **(enter)**
Plot[g[x],{x,.5,2}] **(enter)**

The two graphs are not shown, but will appear in your screen. Then write:

Show[%,%%] **(enter)**

This command will combine the two graphs.

2. **Finding Fixed Points and Periodic Points (Graphically)**

1. ***Finding fixed points***

The fixed points of a function f are the solution of the equation f(x)=x. Thus, they are graphically represented by the intersections of y=x with the graph of f. For example, f(x)=cosx has a fixed point in [0,1]. Recall that by listing both functions we obtain the two graphs in the same plot.

Plot[{Cos[x],x},{x,0,1}] **(enter)**

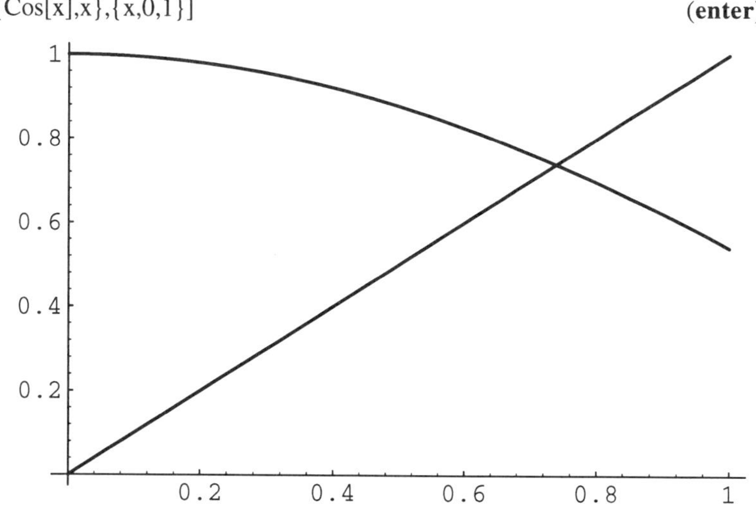

2. ***Finding periodic points***

The periodic orbits of period 2 of f are given by those intersections of the graph of f(f(x)) with the line y=x, which are not on the graph of f(x). Plotting all three functions: f(x), f(f(x)), and x, helps in solving this problem. Suppose that f(x) = 3.2 x(1-x). The second iterate of f is denoted with f[f[x]].

```
f[x_]:=3.2x(1-x)                                  (enter)
Plot[{f[x], f[f[x]],x},{x,0,1}]                   (enter)
```

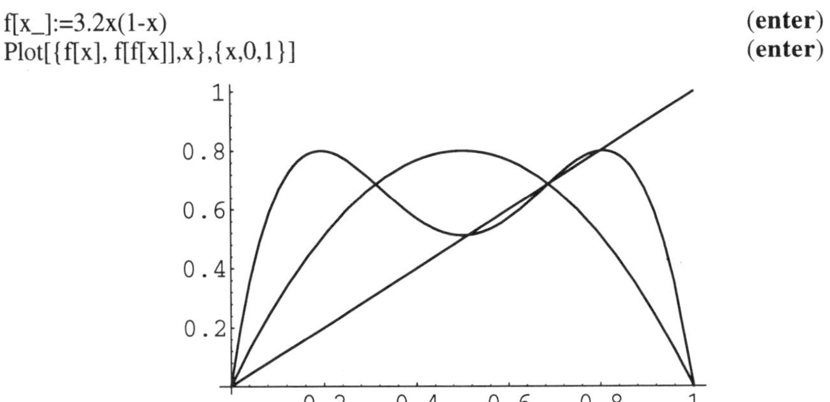

We see that f has two fixed points, which are also fixed for its second iterate. The two points where only the graph of the second iterate crosses the line y=x are periodic points of period 2.

3. ***Different textures to recognize different graphs***

When several graphs are combined we may want to distinguish them with different textures. Here is an example.

```
Plot[{f[x], f[f[x]],x},{x,0,1}, PlotStyle->{{GrayLevel[0]},        (return)
{GrayLevel[0.5]}, {Dashing[{0.01,0.01}]}}]                         (enter)
```

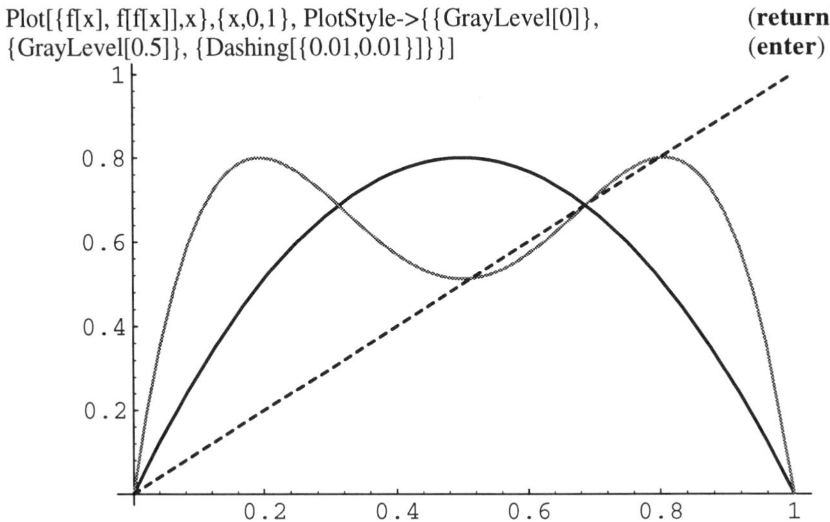

3. **The Cobweb Method**

The method connects the successive points of an orbit on the graph of the function using as bridges the corresponding points on the line y=x. Use the function f(x)=3.83x(1-x), which has a stable periodic orbit of period 3.

f[x_]:=3.83x(1-x) **(enter)**

Plot f in the interval of interest, in this case is [0,1], together with the line y=x.

q1=Plot[{x,x=f[x]},{x,0,1}] **(enter)**

When we hit **enter** we will obtain the graph of f and the line y=x (not shown). We select an initial condition.

x=.3 **(enter)**

We construct how many lines we want in our cobweb diagram. Suppose we want 30.

q2=Graphics[{Line[Flatten[Table[{{x,x},{x,x=f[x]}},{30}],1]]}] **(enter)**

The command Flatten[Table[{{x,x},{x,x=f[x]}},{30}],1] has the purpose of representing the table as a sequence of 30 pairs (x,y). The command Line[.] joins these pairs. The answer of Mathematica to the line above will be simply

Graphics

Now we put everything together.

Show[q1,q2] **(enter)**

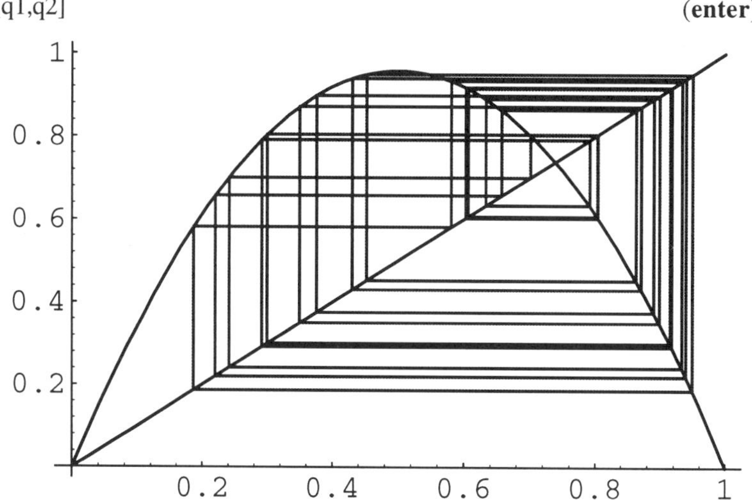

Our next step is to eliminate the **transient** of the orbit. In fact, it appears that the orbit is chaotic, but this impression is incorrect. Put the cursor at the end of the string q2 and press **enter** once. Repeat the procedure. In this way we neglect the first 60 (30 for each time we press **enter**) states of the orbit and we look at the states from 61 to 90. Then,

Show[q1,q2] **(enter)**

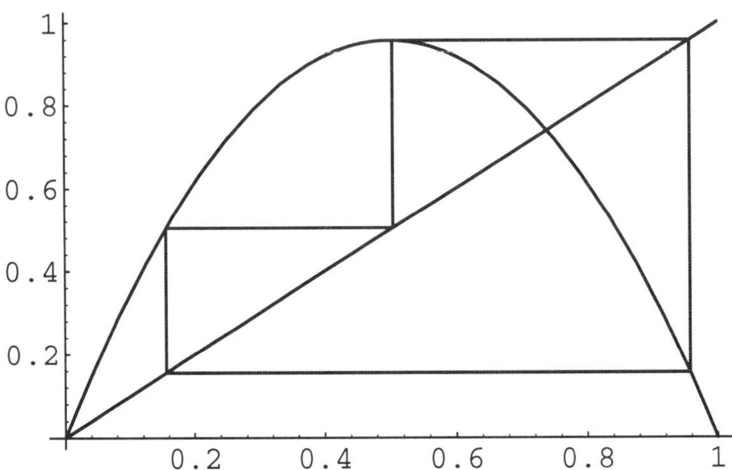

The orbit is asymptotically periodic of period 3. Let us look at an aperiodic orbit. Change the function to g(x)= 4x(1-x) in the same interval [0,1].

```
g[x_]:=4x(1-x)                                                (enter)
p1=Plot[{x,g[x]},{x,0,1}]                                     (enter)
x=.3                                                          (enter)
p2=Graphics[{Line[Flatten[Table[{{x,x},{x,x=g[x]}},{100}],1]]}]  (enter)
Graphics
Show[p1,p2]                                                   (enter)
```

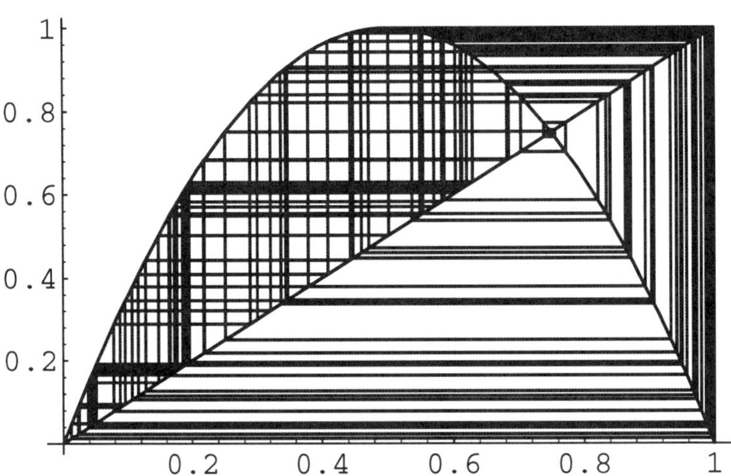

To make sure that the picture is not misleading we repeat the process we did before. Put the cursor at the end of the string p2 and press the **enter** key two or three times. The qualitative features of the picture we obtain do not change. The orbit is aperiodic!

We may want to include the points of the orbit on the graph.

```
x=.3                                                          (enter)
```

```
q3=Graphics[{RGBColor[1,0,0],PointSize[0.02],Table[Point[{x,x=f[x]}],{30}]}]
```
(enter)

Put everything together (use the graphs q1 and q2 produced before and not shown). In your screen the dots will be in red.

```
Show[q1,q2,q3]
```
(enter)

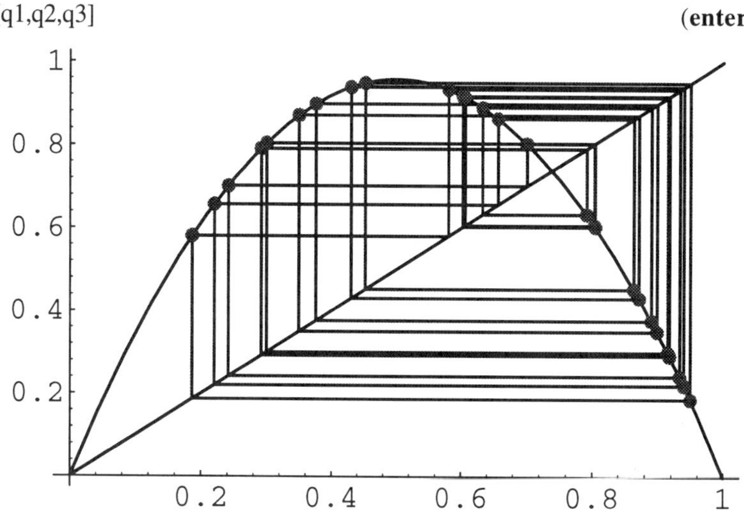

Section 2. **ITERATES AND ORBITS**

1. **Iterates**

1. ***Iterating a function of a single variable***

We want to iterate f(x)=2x(1-x).

```
f[x_]:=2x (1-x); g[x]=f[f[x]]
```
(enter)

The above command iterates the function twice. We get 4(1-x)x(1-2(1-x)x). We can simplify the result by writing

```
Simplify[g[x]]
```
(enter)

The result is $4x\ (1\text{-}3x+4x^2\ \text{-}2x^3)$.

We can factor g(x) with the command

```
Factor[g[x]]
```
(enter)

The result is $4(1\text{-}x)x(1\text{-}2x+2x^2)$.

There is another way to do it, using the command Nest. For example, if we want to iterate f three times, we write

```
g[x]=Nest[f,x,3]
```
(enter)

We get 8(1-x)x(1-2(1-x)x)(1-4(1-x)x(1-2(1-x)x))

Simplify[g[x]] **(enter)**

We obtain 8(1-x)x(1-2x+2x^2)(1-4x+12x^2-16x^3+8x^4)

2. ***Iterating a function depending on a parameter***

Assume that f depends on an independent variable x and on a parameter a.

f[a_,x_]:=a x(1-x) **(enter)**

Notice that the parameter a and the variable x must be separated by a space, otherwise, the two symbols are intended as a new variable ax. Iterate f twice, considering x as the variable, namely replacing x with f(a,x).

g[a,x]=f[a,f[a,x]] **(enter)**

We obtain a^2(1-x)x(1-a(1-x)x)

We can use the Nest command.

Nest[Function[x,f[a,x]],x,2] **(enter)**

We obtain a^2(1-x)x(1-a(1-x)x).

The Nest command must be used with the Function command since f depends on two variables (x and a). Notice the difference with iterating f with respect to a, namely substituting a with ax(1-x). Here is the answer.

Nest[Function[a,f[a,x]],a,2] **(enter)**

We obtain a(1-x)^{2}x^2
We can also say

Nest[Function[x,a x(1-x)],x,3] **(enter)**

We obtain a^3(1-x) x (1-a(1-x)x)(1-a^2(1-x)x(1-a(1-x)x))

3. ***Computing a sequence of iterates***

We want to compute a sequence of iterates (an orbit) starting from a point x. This is really a partial orbit. We choose an integer n and to compute the first n states of the orbit of a given dynamical system starting with x. We first define the function governing the system.

f[x_]:= 3.8 x(1-x) **(enter)**

Then we select an initial condition.

x=.3 **(enter)**

The first 100 states of the orbit are generated below using the command Table. For successive use we name the Table q1. Putting ; after the string instructs the computer not to show the iterates.

q1= Table[x = f[x], {100}]; **(enter)**

We have stored the first 100 states of the orbit starting at .3 for the dynamical system governed by the function f(x)=3.8x(1-x). Suppose that we want to see the iterates from 80 to 90. This can be accomplished using the Range command. The reader may not get the same numbers we include here. Since the system is very sensitive to small changes, the machine used and the accuracy required may alter the result significantly.

q1[[Range[80,90]]] **(enter)**

We obtain {0.596233, 0.914809, 0.296148, 0.792088, 0.6125801, 0.898005, 0.348048, 0.889861, 0.372431, 0.888159, 0.377463, 0.892942, 0.363268}

Suppose that we want to see the last six iterates in a column form. The MatrixForm command will accomplish this task.

MatrixForm[q1[[Range[95,100]]]] **(enter)**
0.790226
0.629922
0.885857
0.384235
0.899074
0.344811

Doing everything as above does not tell us the iterate's number explicitly. We can introduce this important element as follows. We initialize n and x.

x = .3; n = 0 **(enter)**

Then we compute a different table, which includes the iteration number.

q2=Table[{n=n+1,x=f[x]},{100}]; **(enter)**

Suppose that we want to see the iterates from 90 to 100 with their iteration number.

MatrixForm[q2[[Range[90,100],Range[1,2]]]] **(enter)**

95	0.790226
96	0.629922
97	0.885857
98	0.384235
99	0.899074
100	0.344811

The table shows the iterates from 95 to 100 and their iteration number.

2. **Orbits**

1. *Plotting one orbit versus the iteration number*

We want to plot the sequence (n,x_n) to visualize the state of the orbit as a function of n. We compress n dividing it by 10. Hence 10 states of the orbit are condensed on every interval of length 1. We choose f, an initial condition x, and a starting value for n.

```
f[x_]:=3.8x(1-x)                                   (enter)
x=.3; n = 1                                        (enter)
q3=Table[{n/10, x=f[x],n=n+1}, {50}];              (enter)
```

We have a table of 50 triples: $(n/10, x_n, n+1)$. We want to plot the pairs $(n/10, x_n)$. We select these entries using the Range command.

```
p=q3[[Range[1,50],Range[1,2]]];                    (enter)
```

All 50 triples are used (Range[1,50]), but only the first two entries of each triple are selected (Range[1,2]). The table is like a matrix. We select all rows, but only the first and second column. We now plot the list using the ListPlot command.

```
ListPlot[p, PlotStyle->PointSize[0.02], PlotRange->{{0,5},{0,1}}]     (enter)
```

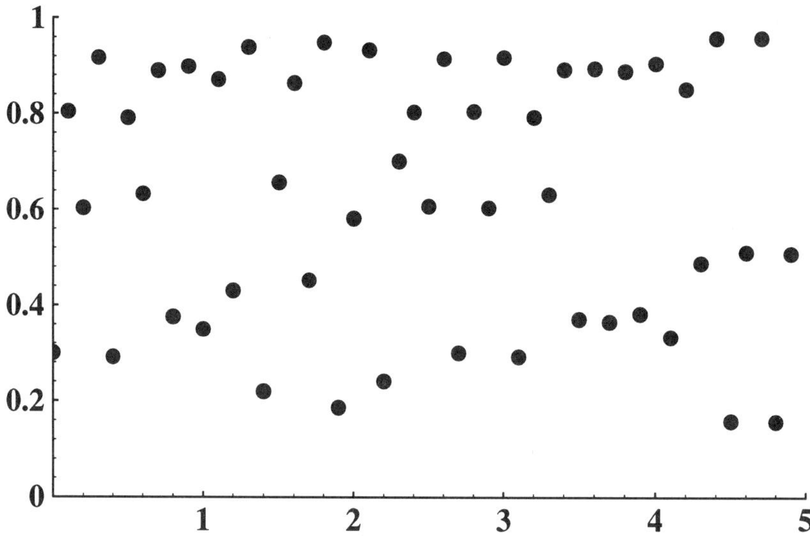

PointSize[0.02] shows the iterates more clearly. PlotRange positions the graph. The part {0,5} is the range of the horizontal coordinate, and the part {0,1} is the range of the vertical coordinate. We could have used n instead of n/10 in q3. In this case the horizontal axis would have had labels from 0 to 50, rather than from 0 to 5. Let us do another example.

```
f[x_]:=3.2 x (1-x); x=.3; n=0                      (enter)
q4=Table[{n/10,x, x=f[x],n=n+1}, {50}];            (enter)
p=q4[[Range[1,50],Range[1,2]]];                    (enter)
q5=ListPlot[p, PlotStyle->PointSize[0.02], PlotRange->{{0,5},{0,1}}] (enter)
```

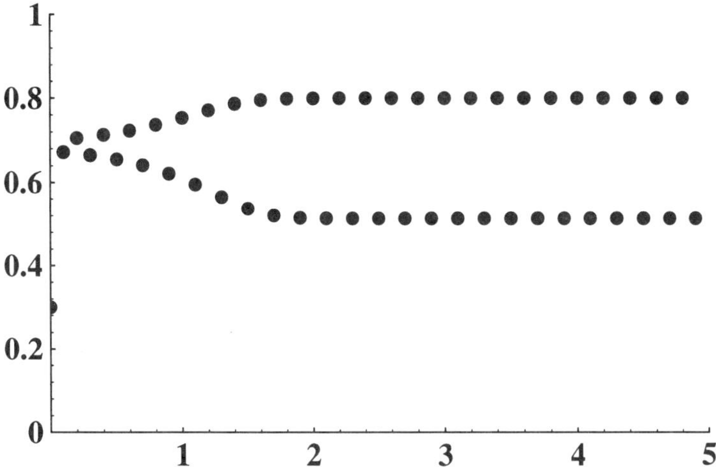

After a transient of about 15 iterations the orbit becomes "periodic" of period 2. The graphic power of these pictures is self-explanatory. We can add an extra feature to it.

q6=ListPlot[p,PlotJoined->True] (**enter**)

The string joins the points of the plot with line segments. We do not show here this intermediate plot . We combine the two plots using the Show command.

Show[q5,q6]

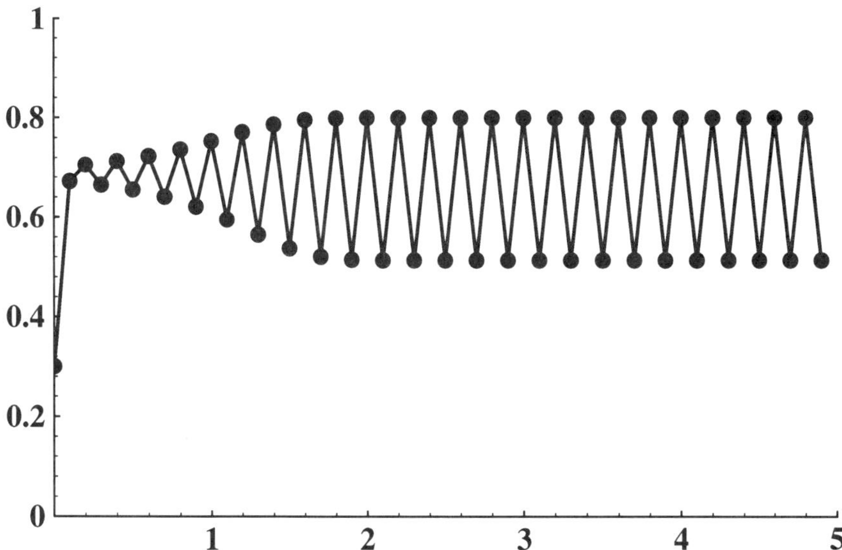

2. ***Plotting one orbit versus another***

The purpose of this activity is to "test" if two orbits tend to travel "together" (or stay at the same distance) or tend to separate, come back, and separate again in a random-like fashion. What is under scrutiny here is the **stability** of the system. We choose two initial conditions close to each other so that the instability character of the system will be evident.

f[x_]:=3.9 x (1-x); x=.3; y=.31 (**enter**)

We construct a table of 200 iterates for each of the two orbits.

q1=Table[{x=f[x],y=f[y]},{200}]; (**enter**)

Use the ListPlot command. The plotted pairs (x_n,y_n) are the corresponding states of the two orbits.

ListPlot[q1,PlotStyle->PointSize[0.015],PlotRange->{{0,1},{0,1}}] (**enter**)

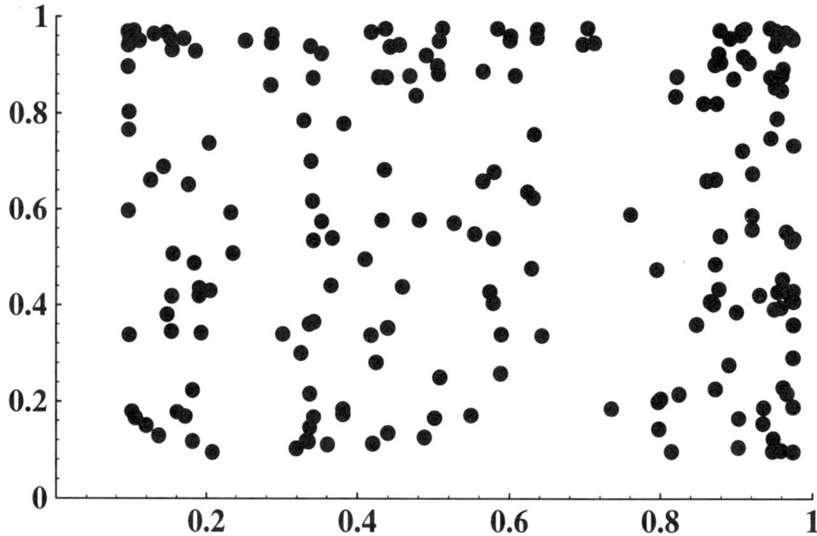

By plotting one orbit versus the other, we see that corresponding states are sometimes close to each other [when (x_n,y_n) is close to the line y=x], and sometimes far from each other [when (x_n,y_n) is close to the left-hand-side top corner or the right-hand-side bottom corner].

3. ***Plotting the distance between two orbits***

The graph above does not tell us the "story" of the two orbits. We have a global view, but we do not follow the evolution. We can obtain information on the evolution by plotting how far the two orbits are at each stage. To accomplish this, let us redefine the initial conditions and initialize the variable n.

x=.3; y=.31; n=0 (**enter**)
q2=Table[{x=f[x],y=f[y],n/10,z=Abs[x-y],n=n+1},{200}]; (**enter**)

The first two entries are the orbits, the third is n/10, the fourth is $|x_n-y_n|$, and the fifth is n+1. We select the third and fourth entries. We also eliminate the first 100. This decision eliminates the transient.

p=q2[[Range[100,200],Range[3,4]]]; (**enter**)

```
q3=ListPlot[p,PlotStyle->PointSize[0.02], PlotRange->{{10,20},{0,1}}]
```
(enter)

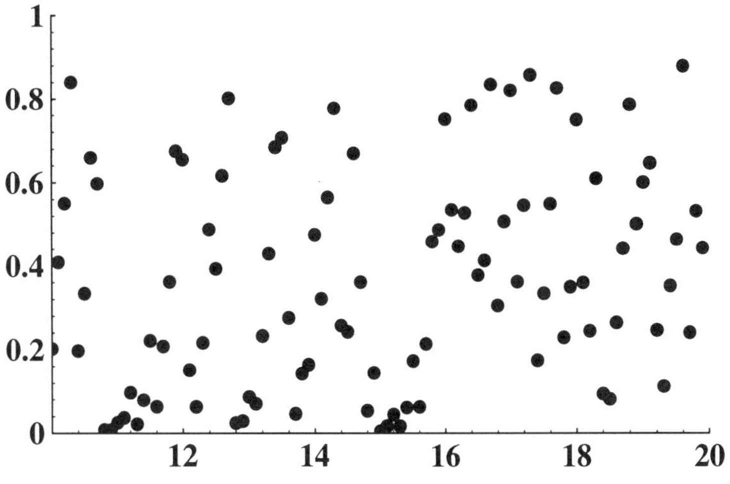

```
q3=ListPlot[p,PlotJoined->True,PlotRange->{{10,20},{0,1}}]
```
(enter)

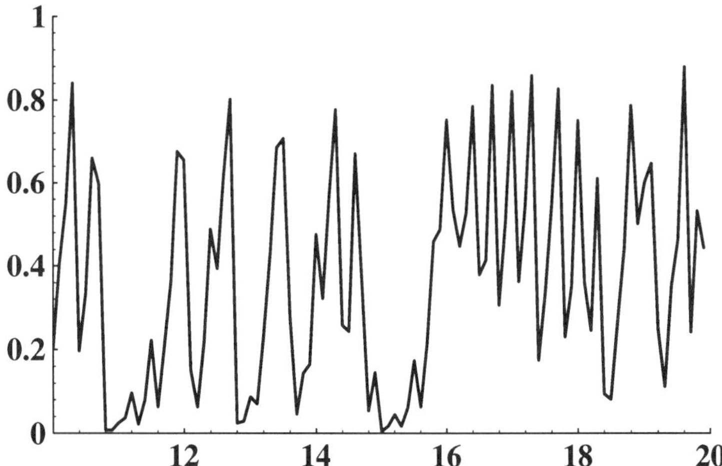

We join the points (n/10, $|x_n - y_n|$) with line segments. The lines are for visual purposes only. They are not part of the evolution of the system. The task is accomplished with the command PlotJoined->True.

We have called the first plot q1 and the second q2. With the Show command we can now see them together. The graph tells us the "story" of the two orbits from the 100th to the 200th positions. We see that the distance between the two orbits is bouncing between 0 and .9 in an apparent erratic manner. This type of information is important for analyzing chaotic behavior.

```
Show[q3,q4]
```
(enter)

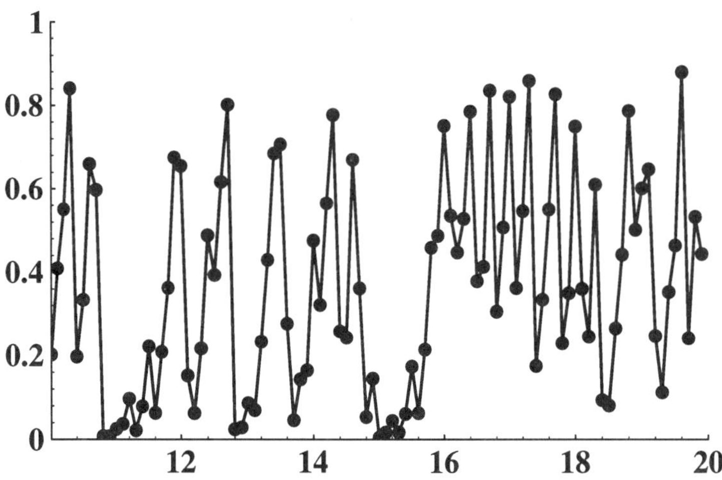

4. ***Using an orbit to reconstruct the graph***

This activity is useful to detect the chaotic behavior of a system governed by a function f. Asymptotically periodic orbits, from which the transient part has been eliminated, will give only finitely many points on the graph of f, whereas aperiodic orbits will give infinitely many, and will possibly reconstruct entire portions of the graph. Let us see the difference between these two situations. First we choose a system with an attracting periodic orbit of period 4.

```
f[x_]:=3.54 x(1-x); x=.3                                    (enter)
q1=Table[{x,x=f[x]}, {200}];                                (enter)
```

Run the table **once more** by putting the cursor at the end of the statement and pressing the enter key. The computer will evaluate 200 iterations using as initial condition x_{200}. Plot the points.

```
q2=ListPlot[q1,PlotStyle->PointSize[0.02], PlotRange->{{0,1},{0,1}}] (enter)
```

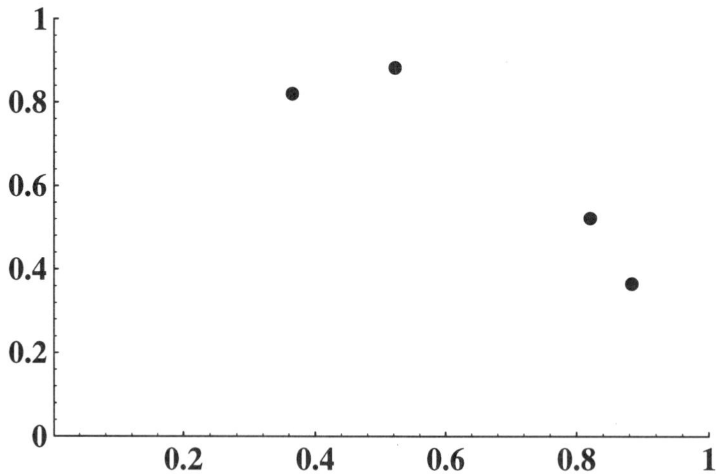

q3=Plot[f[x],{x,0,1}] **(enter)**

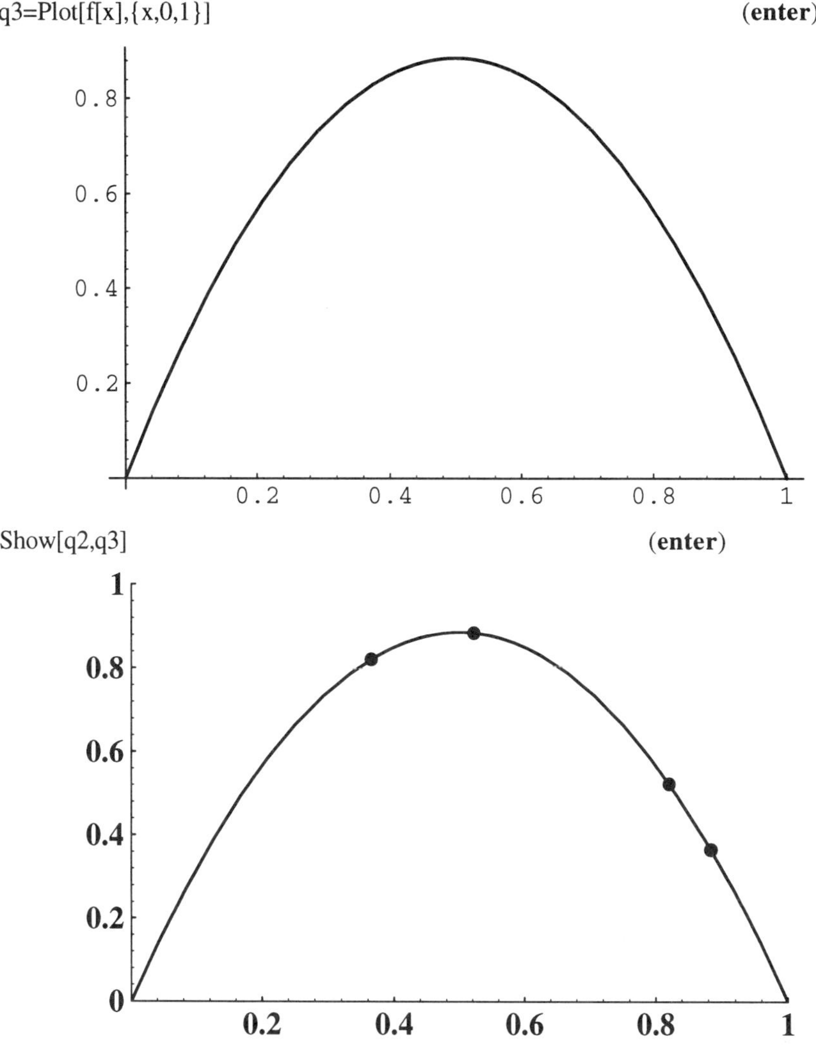

Show[q2,q3] **(enter)**

Try with a function f having an aperiodic orbit, namely an orbit with infinitely many limit points.

f[x_]:=3.9 x(1-x); x=0.3 **(enter)**

Construct a table of 200 iterations. Repeat the previous procedure to eliminate the transient.

r1=Table[{x,x=f[x]}, {200}];

Put the cursor after the semicolon and press **enter** to discard the first 200 iterations and obtain the next 200. Plot the points.

q4=ListPlot[r1, PlotStyle->PointSize[0.01]] (not shown) **(enter)**

q5=Plot[f[x],{x,0,1}] (not shown) **(enter)**
Show[q4,q5] **(enter)**

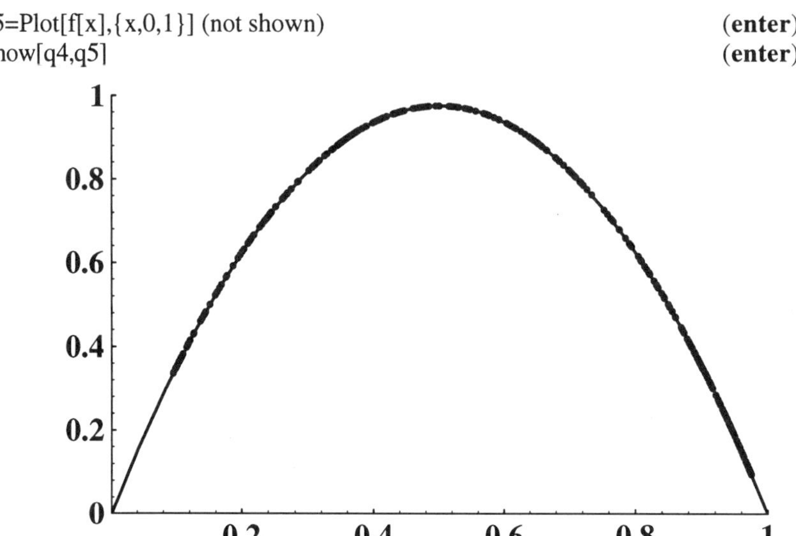

5. ***Density plots (histograms)***

The purpose of this activity is to detect the frequency with which the different intervals of the domain are "visited" by a given orbit. We divide the domain in small intervals (choose a number of intervals to be used), we compute several states of a trajectory, and we construct rectangles whose height is proportional to the number of states found in each interval. We may want to discard the "transient."

f[x_]:=4x(1-x); x=.3 **(enter)**
q1=NestList[f,Nest[f,.3,100],1000]; **(enter)**

The string above computes the first 100 iterates of the orbit of .3, discards them, and then computes the next 1000. Hence its result is the string $\{x_{101},x_{102},...,x_{1101}\}$. 1000 iterates are a bit too few for constructing a density plot, but our goal here is to illustrate the procedure. We compute a table of frequencies of where the 1000 iterates are located. We choose the number of "bins" in which the interval [0,1] is divided, say n=10. This value is too small but it is used here for illustration purposes.

q2=Table[Apply[Plus,Flatten[Table[If[(i-1)/10<=q1[[j]]<i/10,1,0], **(return)**
{j,1,1000}]]], {i,1,10}] **(enter)**
{229,90,67,75,64,65,62,56,88,204}

The string is the answer from Mathematica to q2. The readers may get a different string, depending on the machine used. The accuracy and the round-off error may alter the numbers, since the system is very sensitive to small changes.

The string If[(i-1)/10<=q1[[j]]<i/10,1,0],{j,1,1000}] returns 1 when the inequality is verified and 0 otherwise. The string Apply[Plus,] adds up all the numbers so obtained giving the total number of iterates in the interval [(i-1)/10, i/10). If we do not wish to see the string we simply add the symbol ";" at the end. We now construct the rectangles (i of them) with base the closed interval [(i-1)/10,i/10] and height q2[[i]]/1000. The height of the rectangle is "proportional" to the number of iterates contained in the interval [(i-1)/10,i/10].

```
q3=Graphics[{GrayLevel[0.4],Table[Rectangle[{(i-1)/10,0},        (return)
{i/10,q2[[i]]/1000}], {i,1,10}]}]                                (enter)
Graphics
```

The command Rectangle[{5,0},{6,4}] produces the rectangle with base [5,6] and height 4. The GrayLevel is darker with smaller values. GrayLevel[0] is black. We could use color. Replace GrayLevel[.4] with RGBColor[1,0,0] to get red rectangles.

Now we plot q3. The DefaultFont command selects the font to be used and its size. The command Axes-> tells the computer to print the x- and y-axes.

```
Show[q3, Axes->True, DefaultFont->{"Times",12}]                  (enter)
```

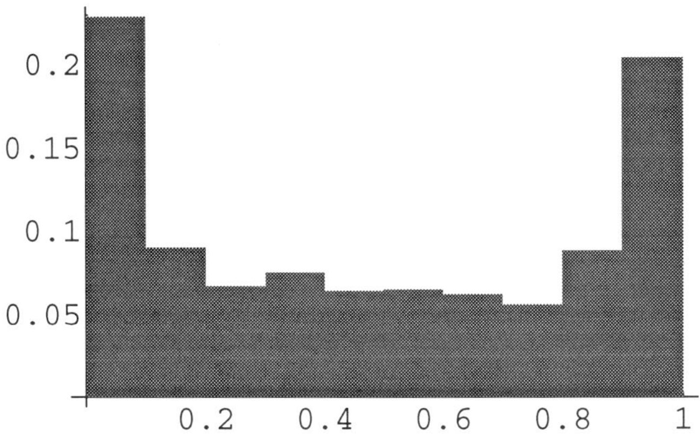

3. Orbits of Two-Dimensional Systems

1. *Plotting both orbits versus the iteration number*

Suppose we have the dynamical system

$$x_{n+1}=1.1x_n-.1x_ny_n,$$
$$y_{n+1}=.9y_n+.05\ y_n(1.1x_n-.1x_ny_n).$$

We want to graph the orbits of this system. There are several things we can do. First plot (n,x_n) and (n,y_n). Then combine them in one plot.

```
f[x_,y_]:=1.1x-.1x y; g[x_,y_]:=.9y+.05 y x(1.1-.1 y)            (enter)
```

Do not forget to separate x and y, otherwise, Mathematica reads xy as a new variable.

```
x=.3;y=.2;n=0                                                    (enter)
q1=Table[{n,u=x,n,y,x=f[x,y],y=g[u,y],n=n+1}, {200}];            (enter)
```

Notice that in y=g[u,y] we have used u in place of x. We need this approach to avoid using the value of x coming from x=f[x,y]. Select the points (n,x_n) and plot them.

```
q2=q1[[Range[1,200],Range[1,2]]];                                (enter)
q2=ListPlot[q2, PlotStyle->PointSize[0.02]]                      (enter)
```

We omit the plot. Do the same for (n,y_n).

```
q3=q1[[Range[1,200],Range[3,4]]];                    (cntcr)
q4=ListPlot[q3,PlotStyle->PointSize[0.01]]           (enter)
```

To distinguish from the previous plot we use smaller dots (PointSize[0.01]). The plot is omitted. Show them together.

```
Show[q3,q4]                                          (enter)
```

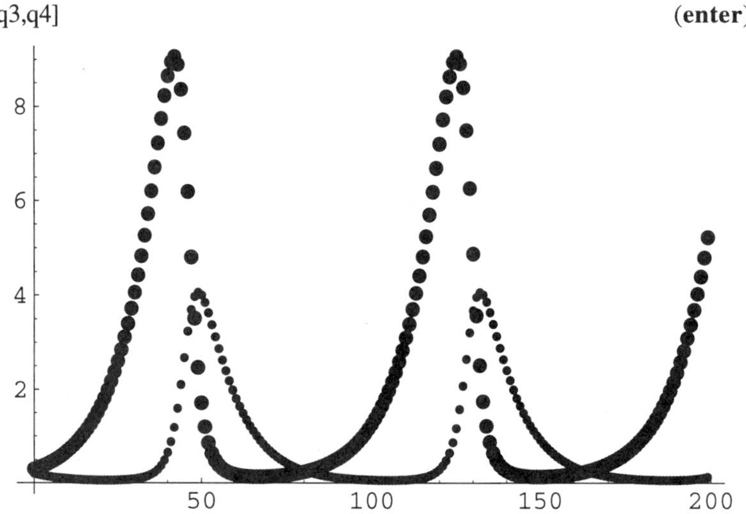

It appears that the two variables oscillate in a periodic fashion.

2. ***Plotting the orbit in the plane***

We refer to the previous table, q1.

```
q5=q1[[range[1,200],Range[5,6]]];                    (enter)
ListPlot[q5,PlotStyle->PointSize[0.02]]              (enter)
```

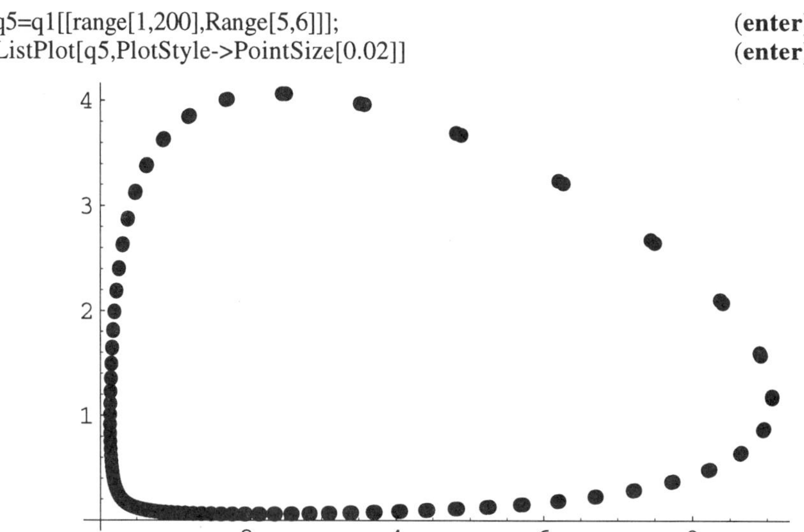

The plot confirms our previous observation on the periodicity of the orbit.

3. ***Instability of orbits of two-dimensional systems***

Now we study a different system to show instability.

f[x_,y_]:=3.9x(1-x-y); g[x_,y_]:=3.8x y (**enter**)

Do not forget to separate x and y, otherwise, Mathematica reads xy as a new variable.

x=0.3;y=0.2;u=0.3001;v=.2001;n=0 (**enter**)
q1=Table[{n, Sqrt[(x-u)^2+(y-v)^2],w=x,z=u,x=f[x,y], (**return**)
y=g[w,y],u=f[u,v],v=g[z,v],n=n+1},{400}]; (**enter**)

Notice that the second entry of the table is the Euclidean distance between $\mathbf{x}_n$ and $\mathbf{y}_n$.

q2=q1[[Range[1,400],Range[1,2]]]; (**enter**)
ListPlot[q2,PlotStyle->PointSize[0.02]] (**enter**)

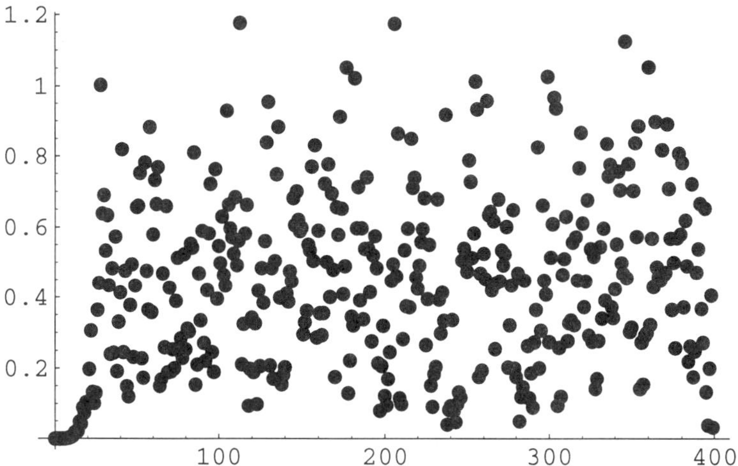

It appears that the system is unstable.

4. **Orbits of Linear Systems**

Suppose we have a linear system of the form $\mathbf{x}_{n+1}$=M$\mathbf{x}_n$+**b** and we want to study its evolution. To make it more concrete assume that

$$M = \begin{pmatrix} -1 & 0 \\ 2 & -1/4 \end{pmatrix}.$$

The eigenvalues are -1 and -1/4. Corresponding eigenvectors are **u**=(-3/8,1), **v**= (0,1). Assume that **b**=(1,1). The fixed point is **c**=(1/2,8/5). We study the conjugate system $\mathbf{y}_{n+1}=M\mathbf{y}_n$ where $\mathbf{y}=\phi(\mathbf{x})=\mathbf{x}-\mathbf{c}$. The orbit of $\mathbf{y}_0$=a**u**+b**v** converges to the periodic orbit of period 2 {a**u**,-a**u**}. Consequently, the orbit of the original system converges to {$\mathbf{x}_s$+a**u**,$\mathbf{x}_s$-a**u**} (see Chapter 4, Section 5). Given the initial condition $\mathbf{x}_0$=(2,3), we have $\mathbf{y}_0$=(1.5,1.4)=-4**u**+5.4**v**. Hence the orbit of $\mathbf{x}_0$ should converge to

$$\{\mathbf{x}_s+4\mathbf{u},\mathbf{x}_s-4\mathbf{u}\}=\{(-1,5.6),(2,-2.4)\}.$$

Here is the program

```
m={{-1,0},{2,-1/4}}                                    (enter)
```

The string above defines the matrix m.

```
x0={2,3}; n=0                                          (enter)
q1=Table[{n=n+1,x0=m.x0+{1,1}},{20}];                  (enter)
q2=MatrixForm[q1[[Range[8,14],Range[1,2]]]             (enter)
```

7	-1	5.59967
8	2	-2.39992
9	-1	5.59998
10	2	-2.39999
11	-1	5.6
12	2	-2.4
13	-1	5.6

As anticipated, the orbit converges to the periodic orbit {(-1,5.6),(2,-2.4)}!

Section 3. **BIFURCATION DIAGRAMS, LYAPUNOV EXPONENTS, AND CORRELATION DIMENSION**

1. **Bifurcation Diagrams**

1. ***Finding fixed points***

Suppose that we want to find the fixed points of the dynamical system (depending on a parameter a) f(a,x)=ax(1-x). We assume that x∈[0,1] and a∈[0,4]. In this case f(a,x)∈[0,1], i.e., f maps the interval [0,1] into itself.

```
f[a_,x_]:=a x(1-x)                                     (enter)
Solve[f[a,x]-x==0,x]                                   (enter)
```

$$\{\{x\to 0\},\{x\to\frac{a-1}{a}\}\}$$

There are two fixed points, or more precisely two "branches" of fixed points, one independent of a, namely $(a,x_1(a)=0)$, and the other dependent on a, namely $(a,x_2(a)=1-1/a)$. The second branch comes to life after a=1, since for a<1 the value of x is negative and we want x∈[0,1].

2. ***Finding periodic points of period*** 2

The fixed points will be included in the answer, since every point that is fixed for f is also fixed for the second iterate f^2. Let us use the same function, f(a,x)= ax(1-x).

```
Solve[f[a,f[a,x]]-x==0,x]                              (enter)
```

$$\{\{x\to 0\},\{x\to\frac{a-1}{a}\},\{x\to\frac{-(-1-a)a-a\sqrt{-3-2a+a^2}}{2a^2}\},\{x\to\frac{-(-1-a)a+a\sqrt{-3-2a+a^2}}{2a^2}\}\}$$

We obtain four points. Only x_3 and x_4 are periodic points of period 2. They form a periodic orbit of period 2. Using the Simplify command we obtain

Simplify[Solve[f[a,f[a,x]]- x==0,x]] **(enter)**

$\{\{x\to 0\},\{x\to \frac{a-1}{a}\},\{x\to \frac{1+a-\sqrt{-3-2a+a^2}}{2a}\}, \{x\to \frac{1+a+\sqrt{-3-2a+a^2}}{2a}\}\}$

The investigation above cannot be done graphically since the function depends on a parameter. The periodic orbit of period 2 starts at a=3 since for a<3 the quantity under the radical sign is negative.

3. ***Stability of fixed points***

Suppose that we want to study the stability of these fixed points and periodic orbits. We first need the derivative of f with respect to x: $f_x(a,x) = a - 2ax$, and we compute it along the first branch (x=0) of fixed points.

```
df[a_,x_]:=a - 2 a x                                    (enter)
df[a,0]                                                 (enter)
a
```

Since a is assumed to be in the interval [0,4], we see that the fixed point $x_s=0$ is a sink as long as a<1 and a source when a>1. Now we compute the derivative along the second branch of fixed points.

```
Simplify[df[a,1- 1/a]]                                  (enter)
2 - a
```

This gives us the value of the derivative along the fixed points of the second branch, namely the branch $x_2(a)=1-1/a$. We see that these points are sinks as long as 1<a<3, and they are sources when a>3. For a<1 the points are discarded.

4. ***Stability of periodic orbits***

We define $x_3(a)$ and $x_4(a)$ to be the two periodic points of period 2 for f(a,x)=ax(1-x). We compute the derivative of the second iterate of f along the orbit $\{x_3(a), x_4(a)\}$. Recall that such a derivative is the product of the derivative of f at $(a,x_3(a))$ and the derivative of f at $(a,x_4(a))$. In fact from $f(a,x_3(a))=x_4(a)$, $f(a,x_4(a))=x_3(a)$ and using the chain rule we obtain $(d/dx)f(a,f(a,x_3(a)))=f_x(a,x_4(a))\, f_x(a,x_3(a))$. We define the periodic orbit and compute the derivative of the second iterate.

```
x3 = (a + 1 + Sqrt[a^2 - 2a - 3])/(2a);                 (return)
x4 = (a + 1 - Sqrt[a^2 - 2a - 3])/(2a)                  (enter)
Simplify[df[a,x3] df[a,x4]]                             (enter)
4 + 2 a - a^2
```

At this point we know that the orbit exists only for a>3. We see that the derivative, as a function of a, is decreasing in the interval [3,4] and it is equal to 1 at a=3. Therefore, we find when the derivative is -1.

```
Simplify[Solve[4 + 2 a - a^2==-1,a]]                    (enter)
```

{{a -> 1 - $\sqrt{6}$}, {a -> 1 + $\sqrt{6}$}}

The value $1\text{-}6^{.5}$ is outside of our domain. Hence we choose $a=1+6^{.5}$. The periodic orbit is a sink for $a \in (1,1+6^{.5})$ and it is a source for $a \in (1+6^{.5},4]$.

5. ***Bifurcation diagram of some branches of periodic orbits***

The following sequence of commands produces the branches of fixed points and periodic orbits of period 2 of f(a,x)=ax(1-x).

```
p1[a_]:=0; p2[a_]:=1 - 1/a                                      (enter)
p3[a_]:=(1 + a + Sqrt[a^2 - 2 a - 3])/(2 a)                     (enter)
p4[a_]:=(1 + a - Sqrt[a^2 - 2 a - 3])/(2 a)                     (enter)
q1=Plot[p1[a],{a,0,4},PlotStyle->Thickness[0.01]]               (enter)
```

We are not showing neither this plot, nor the next two.

```
q2=Plot[p2[a],{a,1,4},PlotStyle->Thickness[0.007]]              (enter)
q3=Plot[{p3[a],p4[a]},{a,3,4},PlotStyle->Thickness[0.006]]      (enter)

Show[q1,q2,q3]                                                  (enter)
```

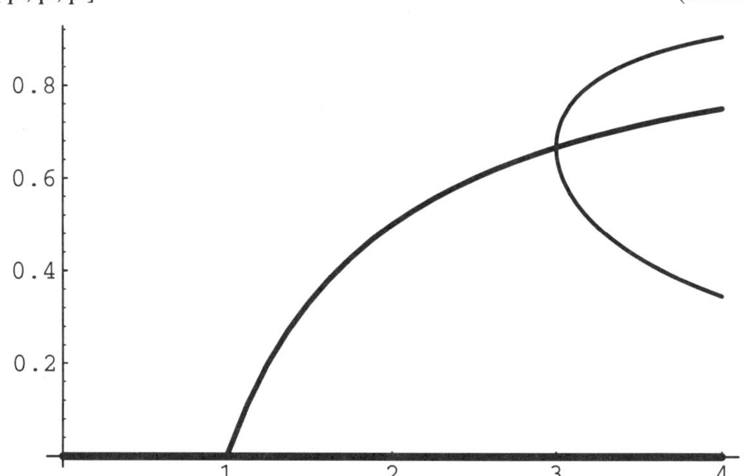

The bifurcation diagram above is far from complete, because it includes only the branches of fixed points and periodic orbits of period 2.

6. ***Bifurcation diagram of the stable portion of all branches***

The following program produces the stable portion of all branches. The idea is first to choose a value of the parameter and an initial condition. Then we iterate the function a certain number of times until we feel that the transient has disappeared. We iterate a few more times to produce the nontransient part of the orbit.

The parameter (named r) ranges between **start** and **end** and its increments are decided by **n**. In the example the parameter starts at 2.5 and ends at 3.5, with increments of 1/600. The program does **init** iterations, discards them, does **final** more and plots them.

It is designed for the quadratic map but can easily be adjusted to other maps. It runs quite slowly. Thus it should be done in successive steps.

Not too many iterations are needed (small **final**) in those intervals of the parameter range where the system has a stable periodic orbit. In those intervals of the parameter range where chaotic behavior is suspected or known, more iterations (large **final**) must be kept in memory.

```
f= r # (1-#) &

OrbitDiagram[start_, finish_, n_, init_, final_]:=                    (return)
Show[Graphics[{PointSize[0.001], Table[
        Map[Point[{r, #}] &, NestList[f, Nest[f, .3, init], final]],  (return)
        {r, finish, start, (start - finish)/n}]}],                     (return)
        PlotRange->{0,1}, AxesOrigin->{start, 0},Axes->True]           (enter)

OrbitDiagram[2.5, 3.5, 1000, 200, 4]                                  (enter)
```

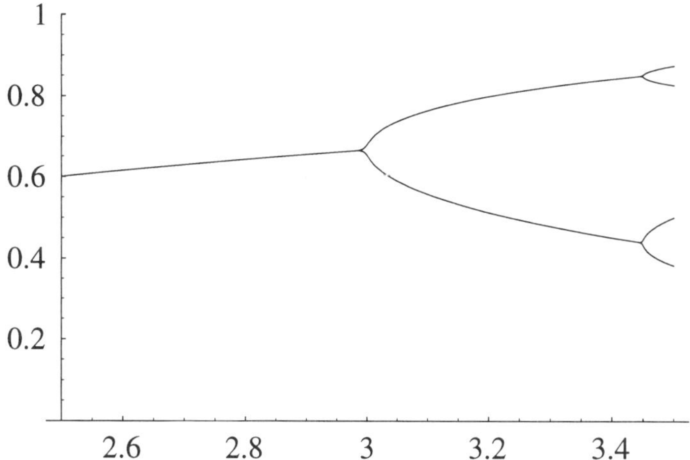

To adjust the program to different functions, such as $g(a,x)=ax^2-1$ we set

```
f=r #^2-1&                                                            (enter)
```

In the OrbitDiagram program we change PlotRange->{-1,1}, AxesOrigin->{start,-1}. We may want to change the initial point to -.7. Everything else is unchanged. However, recall that $a \in [0,2]$.

7. *Branches of fixed points by numerical methods*

This procedure applies also to periodic points. Sometimes the branches are not made of sinks. Therefore, the previous approach will not work. We can use Newton's method to find the branch of fixed points as a function of a and then plot $(a,x_s(a))$. Here is an example.

```
f[a_,x_]: =-Exp[x]+a-1                                                (enter)
```

It can bc shown that for every a the equation f(a,x)=x has one and only one solution x(a). We choose and initial value for a, say a=2. Notice that x(a)>0 when a>2. Hence the fixed point is a source.

a=2; **(enter)**

We construct the following table:

q1=Table[{a,x/.FindRoot[f[a,x]-x==0,{x,0}],a=a+0.01},{100}]; **(enter)**

The command x/.FindRoot[...] reduces the result x(a) of FindRoot[.], to a number that can be plotted as the second coordinates of (a,x(a)). The table produces the solution x(a) of f(a,x)=x for a∈[2,3] with increments in a of .01. We select the first two entries and we plot them.

q2=q1[[Range[1,100],Range[1,2]]]; **(enter)**

ListPlot[q2,PlotStyle->PointSize[0.01]] **(enter)**

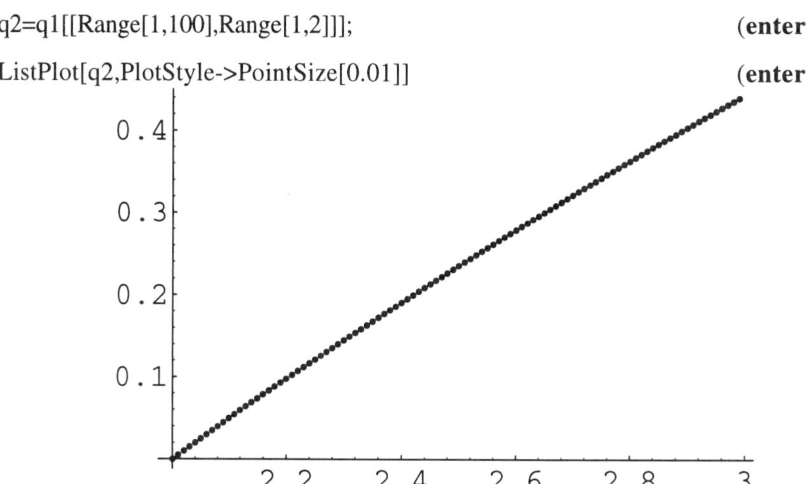

2. **Lyapunov Exponents**

1. ***Lyapunov exponent in* R**

The following sequence of steps is aimed at showing how the Lyapunov exponent changes along a trajectory. The purpose is to test the stability of an orbit.

```
f[x_]:=4 x(1-x); g[x_]:=4 - 8 x; x=.3; n=1; z=0                    (enter)
q=Table[{y=Log[Abs[g[x]]],z=z+y,m=n/100,u=z/n,                     (return)
        n,n=n+1,x=f[x]},{800}];                                    (enter)
```

The table gives the logarithm of the absolute value of the derivative at each point of the orbit (entry 1), the sum of all the logarithms up to the point of interest (entry 2), the iterate number (compressed by 100 for graphing purposes: entry 3), the approximate value of the Lyapunov exponent of the orbit up to the given iteration (entry 4), the iteration number (entry 5), the next iteration number (entry 6) and the next state of the orbit (entry 7).

p=MatrixForm[q[[Range[790,800],Range[4,5]]]] **(enter)**

0.691508 790
0.692345 791
0.693051 792
0.693119 793
0.692969 794
0.693211 795
0.692414 796
0.693094 797
0.693037 798
0.693142 799
0.692889 800

We plot the Lyapunov exponent versus the iteration number.

```
d=q[[Range[1,800],Range[3,4]]];                                   (enter)
ListPlot[d, PlotRange->{Automatic,{0,1}}, AspectRatio->.5]        (enter)
```

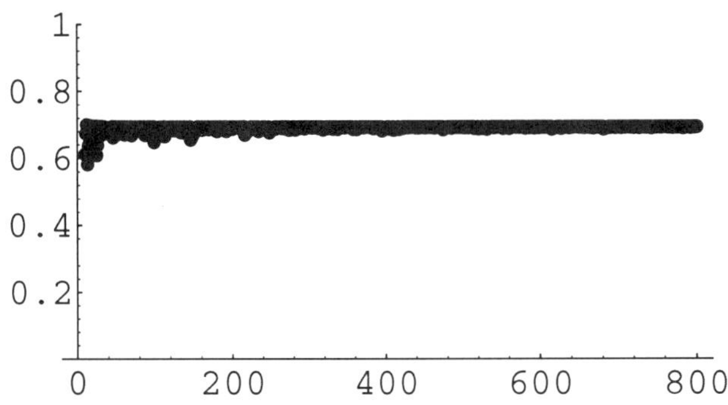

The graph shows that the Lyapunov exponent is positive and larger than .5. This suggests that the orbit is unstable. The same analysis must be repeated for several orbits. If all results agree, we can conclude that the orbits of the system are unstable.

2. *Lyapunov exponent in* $\mathbf{R}^q$

The same strategy can be used for functions of several variables. The only difference is that the absolute value of the derivative at the given point has to be substituted by the spectral radius of the derivative. This can easily be accomplished for dynamical systems in the plane, but becomes a nontrivial numerical task in higher dimension. What it is needed is a fast and reliable method to compute the spectral radius of a matrix. One such method is the **power method**, which works very well as long as there is an eigenvalue which is significantly bigger than all others.

We present here a program for computing the Lyapunov exponent of a plane system. We first define a function $F : \mathbf{R}^2 \to \mathbf{R}^2$, by means of the two component functions $F1(x,y) = 3.8x(1-x-y)$, $F2(x,y) = 3.8xy$.

```
F1[x_,y_]:=3.8 x(1-x-y); F2[x_,y_]:=3.8 x y                       (enter)
```

We define the derivative of F. This is a 2×2 matrix, which we call M and depends on the point (x,y): i.e., we have M=M(x,y).

```
M[x_,y_]:={{3.8(1-2 x-y),-3.8 x},{3.8 y,3.8 x}}                    (enter)
```

We also need the 2×2 identity matrix.

```
q={{1,0},{0,1}}                                                    (enter)
```

Select an initial condition and the starting value of the iteration number.

```
x=.3;y=.4; n=1                                                     (enter)
```

Below is the scheme for approximating the Lyapunov number of the orbit O(.3,.4). Recall that the Lyapunov exponents is the logarithm of the Lyapunov number. The third entry represents the derivative of the nth iterate of F at the initial point of the orbit. By the chain rule this derivative is the product of all matrices representing the derivative of F at all points of the orbit up to $\mathbf{x}_{n-1}$. For example, for n=3, we have $(F^3)'(\mathbf{x}_0)=F'(\mathbf{x}_2)F'(\mathbf{x}_1)F'(\mathbf{x}_0)$. This is the product of three matrices. The order is important, put M[x,y] before q.

The next entry is the iteration number. The fifth entry is the approximate value of the Lyapunov number up to the given iterate. Notice that we compute the eigenvalues of $(F^n)'(.3,.4)$, then their absolute value (or modulus if they are complex), then choose the largest of them (spectral radius), and then compute the nth root. We increase the iteration number by one and compute the next state of the orbit. In F2 we must denote x with u, otherwise, we would be feeding F1(x,y) instead of x into F2.

```
q1=Table[{u=x,y,q=M[x,y].q,n,                                      (return)
(Max[Abs[Eigenvalues[q]]])^(1/n),n=n+1,                            (return)
x=F1[x,y],y=F2[u,y]},{1200}];                                      (enter)
p=MatrixForm[q1[[Range[195,200],Range[5,6]]]]                      (enter)
        1.25484             195
        1.25215             196
        1.26645             197
        1.25675             198
        1.26142             199
        1.25790             200
k=q[[Range[600,1200],Range[4,5]]];                                 (enter)
ListPlot[k,PlotRange->{Automatic,{0,2}},PlotStyle->PointSize[0.02],(return)
Ticks->{Automatic,Range[0,2,.5]}]                                  (enter)
```

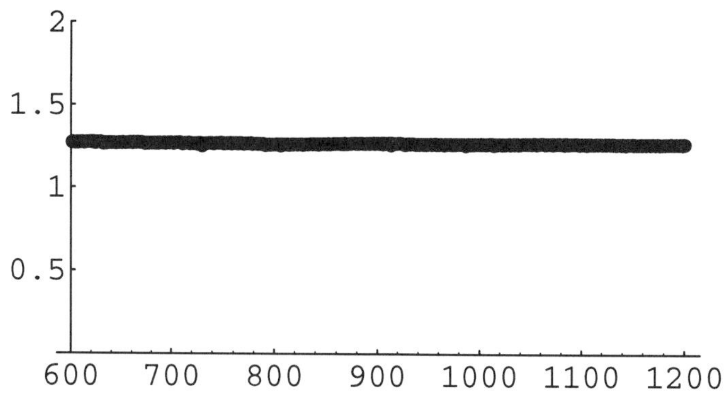

3. Correlation Dimension

1. *Correlation dimension in* R

The correlation dimension is another numerical quantity relevant to chaotic behavior. We describe the general design. We define the unit step function h_r: $h_r(x)=1$ when $x<r$ and $h_r(x)=0$ when $x\geq r$. We choose an initial condition x and we compute a large number of states of O(x). For illustration purposes we have selected the quadratic map f(x)=4x(1-x), with initial condition x=.3, and we compute 50 iterations.

We define two vectors, q and p, with 50 entries equal to the 50 iterations. Next, we selected an initial value of r. Such value cannot be too small or too large. In the main body of the program we compute how many pairs are closer than r. Call this number s. We divide s by the total number of pairs. Assuming N pairs, we are comparing M=N(N-1)/2 pairs. Compute Log[s/M]/Log[r], increase r by a fixed amount, and repeat the procedure. We should be careful not to reach a value of r that is too far from 0.

Explanation of the string: Apply[Plus[Flatten[Table.

Suppose we compute a table of values.

```
b=Table[Abs[i-j], {i,1,5},{j,i+1,5}]                    (enter)
{{1, 2, 3, 4}, {1, 2, 3}, {1, 2}, {1}, {}}
```

We Flatten b.

```
b1=Flatten[b]                                           (enter)
{1, 2, 3, 4, 1, 2, 3, 1, 2, 1}
```

Then we add up all the entries

```
Apply[Plus,b1]                                          (enter)
20
```

In one line only

```
Apply[Plus,Flatten[Table[ Abs[i-j],{i,1,5},{j,i+1,5}]]]  (enter)
20
```

Here is the procedure for the correlation dimension.

```
h[x_]:=1/;x<0                                           (return)
h[x_]:=0/;x>=0                                          (enter)
y=.3                                                    (enter)
q=Table[y=4y (1-y),{50}];                               (enter)
p=q; r=.01;                                             (enter)
t=Table[{s=Apply[Plus,Flatten[Table[ h[Abs[q[[i]]-      (return)
p[[j]]]-r],{i,1,50},{j,i+1,50}]]],u=r,                  (return)
N[Log[s/1225]]/Log[r],r=r+.01},{10}];                   (enter)
```

s represents the total number of pairs that are closer than r; N[Log[s/1225]]/Log[r] represents the approximate value of the correlation dimension for the given value of r.

The first column of the following matrix tells us how many pairs are closer than the numerical value provided in the second column (the value of r). The third column computes the correlation dimension up to that point.

v=MatrixForm[t[[Range[1,10],Range[1,3]]]] **(enter)**

41	0.01	0.737676
62	0.02	0.762665
83	0.03	0.767663
104	0.04	0.766201
123	0.05	0.767262
143	0.06	0.763433
163	0.07	0.758461
184	0.08	0.75058
202	0.09	0.748534
213	0.1	0.759756

We can now have a graphical view.

le=t[[Range[1,10],Range[2,3]]]; **(enter)**
ListPlot[le,PlotRange->{{0,.1},{.6,1}},Ticks->{Range **(return)**
[0,.1,.02], Range[.6,1,.2]}, AxesOrigin->{0,.6}] **(enter)**

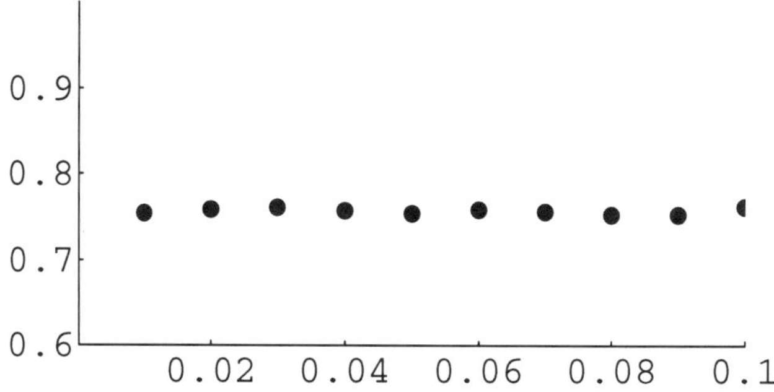

Using the least squares method we can see that the slope of the line is almost 0.

Clear[x]; **(enter)**
Fit[le,{1,x},x] **(enter)**
0.758645 - 0.007672 x

The correlation dimension is about .75. This numerical process is very time consuming. With 50 iterations we have to compare 1225 pairs, i.e., (50×49)/2.

2. ***Two-dimensional example***

We first define the function $F : \mathbf{R}^2 \to \mathbf{R}^2$.

f[x_,y_]:=3.8x(1-x-y); g[x_,y_]:=3.8 x y **(enter)**

We choose an initial condition $\mathbf{x}_0$.

x=.4;y=.3 **(enter)**

We define the function h as before.

```
h[x_]:=1/;x<0                                    (return)
h[x_]:=0/;x>=0                                   (enter)
```

We construct a table of 50 iterations from $\mathbf{x}_0$.

```
q=Table[{u=x,x=f[x,y],y=g[u,y],m={x,y}},{50}];
```

We select the last entry of the above table, which represents the 50 points in $\mathbf{R}^2$.

```
p=q[[Range[1,50],Range[4,4]]];                   (enter)
```

We double it.

```
b=p;                                             (enter)
```

There are 1225 pairs of vectors. Notice that the pairs are of the form p[[i]], b[[j]]. To compute their distance, we use Apply[Plus,...].

```
t=Flatten[Table[Sqrt[Apply[Plus,Flatten[(p[[i]]-b[[j]]))^2]]],      (return)
{i,1,50},{j,i+1,50}]];                                               (enter)
```

To verify how many distances there are, we can ask

```
Dimensions[t]                                    (enter)
{1225}
```

We choose an initial value of r.

```
r=.15                                            (enter)
```

Now comes the big work. In the following line h[t[[k]]] computes the function h with the given r at the entry k of the vector t (which has 1225 entries).

```
v=Table[{s=Apply[Plus,Flatten[Table[h[t[[k]]-r],{k,1,1225}]]],u=r,   (return)
N[Log[s/1225]]/Log[r], r=r+.01},{10}];                                (enter)
```

For every r, starting from r=.15 and ending with r=.24, with increments of .01, count how many distances are smaller than r (denote the number by s), and compute Log[s/1225]/Log[r] for all 10 different values of r.

In the following table the first column tells us how many pairs are closer than r, the second column contains the value of r, and the third column contains the ratio Log[s/1225]/Log[r].

```
w=MatrixForm[v[[Range[1,10],Range[1,3]]]]        (enter)
```

```
114    0.15    1.25163
127    0.16    1.23678
143    0.17    1.21214
163    0.18    1.1762
175    0.19    1.17172
188    0.2     1.16454
204    0.21    1.14861
218    0.22    1.14006
```

```
235     0.23    1.12345
250     0.24    1.1136
```

Let us look at it graphically.

```
le=v[[Range[1,10],Range[2,3]]];                                          (enter)
ListPlot[le,PlotRange->{{.15,.25},{0,2}},                                (return)
Ticks->{Range[.15,.25,.02], Range[0,2,.5]},                              (return)
AxesOrigin->{.15,0}, PlotStyle->PointSize[0.02], AspectRatio->.5]        (enter)
```

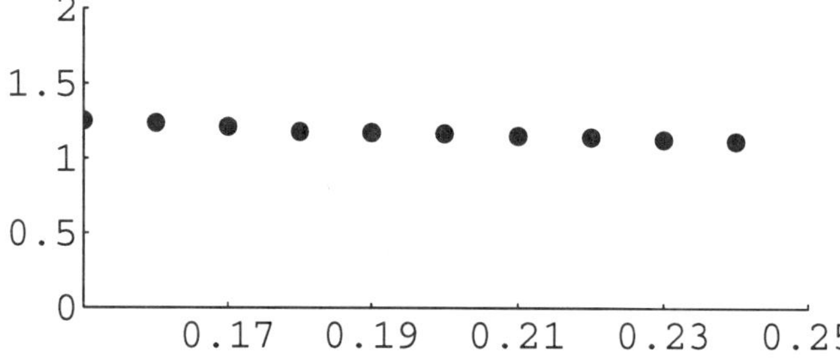

We may have doubts about the true value of the correlation dimension, but it appears that it is not an integer.

Section 4. **ODDS AND ENDS**

1. **Matrices and Vector Operations**

1. ***Applying a map to a vector***

Suppose that

```
f[x_]:=1/;x<0                                                            (return)
f[x_]:=0/;0<=x                                                           (enter)
```

We want to apply this function to each component of a vector x0={x1,x2,x3, x4}. This is accomplished by writing.

```
Map[f,x0]
```

For example, define

```
x0={1.5,-1,-2,2}                                                         (enter)
Map[f,x0]                                                                (enter)
(0,1,1,0}
```

This device may be useful in the study of multidimensional systems. If we say f[x0] we will not get anything. However, for standard functions we will: for example

```
x0^2                                                                     (enter)
{2.25,1,4,4}.
```

2. ***Computing the Euclidean norm of a vector***

Suppose that we want to compute the norm of the vector (2,3,4).

```
x0={2,3,4}                                    (enter)
Sqrt[Plus[x0.x0]]                             (enter)
```
$\sqrt{29}$

3. ***Computing the eigenvalues and eigenvectors of a matrix***
Suppose that we have a matrix m and we want to know its eigenvalues.

```
m={{1,3},{3,1}}                               (enter)
Eigenvalues[m]                                (enter)
{-2,4}
```

Mathematica tries to find the "exact" eigenvalues of m. This operation requires solving equations of degree equal to the size of the matrix and it may give you a very complicated expression. When the matrix has size larger than 2 is better to say

```
N[Eigenvalues[m]]                             (enter)
```

which asks only for the numerical values of the eigenvalues.

Same considerations apply to the search of eigenvectors. We say

```
Eigenvectors[m]                               (enter)
```

The eigenvectors are normally given in the same order as the eigenvalues. Also in this case we may want to say

```
N[Eigenvectors[m]]
{{-1,1},{1,1}}                                (enter)
```

One could try to gain both information at the same time by saying

```
N[Eigensystem[m]]
{{-2,4},{{-1,1},{1,1}}}                       (enter)
```

4. ***Computing the determinant and characteristic polynomial***
The command

```
Det[m]                                        (enter)
```

will compute the determinant of a square matrix m. Assuming that the matrix has size q the command

```
Det[r IdentityMatrix[q]-m]                    (enter)
```

will compute the characteristic polynomial of m.

2. **Solving Equations**

1. ***One equation: trascendental case***
We have seen how to solve equations in some simple polynomial cases (see Section 3, Subsection 1). When the polynomial has degree 3 or higher or the equation is trascendental, we find numerical solutions. We may use Newton's method

or the secant method. For example, suppose that we want to find the solution of cosx=x. We can proceed as follows.

```
FindRoot[ Cos[x]-x==0,{x,1}]                    (enter)
{x -> 0.739085}
```

In fact cos(0.739085)=0.739085.

In the approach above we have used Newton's method starting from x=1. We can use the secant method. Since the root is between 0 and 1, we write

```
FindRoot[Cos[x]-x==0,{x,0,1}]                   (enter)
{x -> 0.739085}
```

For example, we may want to solve x^3-3x -1=0. There is a root between 1 and 2. Hence we try

```
FindRoot[x^3-3x-1==0,{x,1,2}]                   (enter)
{x -> 1.87939}
```

Another root is between -2 and -1.

```
FindRoot[x^3-3x-1==0,{x,-2,-1}]                 (enter)
{x -> -1.53209}
```

Finally, a third root is between -1 and 0.

```
FindRoot[x^3-3x-1==0,{x,-1,0}]                  (enter)
{x -> -0.347296 }
```

2. ***Systems of equations***

To solve a system we can use the following strategy.

```
Solve[{x^2-x-y==0, x-y==0,{x,y}]                (enter)
{{x -> 0, y -> 0},{ x -> 2,y -> 2}}
```

Be careful! If the equations are not that simple, the machine may get stuck! In this case it is better to use the NSolve statement:

```
NSolve[{x^3 - y-1==0,x^3-y^2+1==0},{x,y}]       (enter)
{{y->-1,x->0},{y->-1,x->0},{y->-1,x->0},
{y->2,x->-0.721125-1.29402i},{y->2,x->1.44225},
{y->2,x->-0.721125+1.29402i}
```

The first root has multiplicity 3. The reader should verify that this is indeed the case. Two of the roots are complex conjugate.

3. **Assigning a Numerical Value to the Result of an Operation**

Suppose that we want to solve an equation and we want to assign a numerical value to the result of this operation in a table of numbers we want to use for plotting purposes (see also Section 3, Subsection 1, item 7). For example, suppose that we want to solve the equation $e^{-x}-1+a=x$ numerically for $a\in[-1,1]$ with increments of 0.01. We also want to plot the solution as a function of a.

```
f[a_,x_]:=Exp[-x]-1+a                                        (enter)
a=-1                                                         (enter)
q1=Table[{a,x/.FindRoot[f[a,x]-x==0,{x,0}],a=a+.01},{200}];  (enter)
q2=q1[[Range[1,200],Range[1,2]]];                            (enter)
ListPlot[q2]                                                 (enter)
```

APPENDIX 2

REFERENCES AND PROJECTS

1. References

[Arrowsmith-Place, 1992] Arrowsmith D.K., Place C.M.: Dynamical Systems. Chapman & Hall Mathematics, New York, 1992.

[Bak, 1986] Bak P.: *The devil's staircase.* Phys. Today, **39** (1986), 38-45.

[Banks et al., 1992] Banks J., Brooks J., Cairns G., Davis G., Stacey P.: *On Devaney's definition of chaos.* Am. Math. Monthly, **99** (1992), 332-334.

[Barnsley, 1988] Barnsley M.: Fractals Everywhere. Academic Press, San Diego, CA, 1988.

[Bassein, 1995] Bassein S.: *Order and chaos on your desk.* Am. Math. Monthly, **102** (1995), 409-416.

[Berry-Percival-Weiss, 1987] Berry M.V., Percival I.C., Weiss N.O. (Eds.): Dynamical Chaos. Princeton University Press, Princeton, NJ, 1987.

[Birkhoff, 1927] Birkhoff G.D.: Dynamical Systems. Colloquium Publications, **9**. American Mathematical Society, Providence, RI, 1927.

[Braun, 1982] Braun M.: Differential Equations and Their Applications. Applied Mathematical Sciences **15**. Springer-Verlag, New York, 1982.

[Bunde-Havlin, 1994] Bunde A., Havlin S. (Eds): Fractals in Science. Springer-Verlag, New York, 1994.

[Codenotti-Margara, 1996] Codenotti B., Margara L.: *Transitive cellular automata are sensitive.* Am. Math. Monthly, **103** (1996), 58-61.

[Collet-Eckmann, 1980] Collet P., Eckmann J. P.: Iterated Maps on the Interval as Dynamical Systems. Birkhäuser, Boston, 1980.

[Corless, 1992] Corless R.M.: *Continued fractions and chaos.* Am. Math. Monthly, **99** (1992), 332-334.

[Crannell, 1995] Crannell A.: *The role of transitivity in Devaney's definition of chaos.* Am. Math. Monthly, **102** (1995), 788-793.

[Crannell-Martelli, 1997] Crannell A., Martelli M.: *Periodic orbits from nonperiodic orbits on an interval.* App. Math. Lett., **10** (1997), 45-47.

[Crannell-Martelli, 1998] Crannell A., Martelli M.: *Dynamics of quasicontinuous systems.* Preprint.

[Derrick-Eidswick, 1995] Derrick W., Eidswick J.: *Continued fractions, Chebichev polynomials and chaos.* Am. Math. Monthly, **102** (1992), 337-344.

[Devaney, 1989] Devaney R. L.: An Introduction to Chaotic Dynamical Systems. Second Edition. Addison-Wesley, Reading, MA, 1989.

[Fatou, 1926] Fatou P.: *Sur l'itération des fonctionnes transcendantes entières.* Acta Math., **47** (1926), 337-370.

[Feigenbaum, 1983] Feigenbaum M.J.: *Universal behavior in nonlinear systems.* Physica, **7**D (1983), 16-39.

[Ford, 1989] Ford J.: *What is chaos, that we should be mindful of it?* In The New Physics, P. Davies (Ed.), Cambridge University Press, Cambridge, 1989, 348-371.

[Furi-Martelli, 1991] Furi M., Martelli M.: *On the mean value theorem, inequality, and inclusion.* Am. Math. Monthly, **98** (1991), 840-846.

[Gearhart-Martelli, 1990] Gearhart W., Martelli M.: *A blood cell population model, dynamical diseases and chaos.* UMAP, **11** (1990), 311-338.

[Gleick, 1987] Gleick J.: Chaos: Making a New Science. Viking Penguin, New York, 1987.

[Grassberger-Procaccia, 1983] Grassberger P., Procaccia I.: *Measuring the strangeness of strange attractors.* Physica, **9**D (1983), 189-208.

[Hartman, 1973] Hartman P.: Ordinary Differential Equations. Wiley, New York, 1973.

[Hastings-Sugihara, 1993] Hastings H.M., Sugihara G.: Fractals: A User's Guide for the Natural Sciences. Oxford Science Publications, Oxford University Press, Oxford, 1993.

[Hénon, 1976] Hénon M.: *A two-dimensional mapping with a strange attractor.* Commun. Math. Phys., **50** (1976), 69-78.

[Hofstadter, 1982] Hofstadter D.R.: *Strange attractors: mathematical patterns delicately poised between order and chaos.* Sc. Am., **245**, May 1982, 16-29.

[Hopfield, 1982] Hopfield J.J.: *Neural networks and physical systems with emergent collective computational abilities.* Proc. Natl. Acad. Sci. USA, **79** (1982), 2554-2558.

[Julia, 1918] Julia G.: *Mémoirs sur l'itération des fonctionnes rationnelles.* J. Math., **7** (1918), 47-245.

[Kaplan-Glass, 1995] Kaplan D., Glass L.: Understanding Nonlinear Dynamics. Textbooks in Mathematical Sciences, Springer Verlag, New York, 1995.

[Kennedy-Yorke, 1995] Kennedy J.A., Yorke J.A.: *Bizarre topology is natural in dynamical systems.* Bull. Am. Math. Soc., **32** (1995), 309-316.

[Lasota, 1977] Lasota A.: *Ergodic problems in biology.* Soc. Math. France, Astérisque, **50** (1977), 239-250.

[Lauwerier, 1986a] Lauwerier H.A.: *One-dimensional iterative maps.* In Chaos, A. V. Holden (Ed.), Princeton University Press, Princeton, NJ, 1986, 39-58.

[Lauwerier, 1986b] Lauwerier H.A.: *Two-dimensional iterative maps.* In Chaos, A. V. Holden (Ed), Princeton University Press, Princeton, NJ, 1986, 58-97.

[Li-Yorke, 1975] Li T., Yorke J. A.: *Period three implies chaos.* Am. Math. Monthly, **82** (1975), 985-992.

[Lorenz, 1963] Lorenz E. N.: *Deterministic nonperiodic flow.* J. Atmos. Sci., **20** (1963), 130-141.

[Lyapunov, 1907] Lyapunov A. M.: *Problème générale de la stabilité du mouvement.* Ann. Fac. Sci. Univ. Toulouse, **9**(1907), 203-475. Reproduced in Ann. Math. Study, **17** (1947), Princeton.

[MacEachern-Berliner, 1993] MacEachern S.N., Berliner M.L.: *Aperiodic chaotic orbits.* Am. Math. Monthly, **100** (1993), 237-241.

[Mandelbrot, 1982] Mandelbrot B.B.: The Fractal Geometry of Nature. W. H. Freeman, 1982.

[Marsden-Hoffman, 1993] Marsden J.E., Hoffman M.J.: Elementary Classical Analysis. W.H. Freeman, New York, 1993.

[Marotto, 1979a] Marotto F.R.: *Perturbation of stable and chaotic difference equations*. J. Math. Anal. Appl., **72** (1979), 716-729.

[Marotto, 1979b] Marotto F. R.: *Chaotic behavior in the Hénon mapping.* Commun. Math. Phys., **68** (1979), 187-194.

[Martelli-Dang-Seph, 1998] Martelli M., Dang M., Seph T.: *Defining chaos.* Math. Mag., **71** (1998), 112-122.

[Martelli-Johnston, 1996] Martelli M., Johnston B.: *Global attractivity and forward neural networks*. Appl. Math. Lett., **9** (1996) , 77-83.

[Martelli-Marshall, 1995] Martelli M., Marshall D.: *Stability and attractivity in discrete dynamical systems*. Math. Biosci., **128** (1995), 347-355.

[May, 1976] May R.M.: *Simple mathematical models with very complicated dynamics*. Nature, London, **261** (1976), 459-467.

[May, 1979] May R. M.: *Bifurcation and dynamic complexity in ecological systems.* Ann. N.Y. Acad. Sci., **316** (1979), 517-529.

[May, 1995] May R. M.: *Necessity and chance: deterministic chaos in ecology and evolution*. Bull. Am. Math. Soc., **32** (1995), 291-308.

[Mischaikow-Mrozek, 1995] Mischaikow K., Mrozek M.: *Chaos in the Lorenz equations, a computer assisted proof.* Bull. Am. Math. Soc., **32** (1995), 66-72.

[Moon, 1992] Moon F.C.: Chaotic and Fractal Dynamics. Wiley, New York, 1992.

[Ott-Sauer-Yorke, 1994] Ott E., Sauer T., Yorke J.A.: Coping with Chaos. Series in Nonlinear Science. Wiley, New York, 1994.

[Peitgen-Jürgens-Saupe, 1992a] Peitgen H-O., Jürgens H., Saupe D.: Chaos and Fractals: New Frontiers of Science. Springer-Verlag, New York, 1992.

[Peitgen-Jürgens-Saupe, 1992b] Peitgen H-O., Jürgens H., Saupe D.: Fractals for the Classroom, Part One. Springer-Verlag, New York, 1991.

[Peitgen-Richter, 1986] Peitgen H-O., Richter P.H.: The Beauty of Fractals. Springer-Verlag, New York, 1986.

[Pickett et al., 1996] Pickett C.H., Ball J.C., Casanave K.C., Klonski K.M., Jetter K.M., Bezark L.G., Schoenig S.E.: *Establishment of the ash whitefly parasitoid Encarsia inaron* (Walker) *and its economic benefit to ornamental street trees in California*. Biol. Control, **6** (1996), 260-272.

[Ruelle-Takens, 1971] Ruelle F., Takens F.: *On the nature of turbulence.* Commun. Math. Phys., **20** (1971), 167-192.

[Sarkovskii, 1964] Sarkovskii A.N.: *Coexistence and cycles of a continuous map of a line into itself.* Ukr. Math. Z., **16** (1964), 61-71.

[Stewart, 1989] Stewart I.: Does God Play Dice ? The Mathematics of Chaos. Blackwell, Malden, MA, 1989.

[Touhey, 1997] Touhey P.: *Yet another definition of chaos.* Am. Math. Monthly, **104** (1997), 411-415.

[Vellekoop-Berglund, 1994] Vellekoop M., Berglund R.: *On intervals, transitivity = chaos.* Am. Math. Monthly, **101** (1994), 353-355.

[Volterra, 1931] Volterra V.: Leçons sur la théorie mathématique de la lutte pour la vie. Paris, 1931.

[Walsh, 1996] Walsh J.A.: *Fractals in Linear Algebra.* College Math. J., **27** (1996), 298-304.

[Whittaker, 1991] Whittaker J. V.: *An analytical description of some simple cases of chaotic behavior.* Am. Math. Monthly, **98** (1991), 489-504.

2. Projects

1. ***Analysis of a parasitoid-host interaction model***

This model belongs to the so-called predator-prey models, namely mathematical models of the interaction between two species. Read Chapter 1, Section 4 for additional information on this problem. Read also Chapter 7, Section 2. Part of this project is analyzed in that section. Consider the parasitoid-host interaction model:

$$\begin{cases} x_{n+1} &= a x_n(1 - x_n - y_n) \\ y_{n+1} &= b x_n y_n. \end{cases}$$

Possible topics for investigation.

1. Prove that for $0 \le b \le a \le 4$ every orbit with initial condition (x_0, y_0) in the region, R, bounded by the lines $x = 0$, $y = 0$ and $x + y = 1$ will remain in R forever.
2. Prove that for $0 \le a \le b \le 4$ every orbit with initial condition (x_0, y_0) in the region, R, bounded by the lines $x = 0$, $y = 0$ and $x + y = 1$ will remain in R forever.
3. Find the branches of fixed points of the system.
4. Study the stability of each branch for the particular case $a = b$.

5. Study the stability of each branch in general, namely with a and b selected in the square $[0,4]\times[0,4]$. For each branch determine the region of the square where the parameters ensure stability and the region where instability occurs. Also determine the region where the eigenvalues of the derivative along a branch are real and where they are complex.
6. Study the Hopf bifurcation of the system.
7. Investigate numerically the presence of chaotic behavior for some values of the control parameters, by means of graphical analysis, Lyapunov exponents and correlation dimension.

Additional source: [Lauwerier, 1986b]

2. *Blood-cell population model*

Dynamical systems of the form

$$x_{n+1} = (1 - a)x_n + b x_n^r \exp(-s x_n) \tag{1}$$

have been studied as models of the blood-cell population in humans. The article [Gearhart-Martelli , 1990] has an extensive discussion about these models. Consider the simplified version of (1):

$$x_{n+1} = (1 - a)x_n + 4\, x_n^6 \exp(-2x_n). \tag{2}$$

Possible topics for investigation.

1. Is it true that for all $a \in (0,1)$ the dynamical system has three fixed points?
2. Study the stability of the fixed points. In particular, establish that the origin is always a sink and the smallest positive fixed point is always a source.
3. For which values of $a \in (0,1)$ does F have a maximum point $x_M(a)$ such that $F(a,x_M(a))>x_M(a)$? For these values of a, find an interval that is mapped onto itself by F.
4. Find the value a_0 for which the derivative $F_x(a_0,x_s)=-1$, where x_s is the largest fixed point for F.
5. Establish that for $a>a_0$ the system has a periodic orbit of period 2. Find the orbit numerically for some values of a close to a_0 and analyze the stability numerically.
6. Are there values of a for which the system has a periodic orbit of period 3?
7. Find values of a for which the system appears to be chaotic.
8. Investigate what is happening when a is very close to 1.

Major sources:
- [Gearhart-Martelli, 1990]
- [Lauwerier, 1986a]

3. *Another blood-cell population model*

The following dynamical system has been proposed by Mackey and Glass to model the granulocyte population:

$$x_{n+1} = x_n - d(x_n) + p(x_n), \tag{1}$$

with $d(x)=ax$ representing the cells destroyed and $p(x)=b\theta^m x/(\theta^m+x^m)$ representing the cells produced. The parameters are all positive.

Moreover m is an integer larger than 1; a,b,m are unitless while the dimensions of θ are cells/(kilogram of body weight).

Possible topics for investigation.

1. Show that $p(x)$ has in $(0, \infty)$ a single maximum and a single inflection point and express them as function of the control parameters.
2. Find the fixed points of the dynamical system and study their stability.
3. Let D be the delay time in production. It is estimated that for a healthy person D=.68 day and

$$a = \gamma D, \qquad \gamma = .16 \text{ per day}$$
$$b = \beta D, \qquad \beta = 1.43 \text{ per day}$$
$$\theta = 3.22 \times 10^8 \text{ cells/kg}$$
$$m = 3.$$

Analyze the effects of changing θ while all other parameters are kept constant at the values given above. In particular, does changing θ has any effect in the stability of the fixed points?
4. Find the form of the system obtained from (1) when the variable x is rescaled using the substitution $x=\theta y$. Verify that in this new system y is a dimensionless variable.
5. Assuming γ, β, and m constant (= to the values given above), study the stability of the fixed points of the rescaled system as a function of the delay D. In particular, verify that an orbit of period 2 arises as D is increased.
6. Assume that the parameter $a \in [0,1]$ in the rescaled system does not depend on D, $b=\beta D$ as above and let m=10. Investigate numerically the presence of periodic orbits of period 3 when a is increased.
7. Numerically explore the possibility of chaotic behavior in the same system as in the previous case, using the correlation dimension and Lyapunov exponents.

Major sources:
- [Gearhart-Martelli, 1990]
- [Lauwerier, 1986a]

4. *Fractals and chaos*

Geometric objects having fractional dimension play an increasingly important role in today's research and technology. The purpose of this project is to introduce fractals and discuss some of their properties and their relation to chaotic dynamical systems.

Possible topics for investigation.

1. Introduction of various definitions of dimension, in particular the capacity, the Hausdorff dimension, and the correlation dimension.
2. Introduction of some of the most known fractals. This list should include the Cantor set, the Sierpinski triangle, the Sierpinski carpet, and the Koch snowflake, with a good discussion about their properties. It should also

include a computation of their capacity and Hausdorff dimension with the relevant theorems to be used.

3. Attractors of chaotic dynamical systems frequently have fractional dimension, although this condition is neither necessary nor sufficient for establishing chaotic behavior. In this part of the project some examples of attractors of dynamical systems should be analyzed and their correlation dimension should be established.

4. Add a few comments about the compression of images using fractals.

Major source: [Peitgen-Jürgens-Saupe, 1992a]

5. *Neural networks*

Neural networks, which model the human brain are today in the forefront of research and applications to many different areas of technology. Some neural networks can be written as discrete dynamical systems and their solution to a given problem can be identified with fixed points of the systems.

Possible topics for investigation.

1. Introduction to neural networks and their basic structure. Part of this project can be found in Chapters 1 and 7.

2. Training a neural network with learning rules. Here it would be appropriate to talk about some learning rules, in particular backward propagation, and to give at least one example of training.

3. Develop the mathematical background for the functioning of a neural net. Discuss the contraction case and the boundedness of the orbits of a neural net.

4. Present some cases in which convergence does not occur: for example, the net oscillates or runs into chaotic behavior. Also present some cases in which convergence occurs but the answer provided is not useful.

5. Explore the use of neural networks to predict exchange rates.

Sources:
- The textbook
- Neural Network Approach to Classification Problems, UMAP 1989
- [Codenotti-Margara , 1996]
- [Kaplan-Glass, 1995]

6. *Compression of images using fractals*

Fractals are used to store images more efficiently. For some information on this topic one could consult [Peitgen-Jürgens-Saupe, 1992a]. The authors list articles and books that can be used as a source.

7. *Complete study of a one-parameter family of maps*

Study the one-parameter family of maps

$$f(x,a) = ax+(1-a)x^2.$$

In particular:

1. Find invariant intervals.

2. Find fixed points and periodic orbits of period 2.

3. Study their stability.

4. Find bifurcation points and establish their type.
5. Find periodic orbits of period 3.
6. Find the Lyapunov exponent and correlation dimension on invariant sets where the the system appears to be chaotic.

8. ***Defining chaos***

The article [Martelli-Dang-Seph, 1998] contains some definitions of chaos and points out their advantages and limitations. There are other definitions of chaos, like the one proposed in [Touhey, 1997]. The purpose of this project is to collect all these definitions, explain the relation among them, illustrate their merits, and provide examples that point at their limitations.

9. ***Choose your own project***

For example, the book of Carl Chiarella, The Elements of a Nonlinear Theory of Economic Dynamics (Springer-Verlag), could be used as a source for a project. In Chapter 6 he studies the cobweb model: its instability and the onset of chaos. The chapter is 16 pages long. A team could write a report on it.

APPENDIX 3

ANSWERS TO SELECTED PROBLEMS

CHAPTER 1

Section 1

Examples of Discrete Dynamical Systems

1. 120 months
2. \$9487.64
3. 9.24×10^5
4. 14.404×10^5
5. $x_{s0}=0$, $x_{s1}=2.00445$, $x_{s2}=-2.00445$
6. $x_{s0}=0$, $x_{s1}=0.802771$, $x_{s2}=8.77459$
7. $\mathbf{x}=(2.69179,1.25)$

Definition of Discrete Dynamical Systems

1. State variables: x,y,z. Control parameters: a,b,c.
4. $-a^3x^4+2(ax)^2-a+1$
5. $(ax-bx^2)(a-abx+(bx)^2)$
7. $\begin{cases} x_{n+1} = x_n+y_n \\ y_{n+1} = x_n \end{cases}$
9. $\begin{cases} x_{n+1} = 2x_n - .2z_n w_n \\ y_{n+1} = y_n + .1z_n w_n \\ z_{n+1} = x_n \\ w_{n+1} = y_n \end{cases}$
10. $x_n=a^n x_0$. Since a>1...

Section 2

Stationary States

1. $\frac{2}{3}\times 3+1=3$

3. $a<-\frac{1}{4}$ no fixed points; $a=-\frac{1}{4}$ $x=2$;

$a=0$ $x=1$; all other a's $x=\frac{-1\pm\sqrt{1+4a}}{2a}$.

5. Two. No stationary states.

6. The system

$$\begin{cases} x(1-\cos 1)+y\sin 1 = 1 \\ -x\sin 1+y(1-\cos 1) = 2 \end{cases}$$

has one and only one solution since the determinant of the matrix of its coefficients is $\neq 0$.

9. Define $F(x)=x-\cos x$. Then F is strictly increasing. Apply the intermediate value theorem to F on the interval [0,1].

11. $\|\mathbf{x}_1-\mathbf{x}_s\|=2^{-1}\|\mathbf{x}_0-\mathbf{x}_s\|$

12. $\|\mathbf{x}_1-\mathbf{x}_s\|=2\|\mathbf{x}_0-\mathbf{x}_s\|$

Periodic Orbits

1. $x_0=(1+\sqrt{5})/4$, $x_1=(1-\sqrt{5})/4$

2. $a>3/4$. No.

5. $\mathbf{x}_0=\mathbf{x}_p=F(\mathbf{x}_{p-1})=F^p(\mathbf{x}_0)$. Apply F^p to both sides as many times as needed.

7. F does not change distances.

8. $\frac{d}{dx}F^3(1,a)=-8a^7+24a^6-24a^5+16a^4-8a^3$

10. $\frac{d}{dx}F^3(x_0)=F'(x_2)F'(x_1)F'(x_0)=F'(x_0)F'(x_2)F'(x_1)=\frac{d}{dx}F^3(x_1)\ldots$

11. $x_0=1/7$, $x_1=2/7$, $x_2=4/7$

13. Period 8

Section 3

Limit Points, Limit Sets, and Aperiodic Orbits

1. 0 and 1

2. $-1,0,1$

4. $x_{n+1}=(2/3)^{n+1}(x_0-3)+3$

6. The orbit of $x_0=.10100100010000\ldots$ is aperiodic since each number of the form 2^{-m} for every integer $m\geq 1$ is a limit point of the orbit.

9. $x_0=2/9$, $x_1=4/9$, $x_2=8/9$. Yes.

10. F has a periodic orbit of period 3.

Unstable Orbits and Chaotic Systems

2. $F(-1)=F(1)=-a+1$, and $F(0)=1$. We have $-a+1\in[-1,1]$ for $a\in[0,2]$. Yes.

3. $n=5$

4. $F(-1)=-1$, $F(1)=1$, $F(1/2)=-1$, $F(-1/2)=1$. Yes.

Section 4

Examples

1. F(a,0)=0. Assuming that $x\neq 0$ we obtain $a=4x^5e^{-2x}$. There are two values of x for which this equality is true for every $a\in(0,1)$.

3. (x_1,bx_1), (x_2,bx_2), $x_1=\dfrac{b-1+\sqrt{(b-1)^2+4a}}{2a}$, $x_2=\dfrac{b-1-\sqrt{(b-1)^2+4a}}{2a}$

5. (1,–1), (1,3)

6. Use x,y,z to denote the variables. Then

$x_{n+1} = (1-c_1)x_n + t_{11}f(x_n) + t_{12}f(y_n) + t_{13}f(z_n) + x_I$ etc.

7. $x_{n+1} = (1-c_1)x_n + t_{11}(a_1x_n+b_1) + t_{12}(a_2y_n+b_2) + t_{13}(a_3z_n+b_3) + x_I$, etc.

8. $x(t)=\tan(t+c)$

9. $x(t)=\dfrac{1}{1+ce^{-at}}$

CHAPTER 2

Section 1

The Cobweb Method

1. Go up to the graph, then right to the diagonal... .

2. Go up to the graph, then right to the diagonal... .

Go up to the diagonal, then left to the graph.

3. Period 4

4. x_0 is about .95

Conjugacy

3. Let $G(a,b,y)=y(a-by)$, $F(a,x)=ax(1-x)$, and $\phi(a,b,y)=by/a$. Then $\phi oG=by(1-by/a)$ $=Fo\phi$.

4. $\phi(x)=2x-1$

6. One has two fixed points, the other has three.

7. $F(x)=\cos(x+x_s)-x_s$

8. $\phi^{-1}(y)=\sin\dfrac{\pi y}{2}$ · Domain is [–1,1] .

10. $x_0=-1/7$, $x_1=-5/7$, $x_2=3/7$, $y_0=\sin(-\pi/14)$, $y_1=\sin(-5\pi/14)$, $y_2=\sin(3\pi/14)$. A second periodic orbit of period 3 is $x_0=-7/9$, $x_1=5/9$, $x_2=1/9$.

12. $x_n=(-3/4)^n(x_0-2/7)+2/7$

Section 2

Stationary States and Periodic Orbits

1. $x_{s1}(a,b)=0$, $x_{s2}(a,b)=\sqrt{(1+b)/a}$, $x_{s3}(a,b)=-\sqrt{(1+b)/a}$, $(1+b)/a\geq 0$

3. $x_0=2/17$, $x_1=4/17$, $x_2=8/17$, $x_3=16/17$

4. $x_0=2/33$, $x_1=4/33$, $x_2=8/33$, $x_3=16/33$, $x_4=32/33$

5. $a\in\left(\dfrac{1+\sqrt{5}}{2}, 2\right]$

9. Arctan$x$ is never equal to 2 and F is increasing.
10. –.5,1
12. $x_0=5/3$, $x_1=13/3$

Sinks
1. $a\in(0,2)$
3. $x_{s1}(a)=0$, $a\in(0,2)$; $x_{s2}(a)=1$, $x_{s3}(a)=-1$, $a\in(-1,0)$
4. $x_{s1}(a)=\dfrac{-1-\sqrt{1+4a}}{2a}$, never; $x_{s2}(a)=\dfrac{-1+\sqrt{1+4a}}{2a}$, $a\in(-1/4,3/4)$.
Periodic orbit of period 2 : $a\in(3/4,5/4)$
6. $(3,1+\sqrt{6})$ and $(1-\sqrt{6},-1)$
7. Consider the sequence $n^{-1}\ln 3$. The point is 1.
9. $x_n\to\dfrac{4}{1-r}$
10. $x_n\to\dfrac{a}{1-k}$
11. $|F'(x)|<1$ for $x\in(-\sqrt{(-1+\sqrt{7})/3}, \sqrt{(-1+\sqrt{7})/3})$
12. Follows from $3ax^2(1+x^2)<2+x^2$, which is true for $a>0$ and $x\in(-d,d)$ with

$$d=\left(\frac{1-3a+\sqrt{9a^2+18a+1}}{6a}\right)^{1/2}.$$

14. Apply Theorem 2.7.7 to the second iterate.
15. Theorem 2.2.6 cannot be applied to F since $|F'(2/3)|=1$. Theorem 2.2.8 cannot be applied to F since for $x=2/3+h$ we have $F'(x)=-1-6h$, whose absolute value is larger than 1 for positive h. Since $\frac{d}{dx}F^2(x)=9-72x+162x^2-108x^3$ $=1-54(x-2/3)^2-108(x-2/3)^3$, Theorem 2.2.8 can be applied to the second iterate.

Sources
1. $a\notin[0,2)$
3. $x_{s1}(a)=0$ is a source for $a\notin[0,2]$, $x_{s2}(a)=1$, $x_{s3}(a)=-1$ are sources for $a\notin[-1,0]$.
4. $x_{s1}(a)=\dfrac{-1-\sqrt{1+4a}}{2a}$ is a source for all $a>-1/4$.

$x_{s2}(a)=\dfrac{-1+\sqrt{1+4a}}{2a}$ is a source for $a>3/4$.

The periodic orbit of period 2 is a source for $a>5/4$.
5. Recall that $a<-1$ or $a>3$.

The periodic orbit is a source for $a\notin[1-\sqrt{6},-1]\cup[3,1+\sqrt{6}]$.
7. $a\geq 1/3$
9. Theorem 2.2.9 cannot be applied since F is not differentiable at 0. Theorem 2.2.11 cannot be applied since $|F'(x)|=.5$ for $x<0$ and $|F'(x)|=4$ for $x>0$. The derivative of the second iterate is 2 for every x. Hence, according to the result of Problem 8, the fixed point is a source.

Section 3

Global Sinks

1. Set $F(x)=x-\cos(x+2)$. Apply the intermediate value theorem to F on the interval $[-1,0]$ and use Lemma 2.3.1.

3. Apply Theorem 2.3.2.

5. F is concave up in $[0,1]$. Hence $F(x)<x$. Use Lemma 2.2.3 and the property $F(L(x_0))=L(x_0)$.

7. $x_s\in[2,3]$. Apply Theorem 2.3.2

Section 4

Fold, Transcritical, and Pitchfork Bifurcation

1. Fold bifurcation at $a=-1/4$. Fixed points are $x_1(a)=\frac{-1+\sqrt{1+4a}}{2}$, $x_2(a)=\frac{-1-\sqrt{1+4a}}{2}$.
The first is a sink for $a\in(-1/4, 3/4)$. The second is never a sink.

2. Fixed points $x_1(a)=0$ and $x_2(a)=a-1$. $a=1$ is a transcritical bifurcation point. $x_1(a)$ is a sink for $a\in(-1,1)$. $x_2(a)$ is a sink for $a\in(1,3)$.

4. $a=-1$ is a pitchfork bifurcation point. $x_1(a)=0$ is a sink for $a\in(-1,1)$. $x_2(a)=-\sqrt{1+a}$ and $x_3(a)=-x_2(a)$ are never sinks.

5. $a=-1$ is a fold bifurcation. There are three branches of fixed points $x_1(a)=1-a^2$, $x_2(a)=1-\sqrt{1+a}$ and $x_3(a)=1+\sqrt{1+a}$.

6. $a=0$ is a supercritical pitchfork.

7. $a=0$ is not a bifurcation point.

8. $a=1$ is a transcritical bifurcation point.

Period-Doubling Bifurcation

1. The fixed points are $x_1(a)=(-1+\sqrt{1+4a})/2$, $x_2(a)=(-1+\sqrt{1+4a})/2$. The periodic orbit of period 2 is $x_3(a)=(1+\sqrt{4a-3})/2$, $x_4(a)=(1+\sqrt{4a-3})/2$.
The periodic orbit is a sink for $a\in(3/4, 5/4)$. $a=3/4$ is a period-doubling bifurcation point.

3. The fixed points are $x_1(a)=0$ and $x_2(a)=a-1$. The periodic orbit of period 2 is $x_3(a)=\frac{1+a+\sqrt{a^2-2a-3}}{2}$, $x_4(a)=\frac{1+a-\sqrt{a^2-2a-3}}{2}$. $a=-1$ and $a=3$ are period-doubling bifurcation points. $x_1(a)$ is a sink for $a\in(-1,1)$. $x_2(a)$ is a sink for $a\in(1,3)$. The periodic orbit is a sink for $a\in(1-\sqrt{6},-1)$ and $a\in(3,1+\sqrt{6})$.

5. The fixed points are $x_1(a)=0$, $x_2(a)=-\sqrt{1+a}$, $x_3(a)=\sqrt{1+a}$. The periodic points of period 2 are

$$x_4(a)=\sqrt{a-1},\ x_5(a)=-x_4(a);$$

$$x_6(a)=\frac{\sqrt{a-\sqrt{a^2-4}}}{\sqrt{2}},\ x_7(a)=\frac{\sqrt{a+\sqrt{a^2-4}}}{\sqrt{2}},$$

$$x_8(a)=-x_6(a),\ x_9(a)=-x_7(a).$$

$a=-1$ is a pitchfork, $a=1$ is a period doubling, and $a=2$ is a pitchfork.

7. The fixed points are $x_1(a)=0$, $x_2(a)=1$, $x_3(a)=-1$. A periodic orbit of period 2 is

$x_4(a)=\sqrt{1-2/a}$, $x_5(a)=-\sqrt{1-2/a}$.

The periodic orbit is a sink for $a\in(2, 3)$ and a source for $a\in(3,\infty)$.

Two additional periodic orbits of period 2 are

$$x_6(a)=\sqrt{\frac{a-1-\sqrt{a^2-2a-3}}{2a}},\quad x_7(a)=\sqrt{\frac{a-1+\sqrt{a^2-2a-3}}{2a}},$$

$x_8(a)=-x_6(a)$, $x_9(a)=-x_7(a)$.

$a=-1$ is a subcritical period-doubling, $a=2$ is a supercritical period-doubling, and $a=3$ is a pitchfork.

9. Yes, supercritical

Bifurcation: A Theoretical Viewpoint

1. The derivative of the second iterate is always positive...

3. $a=\sqrt{3}$ is a supercritical period-doubling, $a=-\sqrt{3}$ is a subcritical period-doubling.

5. $F_a(2,0)=0$, $F_{xa}(2,0)=1$, $F_{xx}(2,0)=0$, $F_{xxx}(2,0)=2$.

7. $a=2$ is a supercritical period-doubling.

Section 5

Conjugacy and Chaos

1. $x_0=2/9$, $x_1=4/9$, $x_2=8/9$

3. The conjugacy is $\phi(x)=-2x+1$

5. $y_0=\sin(-5\pi/14)$, $y_1=\sin(-\pi/14)$, $y_2=\sin(3\pi/14)$

7. 5/26, 15/26, 19/26. No, the map is not continuous.

9. $$G(x)=\begin{cases} 3x & 0\le x<1/3 \\ -3x+2 & 1/3\le x<2/3 \\ 3x-2 & 2/3\le x\le 1 \end{cases}$$

CHAPTER 3

Section 1

Norms and Sets

1. $|x|+|y|=1$

3. (4,2,–1), (–2,2,–1), (1,5,–1), (1,–1,–1)

5. Choose a point $\mathbf{x}_0$ on ∂A which is not in A. For every positive integer n find a point $\mathbf{x}_n$ of A which is at distance less than 1/n from $\mathbf{x}_0$.

7. Let $\mathbf{x}$ be a point in the interior of A. Then $\mathbf{x}$ does not belong to ∂A. Hence, there is r>0 such that ...

9. Let r>0 be such that $B(\mathbf{0},r)$ contains the bounded set. Select $\mathbf{x}_n\in S(\mathbf{0},r+n)$.

11. $\max\{|x_i+y_i| : i=1,2,...,q\}\le\max\{|x_i| : i=1,2,...,q\}+\max\{|y_i| : i=1,2,...,q\}$.

13. If $\|\mathbf{x}\|\le 1$, then $\max\{|x|,|y|\}\le 1$.

Continuity

1. $F_1(\mathbf{x})=F_1(x,y)=x-y^2$, $F_2(\mathbf{x})=F_2(x,y)=x^2-y$. Yes.

3. $F_1(x,y,z)=x-y$, $F_2(x,y,z)=xy-z^2$, $F_3(x,y,z)=\sin(x+z)$

4. The function is defined at every point where $x+y$ is not an odd multiple of $\pi/2$. At every point of its domain the function is continuous.

6. Since $F(\mathbf{x}_n)=\mathbf{x}_n\bullet\mathbf{x}_0$ and $|F(\mathbf{x}_n)-F(\mathbf{x}_0)|=|\mathbf{x}_n\bullet\mathbf{x}_0-\mathbf{x}\bullet\mathbf{x}_0|=|(\mathbf{x}_n-\mathbf{x})\bullet\mathbf{x}_0|\le\|\mathbf{x}_n-\mathbf{x}\|\ \|\mathbf{x}_0\|$, we see that F is continuous.

7. Let $\mathbf{x}_M=\mathbf{x}_0/\|\mathbf{x}_0\|$ and $\mathbf{x}_m=-\mathbf{x}_0/\|\mathbf{x}_0\|$. Both belong to $D(\mathbf{0},1)$ and $F(\mathbf{x}_m)=-\|\mathbf{x}_0\|$, $F(\mathbf{x}_M)=\|\mathbf{x}_0\|$. Since $|F(\mathbf{x})|\le\|\mathbf{x}\|\ \|\mathbf{x}_0\|\le\|\mathbf{x}_0\|$, we see that the maximum and minimum value of F are reached at $\mathbf{x}_M$ and $\mathbf{x}_m$, respectively.

9. F is continuous in [0,1] and since $\sin x\le x$ we have $F(x)\le 1$. Hence its maximum value is at $x=0$. Moreover, the derivative of F is negative in (0,1). Hence F is decreasing and its minimum value is at $x=1$.

11. Let $x=\cos t$, $y=\sin t$. We obtain $1-\cos t\sin t$, $t\in[0,2\pi]$. The derivative is $-\cos 2t$, which is 0 for $t=\pi/4$, $t=3\pi/4$, $t=5\pi/4$ and $t=7\pi/4$. We obtain four points. Two are minimum points and two are maximum points.

13. The two constants are A=q and a=1.

Section 2

Operator Norm

1. $M\mathbf{x}=x(2,3,2)+y(2,-2,3)+z(-1,0,-2)=(2x+2y-z,3x-2y,2x+3y-2z)$

3. Let $\|\mathbf{x}\|=1$, i.e., $x^2+y^2=1$. Since $M\mathbf{x}=(2x+y,3y)$ we obtain $\|M\mathbf{x}\|^2=4+6y^2+4xy$. Set $x=\cos t$ and $y=\sin t$. Then we are looking for the maximum value of the function $6\sin^2 t+4\sin t\cos t$... . $\|M\|\le 3.32$ and $\|M^2\|\le\sqrt{16+95}\approx 10.54$.

5. $\|M\mathbf{x}\|_1=|2x+y|+|3y|$. For $x=0$ and $y=1$ this value is 4. Since $|2x+y|\le 2|x|+|y|$ and $|x|+|y|=1$ we see that 4 cannot be exceeded. Hence $\|M\|_1=4$.
$\|M\mathbf{x}\|_\infty=\max\{|2x+y|,3|y|\}\le 3$. Since $M(1,1)=3$...

7. M is a rotation of $\pi/3$ combined with a compression of 1/2. Hence its operator norm in the Euclidean norm is 1/2.

9. The spectral radius is 3.

11. $1+2\sqrt{7}$; $\|M\|_\infty=14$

Derivative and Mean Value Inequality

1. $F'(\mathbf{x})=\begin{pmatrix}1 & -2y\\ 2x & -1\end{pmatrix}$ which is continuous. In fact, all its entries are polynomials in the variables x and y. $F'(\mathbf{0})=\begin{pmatrix}1 & 0\\ 0 & -1\end{pmatrix}$. The first-order approximation is $(0,0)+F'(\mathbf{0})(x,y)=(x,-y)$.

3. $F'(\mathbf{x})=\begin{pmatrix}1/y & -x/y^2\\ -y/x^2 & 1/x\end{pmatrix}$, $F'(2,3)=\begin{pmatrix}1/3 & -2/9\\ -3/4 & 1/2\end{pmatrix}$, $F(2,3)=(2/3,3/2)$. Hence $(2/3,3/2)+F'(2,3)(x-2,y-3)=\left(\frac{1}{9}(6+3x-2y), \frac{1}{4}(6-3x+2y)\right)$.

5. $F(0,1)=(-1,-1)$ and $F(1,2)=(-3,-1)$. Hence $\|F(1,2)-F(0,1)\|_\infty=2$. Since $\|(1,2)-(0,1)\|_\infty=1$ we must find $\mathbf{c}$ such that $\|F'(\mathbf{c})\|_\infty\ge 2$. Consider $\mathbf{c}=(.5,1.5)$.

7. We have $\|F(1,2)-F(0,1)\|_1=2$ and $\|(1,2)-(0,1)\|_1=2$. Hence we need $\|F'(\mathbf{c})\|_1 \geq 1$. Let $\mathbf{c} = (.5,1.5)$.

CHAPTER 4

Section 1

Orbits of Linear Processes

1. $M=\begin{pmatrix} 3 & -2 \\ 5 & 7 \end{pmatrix}$

3. $M=\begin{pmatrix} 3 & -2 \\ 2 & -1 \end{pmatrix}$

5. No

7. The eigenvalues of M are $2-\sqrt{3}$ and $2+\sqrt{3}$ with corresponding eigenvectors

$$\mathbf{u}=(-1-\sqrt{3},1),\ \mathbf{v}=(-1+\sqrt{3},1).\quad \mathbf{x}_n=(2-\sqrt{3}\,)^n\frac{3-2\sqrt{3}}{6}\mathbf{u} + (2+\sqrt{3}\,)^n\frac{2+\sqrt{3}}{2\sqrt{3}}\mathbf{v}.$$

9. The eigenvalues of M are .5, 1, 2 with corresponding eigenvectors $\mathbf{u}=(1,1,0)$, $\mathbf{v}=(0,1,0)$, and $\mathbf{w}=(2,0,1)$. $\mathbf{x}_n=(.5)^n3\mathbf{u}-2\mathbf{v}-2^n\mathbf{w}$.

Section 2

Stability and Instability of the Origin

1. Every orbit goes to $\mathbf{0}$, since the eigenvalues are $\pm 1/\sqrt{2}$.
3. Every orbit goes to $\mathbf{0}$, since the eigenvalues are 1/3 and 1/2.
5. The eigenvalues are 2 and $(8\pm 12i)/13$. Every nonstationary orbit goes to ∞.
7. $\mathbf{x}_n=2^{-n}((3/2)\mathbf{x}_1-\mathbf{x}_2)+3^{-n}(1/2)\mathbf{x}_3$, with $\mathbf{x}_1=(1,0,1)$, $\mathbf{x}_2=(0,1,0)$, and $\mathbf{x}_3= (-1,0,1)$.

Section 3

Spectral Decomposition Theorem

1. $\mathbf{x}_n=\frac{1}{3}[(-1)^n (-x_0+y_0)\mathbf{u}+2^n(x_0+2y_0)\mathbf{v}]$ with $\mathbf{u}=(-2,1)$ and $\mathbf{v}=(1,1)$

3. $\mathbf{x}_n=2^{-n-1}[2z_0\mathbf{e}_3+(y_0-x_0)\mathbf{x}_2]+2^{n-1}(x_0+y_0)\mathbf{x}_3$ with $\mathbf{x}_2=(-1,1,0)$ and $\mathbf{x}_3=(1,1,0)$

5. $\mathbf{x}_n=M^n\mathbf{x}_0=(2^nx_0+n2^{n-1}y_0)\mathbf{e}_1+2^ny_0\mathbf{e}_2$

7. $\mathbf{x}_n=(2n+1)2^{-n}\mathbf{e}_3+2^{-n}\mathbf{e}_2$. The orbit goes to $\mathbf{0}$.

9. $\mathbf{x}_n=(1/15)[(-5x_0+20y_0+10z_0)2^{-n}\mathbf{w} + 3z_0M^n\mathbf{u}+(5x_0-5y_0-4z_0)M^n\mathbf{v}]$, $\mathbf{w}=(1,1,0)$, $\mathbf{u}=(2,-2,5)$, and $\mathbf{v}=(4,1,0)$. This can be evaluated using (4.3.14). In our case $r=\sqrt{13}$ and $\theta=\arctan(1.5)$. With $x_0=y_0=z_0=1$ we obtain $\mathbf{x}_n=(5/3)2^{-n}\mathbf{w}+(13^{n/2}/5)[(\cos n\theta-(4/3)\sin n\theta)\mathbf{u}-(\sin n\theta+(4/3)\cos n\theta)\mathbf{v}]$.

Section 4

The Origin as a Saddle Point

1. S=Span{(-5/4,1)}, **U**=Span{$\mathbf{e}_2$}

3. S=Span{$\mathbf{e}_1$}, **U**=Span{(4,–3,3), (4,0,5)}

5. U=Span{$\mathbf{e}_1$}, **S**=Span{(-3/5,0,1), (1/5,1/2,0)}

Section 5

Eigenvalues with Modulus 1 (*)

1. The eigenvalues are 1/2 and 1 with corresponding eigenvectors **u**=(0,1) and **v**=(1,4). Given an initial condition $\mathbf{x}_0$=a**u**+b**v**, the orbit converges to b**v**.

3. The eigenvalues are 1/2 and 1 (double). Corresponding eigenvectors are **x**=(–2,0,1) (corresponding to 1/2) and $\mathbf{e}_1$, $\mathbf{e}_2$. Given the initial condition $\mathbf{x}_0$=a**x**+b$\mathbf{e}_1$ +c$\mathbf{e}_2$, the orbit converges to b$\mathbf{e}_1$+c$\mathbf{e}_2$.

5. The eigenvalues are –1 and 1 with corresponding eigenvectors **u**=(–2,1) and $\mathbf{e}_2$. The orbit of $\mathbf{x}_0$=a**u**+b$\mathbf{e}_2$ is periodic of period 2.

7. The eigenvalues are 1/4 and 1 (double). An eigenvector corresponding to 1/4 is **u**=(1,–1,1) and an eigenvector corresponding to 1 is **v**=(2,0,1). Every orbit in the one-dimensional subspace spanned by **u** goes to **0**. The generalized kernel of I–M is spanned by the vectors **v** and **z**=(–1,1,0). Given a point $\mathbf{x}_0$=a**u**+b**v**+c**z** with c≠0 the orbit O($\mathbf{x}_0$) goes to infinity. When c=0 the orbit converges to b**v**.

Section 6

Affine Systems

1. **c**=(2,–1). The eigenvalues of M are 1/2 and 3 with corresponding eigenvectors **u**=(5,–4) and $\mathbf{e}_2$=(0,1). The stable linear variety is **T** ={**c**+Span{**u**}}={(2+5a,–1–4a): a∈**R**}, and the unstable is **V**={**c**+Span{$\mathbf{e}_2$}}={(2,–1+b): b∈**R**}.

3. **c**=(–6,–2). The eigenvalues of M are 1/2 and 2 with corresponding eigenvectors $\mathbf{e}_1$ and **u**=(4,3). The stable linear variety is **T**={**c**+Span{$\mathbf{e}_1$}}={(–6+a,–2): a∈**R**} and the unstable is **V**={**c**+Span{**u**}}={(–6+4b,–2+3b): b∈**R**}.

5. The eigenvalues of M are 1,–.5 with corresponding eigenvectors **u**=(1,–1) and **v**=(2,–1). The orbit of every point, including (1,3) goes to infinity.

7. The two systems are conjugate by P. Yes.

9. Every orbit goes to infinity.

CHAPTER 5

Section 1

Bounded Invariant Sets

1. F is a contraction with constant .6. R≥12.5.

2. Verify that |F(n+1/n)-F(n)| cannot be made smaller than k/n for every n, with k∈[0,1).

3. ||M(0,1)||>1. p=2 is OK.

4. $||F(\mathbf{x})-\mathbf{x}_s||=||F(\mathbf{x})-F(\mathbf{x}_s)||\leq k||\mathbf{x}-\mathbf{x}_s||$

5. $570x^2+57y^2+152u^2+5280u=0$. The fixed points

$$(0,0,-660/19),\ (2(902/57)^{.5},2(902/57)^{.5},-11),$$
$$(-2(902/57)^{.5},-2(902/57)^{.5},-11)$$

belong to the ellipsoid.

7. $\|F(\mathbf{x})\|^2/\|\mathbf{x}\|^2\le\dfrac{\sin^2\theta}{1+(\rho\cos\theta)^2}+\dfrac{\cos^2\theta}{1+(\rho\sin\theta)^2}\le\dfrac{1+\rho^2(\sin^4\theta+\cos^4\theta)}{1+\rho^2+(\rho^2\cos\theta\sin\theta)^2}\le1.$

9. $\|M\mathbf{x}-\mathbf{0}\|=\|M\mathbf{x}-M\mathbf{0}\|\le\|M\|\ \|\mathbf{x}\|\le\|\mathbf{x}\|$

10. $\|T\Phi(\mathbf{x})\|\le\|T\|\ \|\Phi(\mathbf{x})\|\le\|T\|\ k$. Hence $\dfrac{\|T\Phi(\mathbf{x})\|}{\|\mathbf{x}\|}\to0$ as $\|\mathbf{x}\|\to\infty$. Choose a small positive number a such that $\|C\|+a<1$ and select a positive number r_0 such that $\dfrac{\|T\Phi(\mathbf{x})\|}{\|\mathbf{x}\|}\le a$ for every $\|\mathbf{x}\|\ge r_0$.

Section 2

Global Stability of Fixed Points

1. $\|F(\mathbf{x})-F(\mathbf{y})\|\le k\|\mathbf{x}-\mathbf{y}\|<\|\mathbf{x}-\mathbf{y}\|$

2. $F(F^p(\mathbf{x}_0))=F(\mathbf{x}_0)$. The left-hand side can be rewritten as $F^p(F(\mathbf{x}_0))$. Thus $F(\mathbf{x}_0)$ is a fixed point for F^p. Since F^p has only one fixed point...

3. The orbit starting at $\mathbf{x}_0$ can be partitioned into the following subsequences $\mathbf{x}_0, \mathbf{x}_p,\mathbf{x}_{2p},\ldots;\ \mathbf{x}_1,\mathbf{x}_{p+1},\mathbf{x}_{2p+1},\ldots;\ldots;\ \mathbf{x}_{p-1},\mathbf{x}_{2p-1},\ldots$. All these are sequences of iterates of F^p. Hence they converge to the unique fixed point.

4. We have $\|\mathbf{x}_2-\mathbf{x}_1\|=\|F(\mathbf{x}_1)-F(\mathbf{x}_0)\|\le k\|\mathbf{x}_1-\mathbf{x}_0\|$, $\|\mathbf{x}_3-\mathbf{x}_2\|=\|F(\mathbf{x}_2)-F(\mathbf{x}_1)\|\le k\|\mathbf{x}_2-\mathbf{x}_1\|\le k^2\|\mathbf{x}_1-\mathbf{x}_0\|$. In general we obtain $\|\mathbf{x}_{n+1}-\mathbf{x}_n\|\le k^n\|\mathbf{x}_1-\mathbf{x}_0\|$. Since $k\in[0,1)$ the result follows.

5. Since

$$\|\mathbf{x}_0-\mathbf{x}_s\|\le\|\mathbf{x}_0-\mathbf{x}_1\|+\|\mathbf{x}_1-\mathbf{x}_s\|\le\|\mathbf{x}_0-\mathbf{x}_1\|+k\|\mathbf{x}_0-\mathbf{x}_s\|,$$

we obtain $\|\mathbf{x}_0-\mathbf{x}_s\|\le\dfrac{1}{1-k}\|\mathbf{x}_0-\mathbf{x}_1\|$. Replace this quantity in the inequality

$$\|\mathbf{x}_n-\mathbf{x}_s\|\le k^n\|\mathbf{x}_0-\mathbf{x}_s\|.$$

7. Since M^p is a contraction every orbit of the dynamical system governed by M converges to **0**, which is the only fixed point of M.

8. The map is of triangular type and the conditions (5.2.5) are verified.

9. $y=\cos y$ has one and only one solution. Replace in the first equation.

11. No.

Section 3

Sinks

1. $\mathbf{x}=(-1,2,1)$ is a fixed point. It is not a sink.

3. $a\in(-.9,.9)$

5. Consider the point $\mathbf{x}_0=(\varepsilon,0)$. Then $\mathbf{x}_1=(1.4\varepsilon,0)$, $\mathbf{x}_2=((1.4)^2\varepsilon,0)$ ... .

7. $(\pi,\pi)\to(-\pi,\pi)\to(\pi,\pi)$. Starting at the point $(\pi+.01,\ \pi-.01)$ and after 80 iterations we reach the fixed point $(.175832,0)$.

Section 4

Repellers and Saddles

1. The eigenvalues of the derivative at the origin are $\pm\sqrt{6}$.

3. λ_1=.97334, $\lambda_2=1+.005(-11-\sqrt{81+40r})$, $\lambda_3=1+.005(-11+\sqrt{81+40r})$.
Hence, for r>1 we have $\lambda_3>1$.

4. (1,1) and (–1,–1). Always saddles.

5. The fixed points are $(1/\sqrt{a}, 1/\sqrt{a})$ and $(-1/\sqrt{a}, -1/\sqrt{a})$. They are saddles.

7. $x_1=2x-y^3$, $y_1=y/2$;
$x_2=2x_1-y_1^3=2(2x-y^3(1+1/4^2))$, $y_2=y/4$;
$x_3=2x_2-y_2^3=2^2(2x-y^3(1+1/4^2+1/4^4))$, $y_3=y/2^3$;
...
$x_{n+1}=2^n(2x-y^3(1+1/4^2+1/4^4+...+1/4^{2n}))$, $y_{n+1}=y/2^{n+1}$.

9. $\lambda_1=1-hb$ and

$$\lambda_2=1-h(a+1+\sqrt{(a-1)^2+4ar})/2$$

belong to the interval (0,1) for h sufficiently small. The third eigenvalue

$$\lambda_3=1-h(a+1-\sqrt{(a-1)^2+4ar})/2$$

is also in the same interval as long as $a+1-\sqrt{(a-1)^2+4ar}>0$. This requires that r<1. For r>1 the difference is negative and $\lambda_3>1$.

11. The inverse of F is the map $G(u,v)=(u/2, 2v-7u^2/2)$. The stable manifold of the origin is the y-axis. The unstable manifold is the parabola $y=2x^2$.

Section 5

Bifurcation

1. $x(a)=y(a)=0$ if and only if a=1.
$a(1/a-1)+a(1/a-1)^2 = 1/a-1 = x(a)$; $a(1-1/a+2/a-2+(1/a-1)^2=1/a-1= y(a)$.

3. $\mathbf{x}_1(a)=\mathbf{0}$, $\mathbf{x}_2(a)=(1-1/a,0)$, $\mathbf{x}_3(a)=(1/2a,1-1.5/a)$.
$\mathbf{0}$ is a sink for $|a|<1$ and a saddle for $|a|>1$.
$\mathbf{x}_2(a)$ is a sink for $a\in(1,1.5)$, a saddle for $a\in(-1/2,1)\cup(1.5,3)$, and a source for a>3 or a<–1/2.
$\mathbf{x}_3(a)$ is a sink for $a\in(1.5,2.5)$, a saddle for $a\in(-4.5,1.5)$, and a source for a>2.5 or a<–4.5.

5. $\mathbf{x}_3(a)$ is a sink for $a\in(2,3)$, a saddle for $a\in(0,2)$ and a source for a<0 or a>3.

CHAPTER 6

Section 1

Attractors

1. The domain of attraction of the origin include(but is not limited to) the set
$[0,1/4)\cup(7/8,1]\cup(11/24,1/2]\cup[1/2,9/16)...$.

3. F has two fixed points $2(1-\sqrt{2})$ and $2(1+\sqrt{2})$. Every orbit of $x_0 \in (-2(1+\sqrt{2}), 2(1+\sqrt{2}))$ converges to $2(1-\sqrt{2})$. An invariant interval is $[-2(1+\sqrt{2}), 2(1+\sqrt{2})]$.

5. $L(\mathbf{x}_0)$ is a periodic orbit of period 2.

7. Since $|h|<1/2$ we obtain $2h^2<|h|$. Hence x_1 is closer to 1/2 than x_0 and the sequence of iterates converges to 1/2.

Section 2

Chaotic Dynamical Systems

1. Consider (ρ,θ) and $(\rho+\pi/n,\theta)$ for n sufficiently large. Prove that after n iterations the two points are at least as far as 2.

3. Let [a,b] be any subinterval of [0,1] with a<b. Select an irrational number $x \in (a,b)$. Suppose that in base 2 we have $x=.a_1a_2a_3....$. Construct the following sequence of rational numbers

$$x_0=.a_1a_2a_1a_2a_1a_2... , \quad x_2=.a_1a_2a_3a_4a_1a_2a_3a_4... ,$$
$$x_3=.a_1a_2a_3a_4a_5a_6a_7a_8a_1a_2a_3a_4a_5a_6a_7a_8.... \ .$$

4. $F(\phi(x))=-2\cos^2 2\pi x+1$, $\phi(B(x))=-\cos 2\pi(2x-[2x])=-\cos 4\pi x=-2\cos^2 2\pi x+1$

5. Assume that $O(x_0)$, $x_0 \in [0,1]$, is such that $L(x_0)=[0,1]$ and let $z \in [-1,1]$. Since ϕ is onto, there exists $w \in [0,1]$ such that $\phi(w)=z$. Moreover, w is a limit point of $O(x_0)$. Hence we can find a subsequence of $O(x_0)$ which converges to w. The corresponding subsequence of $z_0=\phi(x_0)$ converges to z.

7. When $x_0 \in (0,1)$ we have $x_0^3-x_0 \in (-1,0)$ and $\exp(x_0^3-x_0)<1$. Hence $x_1=F(x_0)<x_0$ and the sequence converges to 0. Assume that $x_0 \in (-1,0)$. Then $x_0^3-x_0>0$ and $\exp(x_0^3-x_0)>1$. Then $x_1<x_0$ and this time the sequence of iterates will converge to -1.

8. $[0,5/16]\cup[7/16,9/16]\cup[11/16,1]$

9. Consider the sequence $(4^n+1)/(3\times 4^n)$

11. Show that $\phi(T(x))=L(\phi(x))$ and use Problem 10.

Section 3

Fractal Dimension

1. The correlation dimension is 0.

2. The correlation dimension is about .73.

3. The correlation dimension is about .9.

5. The conjugacy is $\phi(x)=(x+1)/2$ since $\phi(T(x))=(4x^3-3x+1)/2 =2x^3-1.5x+.5$, and $Q(\phi(x))=16((x+1)/3)^3-24((x+1)/2)^2+9(x+1)/2 =2x^3-1.5x+.5$.

6. The total area removed is$(A/4)(1+3/4+(3/4)^2+...)=A$.

7. We need three segments of length 1 to cover the perimeter of the original triangle.
We need $12=4\times 3$ segments of length 1/3 to cover the first stage.
We need $4^2\times 3$ segments of length 1/9 to cover the second stage.
In general, we need $4^n\times 3$ segments of length $1/3^n$ to cover the nth stage.
The area inside is 8A/5 where A is the area of the triangle.

Section 4

Lyapunov Exponents

1. Notice that the derivative of F at x_1 is 4 and the derivative of F at x_2 is 3/5.

Thus, for n even, n=2m we have $(d/dx)F^n(x_1)=4^m(3/5)^m=2^n\,(3/5)^m$. When n is odd, n=2m+1 we have $(d/dx)F^n(x_1) = 4^{m+1}(3/5)^m=2^{n+2}(3/5)^m$.

2. The Lyapunov number of the orbit is 2.

3. Both the Lyapunov k-number and the Lyapunov number are 2.

4. Depends on the equalities $(d/dx)F^p(x_i)= (d/dx)F^p(x_j)$ for all i,j=1,2,...,p and $(d/dx)F^{pm}(x_i)= (d/dx)F^{pm}(x_j)$.

5. Apply the previous result (Problem 4) with k=2,3,...,p.

7. The Lyapunov number is $\sqrt{2}-1$.

9. By using the conjugacy we should find that the Lyapunov exponent of the orbit is ln2.

10. The spectral radius of M is 6.

Let us start with the point $\mathbf{x}_0$=(1,0,0). The following table shows the behavior of the sequence (6.4.10) for the iterations 10,...,15. We see that the sequence converges very quickly to 6.

10	6.03491240159192532
11	6.02323191953049441
12	6.01546618940770727
13	6.01030023635767829
14	6.00686183612970836
15	6.00457224180906834

CHAPTER 7

Section 1

Blood-Cell Population Model

1. $x_0=p(m-1)^{-1/m}$. It is a maximum point since the first derivative is positive before x_0 and negative after.

$x_1=p((m+1)/(m-1))^{1/m}$. The function is concave down before x_1 and concave up after.

3. a is approximately .11. For all a≥.11 the minimum point comes after the second positive fixed point.

In the following table we compare the numerical values of $F(a,x_m)$ with x_{s2}. The top row are the values of a and the bottom row are the differences $F(a,x_m)-x_{s2}$. We see that at a=.84 we have a change is sign:

.81	.82	.83	.84	.85
.03	.02	.004	–.01	–.02

When this is happening the blood-cell population will reach the value corresponding to x_m and then, the next time around, will fall below the critical value x_{s2}. The individual cannot recover without external help.

4. The following table shows that for $a \geq .86$ we have both $F(a,x_M) > x_m$ and $F(a,F(a,x_M)) < x_{s2}$. The top row are the values of a, the middle are the differences $F(a,x_M) - x_m$, and the bottom row are the difference $F(a,F(a,x_M)) - x_{s2}$.

.83	.84	.85	.86	.87	.88
.306	.295	.282	.270	.256	.243
.038	.022	.005	–.011	–.027	–.044

7. Assume $b>0$, $s>r>1$. The origin is a sink. Assume that $b(r/se)^r > r/s$. Then there are two additional fixed points. The smallest is a source.

Section 2

Predator-Prey Models

1. Assume that the harvesting activity takes the form $-hex$ in the first equation and $-hey$ in the second. The new stationary state is $((c+e)/d, (a(d-c)-e(a+d))/bd)$. Therefore, we see that the harvesting activity is in favor of the prey.

3. Let $M = \begin{pmatrix} a & b \\ c & d \end{pmatrix}$. The assumptions are $a+d<0$ and $ad-bc>0$. The eigenvalues are the solutions of the quadratic equation $r^2-(a+d)r+ad-bc=0$. Since $a+d<0$ and $ad-bc>0$ each eigenvalue has negative real part.

5. Both fixed points (0,0) and (1/2,0) are sinks. Numerical investigation supports this conclusion. Orbits starting close to the two fixed points converge to them.

7. We have $(d/da)(a-2)^{.5} = 0.5(a-2)^{-.5}$, which equals 1/2 when a=3.

Section 3

Lorenz Model of Atmospheric Behavior

1. The matrix M_2 has the form $\begin{pmatrix} -a & a & 0 \\ 1 & -1 & \alpha \\ -\alpha & -\alpha & -b \end{pmatrix}$. A simple computation gives the same characteristic polynomial of M_1.

3. In terms of x,y,z, the fixed points are (0,0,0), $(72^{.5},72^{.5},27)$, $(-72^{.5},-72^{.5},27)$. The equation of the ellipsoid is $30x^2+3y^2+8z^2-304z=0$. The tangent plane at the origin is $z=0$.

4. The eigenvector corresponding to $1-bh=.97333$ is (0,0,1), which is perpendicular to the tangent plane. The remaining two eigenvectors are (1,2.18277,0) and (–1,1.28277,0). Hence they are both in the tangent plane.

5. No.

Section 4

Neural Networks

1. We have $F(\mathbf{x}) = \alpha(x_1,\ldots,x_N) + \beta(1,\ldots,1) = \alpha\mathbf{x}+\mathbf{b}$. Hence $TF(\mathbf{x}) = \alpha T\mathbf{x} + T\mathbf{b}$. The system becomes $\mathbf{x}_{n+1} = (I-C+\alpha T)\mathbf{x} + \mathbf{x}_J = B\mathbf{x}_n + \mathbf{x}_J$, where $B = I-C+\alpha T$ and $\mathbf{x}_J = \mathbf{x}_I + T\mathbf{b}$.

3. T_{13} is the 0 matrix since it represents the action of the output neurons on the input neurons. Most likely, they are not directly linked. The same reasoning applies to T_{31}. If no neuron has an influence on itself, the main diagonal of the matrix T is 0.

5. Notice that I–C is the 0 matrix. The output neuron is always 1. The input neuron is periodic of period 2. It starts with .5, becomes 1.5, etc. The two hidden neurons are also periodic of period 2. The first of them starts at 0, becomes 1, etc. The second start at 1, becomes 0, etc. Here is the entire situation:

n=0	1	2	3	4	...
(.5,0,1,1)	(1.5,1,0,1)	(.5,0,1,1)	(1.5,1,0,1)	(.5,0,1,1)	....

INDEX

WILEY-INTERSCIENCE SERIES IN DISCRETE MATHEMATICS AND OPTIMIZATION

AARTS AND KORST • Simulated Annealing and Boltzmann Machines: A Stochastic Approach to Combinatorial Optimization and Neural Computing
AARTS AND LENSTRA • Local Search in Combinatorial Optimization
ALON, SPENCER, AND ERDŐS • The Probabilistic Method
ANDERSON AND NASH • Linear Programming in Infinite-Dimensional Spaces: Theory and Application
AZENCOTT • Simulated Annealing: Parallelization Techniques
BARTHÉLEMY AND GUÉNOCHE • Trees and Proximity Representations
BAZARRA, JARVIS, AND SHERALI • Linear Programming and Network Flows
CHANDRU AND HOOKER • Optimization Methods for Logical Inference
CHONG AND ZAK • An Introduction to Optimization
COFFMAN AND LUEKER • Probabilistic Analysis of Packing and Partitioning Algorithms
COOK, CUNNINGHAM, PULLEYBLANK, AND SCHRIJVER • Combinatorial Optimization
DASKIN • Network and Discrete Location: Modes, Algorithms and Applications
DINITZ AND STINSON • Contemporary Design Theory: A Collection of Surveys
ERICKSON • Introduction to Combinatorics
GLOVER, KLINGHAM, AND PHILLIPS • Network Models in Optimization and Their Practical Problems
GOLSHTEIN AND TRETYAKOV • Modified Lagrangians and Monotone Maps in Optimization
GONDRAN AND MINOUX • Graphs and Algorithms *(Translated by S. Vajdā)*
GRAHAM, ROTHSCHILD, AND SPENCER • Ramsey Theory, Second Edition
GROSS AND TUCKER • Topological Graph Theory
HALL • Combinatorial Theory, Second Edition
JENSEN AND TOFT • Graph Coloring Problems
KAPLAN • Maxima and Minima with Applications: Practical Optimization and Duality
LAWLER, LENSTRA, RINNOOY KAN, AND SHMOYS, Editors • The Traveling Salesman Problem: A Guided Tour of Combinatorial Optimization
LAYWINE AND MULLEN • Discrete Mathematics Using Latin Squares
LEVITIN • Perturbation Theory in Mathematical Programming Applications
MAHMOUD • Evolution of Random Search Trees
MARTELLI • Introduction to Discrete Dynamical Systems and Chaos
MARTELLO AND TOTH • Knapsack Problems: Algorithms and Computer Implementations
McALOON AND TRETKOFF • Optimization and Computational Logic
MINC • Nonnegative Matrices
MINOUX • Mathematical Programming: Theory and Algorithms *(Translated by S. Vajdā)*
MIRCHANDANI AND FRANCIS, Editors • Discrete Location Theory
NEMHAUSER AND WOLSEY • Integer and Combinatorial Optimization
NEMIROVSKY AND YUDIN • Problem Complexity and Method Efficiency in Optimization *(Translated bv E. R. Dawson)*
PACH AND AGARWAL • Combinatorial Geometry

PLESS • Introduction to the Theory of Error-Correcting Codes, Third Edition
ROOS AND VIAL • Ph. Theory and Algorithms for Linear Optimization: An Interior Point Approach
SCHEINERMAN AND ULLMAN • Fractional Graph Theory: A Rational Approach to the Theory of Graphs
SCHRIJVER • Theory of Linear and Integer Programming
TOMESCU • Problems in Combinatorics and Graph Theory *(Translated by R. A. Melter)*
TUCKER • Applied Combinatorics, Second Edition
WOLSEY • Integer Programming
YE • Interior Point Algorithms: Theory and Analysis